An alternative, but not mutually exclusive, explanation is that cannibalism represents a form of parental allocation of resources to offspring. It seems likely that this form of cannibalism was the ancestral behaviour leading to the production of trophic or nurse eggs. The parent that produces a clutch which is partly consumed by her offspring is providing nutrition. This may be an alternative to producing fewer but larger eggs, and may occur in species in which the size of egg is constrained by adult female morphology, but juvenile survival is enhanced if food resources are readily available at hatching. Comparative studies across several taxa reveal that reproductive output is allocated along a continuum from a few large to many small eggs (e.g. Elgar 1990), and it would be interesting to know if deviations from this general pattern were associated with sibling cannibalism.

Social factors favouring cannibalism

Competition among individuals as a result of social structure may also be an important factor leading to cannibalism. This form of cannibalism effectively represents the outcome of the conflict for reproductive dominance, either for fertilizing opportunities or for access to resources necessary for offspring development.

The conflict between males for mating opportunities may result in cannibalism if females are more likely to re-mate or become sexually receptive more quickly after losing their offspring. For example, in some mammals a female comes into oestrous more quickly if her infants have been killed than if they have been successfully weaned (e.g. Hausfater and Hrdy 1984; see Elwood, Hiraiwa-Hasegawa, this volume). By killing a female's brood, a male might increase his lifetime mating opportunities at the expense of another male. Some forms of cannibalism in fish with male parental care may have similar explanations, except that the competition is between females for mating opportunities with males. In these species, a female that raids a male's nest and eats his eggs is more likely to mate with the male than if she only attempted to court him. Such a strategy would also be favoured if the survivorship of eggs laid in nests that already contained eggs was lower than in nests that did not.

Cannibalism as a means of increasing access to resources for reproduction is found in many taxa, including acorn woodpeckers, but is particularly prevalent in certain insects. In the group-breeding acorn woodpecker there is intense competition among breeding females to have their eggs incubated rather than those of their co-breeders. In groups containing joint-nesting females, the first female to lay eggs invariably has several removed from the nest by her co-breeders. These eggs are then eaten by any or all of the group. Removing the eggs provides additional room in the nest for other eggs laid by the co-breeders. Cannibalism occurs in a similar context in facultatively-eusocial carpenter and euglossine bees. In these taxa, one founding female will cannibalize the eggs of a co-founding female, thereby gaining access to oviposition sites, and ultimately leading to her reproductive dominance (see Kukuk). Cannibalism as a means of establishing reproductive dominance is particularly prevalent in the social insects, and usually reflects the conflicts of interest between queen and workers, and among the workers themselves.

The above arguments are frequently used to explain infanticidal behaviour (see Hausfater and Hrdy 1984), which is associated with cannibalism in many species. However, explanations of infanticide and cannibalism are not necessarily equivalent, and the two behaviours may have evolved for different reasons. For example, selection may have favoured cannibalism of young, and therefore infanticide is a necessary prerequisite. However, even if selection has favoured infanticide, cannibalism is not necessarily expected because it represents a subsequent option for the killer. For example, in addition to the advantages to be gained from infanticide for female prairie dogs, infanticidal females may also benefit nutritionally by subsequent cannibalism (Hoogland *et al.* 1989; see also Elwood). In some species, the option of cannibalism after infanticide may not be available, as a result of morphological or physiological constraints.

The costs of cannibalism

There are costs of cannibalism. The most obvious is that an individual which attempts to kill and consume an individual of roughly the same size and predatory ability, runs a high risk of being prey to its potential victim. This problem is obviously relevant to foraging theory that incorporates the risk of predation. It is not surprising, therefore, that cannibalism involves an asymmetry, with victim being more vulvernable to attack than the cannibal. Risk of infection from parasites, viruses, and other diseases may also represent a potentially important cost of cannibalism, although it is one that has been largely ignored for non-human animals. The opportunity for infection, particularly by host-specific parasites that inhabit the digestive system, would be greatly enhanced by cannibalism.

Possibly the most intriguing cost is the potential loss of fitness when a cannibal kills and consumes a genetically related individual. Clearly, this potential loss of fitness will depend upon the degree of relatedness between cannibal and victim. The importance of the genetic relationship between cannibal and victim has led to the distinction between heterocannibalism, where the two are assumed to be completely unrelated; sibling cannibalism, between offspring that share a parent; and filial or parental cannibalism, between parent and offspring. Although it is convenient to distinguish simply between kin and non-kin cannibalism, the dichotomy may be misleading for two reasons. First, genetic relatedness is a continuous variable, so the point at which two individuals of low relatedness are regarded as kin is more or less arbitrary. Second, the genetic relationship between kin can be low or quite ambiguous under certain mating systems. This is particularly apparent in polygynous mating systems, but even in monogamous mating systems the relatedness between mother and offspring, or even between offspring, can still be ambiguous if egg dumping or brood parasitism occurs. Thus, the costs, in terms of loss of inclusive fitness, of killing and consuming a member of a brood will differ according to who is cannibal and who is victim. This may result in interesting conflicts of interests between the victim, its mother, apparent father, and their offspring.

The potential loss of inclusive fitness as a result of cannibalizing related individuals suggests that cannibalism would be directed away from related individuals in circumstances where it is principally a mechanism of acquiring food resources. The evidence for this is equivocal. Individuals of several taxa apparently avoid killing and eating kin, these include fish (see FitzGerald and Whoriskey), rodents (see Elwood), gerrid waterstriders (Nummelin 1989), and ascidian larvae (Young 1988). In contrast, the available evidence indicates that cannibalistic juvenile gastropods (see Baur) and amphibians (see Crump) do not discriminate between related and unrelated victims. Wild-type amoebae of the predatory slime mould exhibit self/non-self recognition, but it is not known whether ciliates in general are able to distinguish between prey ciliates and cells of their own species (see Waddell). The degree to which cannibalism in social insects and beetles is directed away from kin is not known and certainly requires detailed investigation.

Cannibalism may also incur a cost in those species that benefit from social behaviours. For example, there is a positive correlation between larval group size and survivorship in the willow leaf beetle (*Plagiodera versicolora*), possibly as a result of coordinated feeding, group defensive displays and synchronous moulting (see Stevens). Larval cannibalism occurs in this species, accounting for nearly 50 per cent of the mortality of the hatching individuals. Such a group-related cost may diminish the survival and nutritional benefits that accrue to the cannibalistic individuals. Although larval willow leaf beetle groups usually consist of individuals from a clutch laid by one female and are therefore siblings or half-siblings, this group-related cost could apply equally to other species which form aggregations of non-related individuals.

An important cost for pre-mating sexually cannibalistic females is the risk of remaining virgin (see Elgar). Such a cost implies two predictions: females may be less likely to attempt to cannibalize a courting male if their expected encounter rate with males is low, and females that are still virgin toward the end of the mating season may be less cannibalistic than virgin females at the start of the mating season.

CONSEQUENCES OF CANNIBALISM

Whatever the selective forces responsible for the evolution of cannibalism, it has important life history consequences. It is responsible for a variety of morphological and behavioural characteristics and has broad implications for the population dynamics of many cannibalistic species.

Cannibalistic polyphenisms

Cannibals of several species from quite different taxa are phenotypically distinct from non-cannibals and represent cannibalistic polyphenisms. The cannibal morphs of slime moulds are very much larger than their non-cannibalistic counterparts, and are usually referred to as 'giant cells'. Cannibalistic polyphenisms are widespread among anurans, where certain larvae develop distinctive morphologies such as enlarged heads with wide mouths and elongate vomerine teeth. Behavioural cannibalistic polyphenisms are well known in birds, particularly in gulls and skuas. In these species, some individuals are specialist cannibals and may account for most of the cannibalistic attacks within colonies.

The degree to which these cannibalistic polyphenisms are determined by genetical or environmental factors is largely unknown. In lower eukaryotes, the formation of giant cells occurs under conditions of starvation, and environmental factors have an important influence on the formation of cannibal morphs in anurans. Dietary and other environmental factors may influence the extent of cannibalistic specialization in birds, although there is some evidence of a genetical basis in some species. There is clearly considerable scope for further studies in this area, particularly where polyphenisms appear to result from predictable variation in factors causing cannibalism. Such systems will provide some of the best models for studying how natural selection influences cannibalistic behaviour.

Behavioural

Cannibalism is a particularly anti-social form of behaviour. It is perhaps not surprising that the relatively small number of species of social spiders is frequently explained by the comparatively widespread occurrence of cannibalism among spiders (e.g. Buskirk 1981; Rypstra 1986). The argument is one of irreversible evolution (Bull and Charnov 1985); it may be impossible for gregarious behaviour to evolve within a population in which cannibalism between adults or juveniles is common. Comparative studies of ascidian larvae lend some support to this idea; gregarious species reject their own eggs and larvae as food, but non-gregarious species will consume conspecific eggs and larvae (Young 1988).

Although cannibalism between adult social spiders is rare, cannibalism of eggs is not, which may reflect the conflicts of interest between individuals in an apparently cooperative group. This conflict of interest, manifested by oophagy, may have been an important factor in the evolution of subsocial behaviour. This is very clearly demonstrated in the behaviour of foundress carpenter and euglossine bees in which subordinate females are apparently manipulated by egg cannibalism into helping care for the offspring of the dominant female.

In some species, sexual cannibalism may have been responsible for the evolution of certain characteristics of courtship behaviour. It is sometimes argued that courtship behaviour in cannibalistic species allows the male to alert the female to his presence, thereby indicating to her that he is a potential partner rather than a prey item. This argument supposes that females are unable to distinguish between males and prey items, which seems unlikely. Nevertheless, there is some evidence that certain components of courtship behaviour are effective in reducing the risk of cannibalism. Other comparative evidence for orb-weaving spiders suggests that sexual cannibalism has been influential in the location of courtship, which has led to differences in the degree of sexual dimorphism.

Cannibalism may have been a contributing selective force in the evolution of parental behaviour. For example, herring gulls spend considerable time defending chicks from cannibalistic attacks by other gulls, male sticklebacks defend their nests from raiding females, and certain parental defensive behaviours in rodents are directed toward potential cannibals. However, in all of these cases, the offspring are also potential prey to other predators. Thus, this behaviour may be a response to the threat of predation generally rather than cannibalism specifically. For some taxa, it may be impossible to establish the relative importance of cannibals compared with other predators in favouring parental behaviour. Nevertheless, cannibalism is likely to have been an important selective force in the evolution of parental care in certain subsocial insects in which cannibalism is a major source of juvenile mortality.

Demographic and life history

Numerous studies have shown that cannibalism plays an important role in affecting population dynamics (e.g. Fox 1975; Polis 1981). However, it is not clear whether cannibalism is always density dependent, or whether it generally stabilizes or destabilizes populations. Theoretical studies that attempt to model the effects of cannibalism on population dynamics have focused on population-state variables, such as mortality, fertility, and migration rates. However, Dong and Polis argue that these models have limited value, principally because individuals differ within a population. Instead, they advocate the use of models that examine individual strategies of cannibalism, and then examine the population consequences of these strategies. Such a procedure seems sensible, because the most common feature of cannibalistic behaviour is that it represents the outcome of a basically asymmetric contest.

An aspect of the biology of cannibalism that seems to have been largely neglected is its consequences for life-history variation. This may be particularly important for cannibalism among the juvenile stages, for example in amphibians, gastropods and certain insects. In particular, selection may favour shorter developmental times if smaller individuals are more likely to fall victim to larger individuals.

CONCLUSION

Despite its widespread taxonomic distribution, and the diverse contexts in which it occurs, the biology of cannibalism still remains a relatively neglected topic in evolutionary biology. In some species, cannibalism is unlikely to represent more than opportunistic carnivory. In other species, particularly those in which there are cannibalistic polyphenisms, cannibalism is clearly of evolutionary significance. In most contexts, cannibalism is a clear manifestation of the conflicts of interest between individuals, and consequently may be a particularly useful model behaviour for studies of social species.

The chapters in this volume have a common theme, that of attempting to explain why cannibalism occurs, and in describing the consequences of cannibalism. The presence and frequency of cannibalism in various taxa clearly depends on diverse morphological and behavioural factors, combined with the opportunity for cannibalism. Specifying these factors and the environmental circumstances favouring cannibalism may lead to a general theory predicting the distribution, forms and consequences of cannibalistic behaviour. Perhaps the lack of an obvious general explanation at this time should not be surprising given the variety of contexts in which cannibalism occurs, and that unlike predation, it represents more than just the acquisition of food resources. Nevertheless, in the process of answering these questions, the chapters have revealed numerous avenues for future research, both in attempting to explain why cannibalism occurs and by using it as an obvious manifestation of conflicts of interest. Perhaps the most promising approach to answering these questions is by comparing clades and species, where the putative causes and effects of cannibalism vary.

ACKNOWLEDGEMENTS

We thank Theo Evans, Sally Troy and Nina Wedell for their comments.

REFERENCES

Abler, T. S. (1988). A mythical myth: comments on Sanday's explanation of Iroquoian cannibalism. *American Anthropologist*, **90**, 967–9.

Alexander, R.D. (1974). The evolution of social behaviour. *Annual Review of Ecology and Systematics*, **5**, 325–83.

Arens, W. (1979). *The man-eating myth: anthropology and anthropophagy*. Oxford University Press.

Armstrong, J. T. (1964). The population dynamics of the planarian, *Dugesia tigrina*. *Ecology*, **45**, 361–5.

Banks, C. J. (1968). Cannibalism and predation by aphids. *Nature*, **218**, 491.

Bull, J. J. and Charnov, E. L. (1985). On irreversible evolution. *Evolution*, **39**, 1149–55.

Buskirk, R. E. (1981). Sociality in the arachnida. In *Social insects*, Vol II (ed. H. R. Hermann), pp. 281–367. Academic Press, London.

Cottrell, C. B. (1984). Aphytophagy in butterflies: its relationship to myrmecophily. *Zoological Journal of the Linnean Society*, **79**, 1–57.

Crowley, P. H., Dillon, P. M., Johnson, D. M. and Watson, C. N. (1987). Intraspecific interference among larvae in a semivoltine dragonfly population. *Oecologia*, **71**, 447–56.

de Montellano, B. R. O. (1978). Aztec cannibalism: an economic necessity? *Science*, **200**, 611–17.

Duelli, P. (1981). Is larval cannibalism in lacewings (*Chrysopa carnea*) adaptive (Neuroptera: Chrysopidae)? *Research on Population Biology*, **23**, 193–209.

Elgar, M. A. (1990). Evolutionary compromise between a few large and many small eggs: comparative evidence in teleost fish. *Oikos*, 59, 283–7.

Fox, L. R. (1975). Cannibalism in natural populations. *Annual Review of Ecology and Systematics*, **6**, 87–106.

Gilbert, J. J. (1973). The adaptive significance of polymorphism in the rotifer *Asplanchna*. Humps in males and females. *Oecologia*, **13**, 135–46.

Harner, M. (1977). The ecological basis for Aztec sacrifice. *American Ethnologist*, **4**, 117–35.

Harris, M. (1977). *Of cannibals and kings*. Random House, New York.

Hausfater, G. and Hrdy, S. B. (1984). *Infanticide in animals and man: comparative and evolutionary perspectives*. Aldine Press, New York.

Hoogland, J. L., Tamarin, R. H. and Levy, C. K. (1989). Communal nursing in prairie dogs. *Behavioral Ecology and Sociobiology*, **24**, 91–95.

Houston, A.I., Clarke, C.W., McNamara, J.M. and Mangel, M. (1988). Dynamic models in behavioural and evolutionary ecology. *Nature*, **332**, 29–34.

Jasienski, M. and Jasienska G. (1988). Parental correlates of the variability in body mass at hatching in *Arctia caja* larvae (Lepidoptera: Arctiidae). *Entomological Genetics*, **13**, 87–93.

Kurihara, Y. and Okamoto, K. (1987). Cannibalism in a grapsid crab, *Hemigrapsus penicillatus*. *Marine Ecology Progress Series*, **41**, 123–7.

Lipinski, M. R. and Linkowski, T. B. (1988). Food of the squid *Ommastrephes bartramii* (Lesueur, 1821) from the Southwest Atlantic Ocean. *South African Journal of Marine Science*, **6**, 43–6.

Mecom, J. O. (1972). Feeding habits of Trichoptera in a mountain stream. *Oikos*, **23**, 401–7.

Newman, J. A. and Elgar, M. A. (1992). Sexual cannibalism in orb-weaving spiders: an economic model. *American Naturalist*. (In press).

Nummelin, M. (1989). Cannibalism in waterstriders (Heteroptera: Gerridae): is there kin recognition? *Oikos*, **56**, 87–90.

Orr, B., Murdoch, W., and Bence, J. (1990). Population regulation, convergence and cannibalism in *Notonecta* (Hemiptera). *Ecology*, **71**, 68–82.

Polis, G. (1981). The evolution and dynamics of intraspecific predation. *Annual Review of Ecology and Systematics*, **12**, 225–51.

Polis, G. and Myers, C. A. (1985). A survey of intraspecific predation among reptiles and amphibians. *Journal of Herpetology*, **19**, 99–107.

Rinkevich, B. and Weissman, I. L. (1987). The fate of *Botryllus* (Ascidiacea) larvae cosettled with parental colonies: beneficial or deleterious consequences. *Biological Bulletin*, **173**, 474–88.

Rypstra, A. L. (1986). High prey abundance and a reduction in cannibalism: the first step to sociality in spiders (Arachnida). *Journal of Arachnology*, **14**, 193–200.

Sanday, P. R. (1986). *Divine hunger: cannibalism as a cultural system*. Cambridge University Press.

Villa, P., Bouville, C., Courtin, J., Helmer, D., Mahieu, E., Shipman, P., Belluomini, G. and Branca, M. (1986). Cannibalism in the Neolithic. *Science*, **233**, 431–7.

Wourms, J. (1977). Reproduction and development in chondrichthyan fishes. *American Zoologist*, **17**, 379–410.

Young, C. M. (1988). Ascidian cannibalism correlates with larval behavior and adult distribution. *Journal of Experimental Marine Biology and Ecology*, **117**, 9–26.

2

The dynamics of cannibalistic populations: a foraging perspective

Quan Dong and Gary A. Polis

For many species, cannibalism can strongly affect population dynamics and mortality rates. However, it is less clear if and how it can stablize population size. If cannibalism is density-dependent, as has been shown for over 40 studies (see Fox 1975; Polis 1981 for reviews), it can potentially regulate population size (Polis 1981; Jude *et al.* 1987; Saila *et al.* 1987). However, other models (Ricker 1962; Hastings and Constantino 1987) and empirical studies (Radovich 1962; Mertz 1969; Mertz 1972; Barber 1979*a*; Barber 1979*b*; DeAngelis *et al.* 1979; Higgins 1990; Orr *et al.* 1990) show that cannibalism may also destabilize populations, often by causing cycles.

Population dynamics are explicitly related to individual behaviour and vice versa (Hassell and May 1985; Koehl 1989). The foraging biology of individuals directly affects population dynamics because it connects the characteristics of food resources with the survival and reproduction of individuals. Thus, the use of foraging theory in population studies can provide great insight into what were previously considered separate areas (Schoener 1987). This chapter examines the dynamics of cannibalistic populations in terms of the foraging biology of individuals, and outlines the ways by which the application of foraging theory can contribute to our understanding of cannibalism.

The study of foraging biology has accumulated a series of theoretically based and empirically founded 'rules' that predict, with fair accuracy, 'decisions' made by foragers faced with a variety of prey, foraging habitats and hunting tactics. This work is embodied within the framework of 'optimal foraging theory' (Pyke 1984; Stephens and Krebs 1986). Since cannibalism can be regarded as a special type of foraging behaviour, this theory can be used to analyse and predict cannibalistic behaviour, including all the tactics and decisions of the cannibals, and those used by potential victims to avoid cannibalism.

The major interests of this chapter are to demonstrate how foraging models are heuristic to cannibalistic foraging and how studies of cannibalistic foraging contribute to understanding population dynamics. In particular, the following questions are addressed: what aspects of foraging theory can be applied to cannibalism; what are the strengths and limitations of this approach; what type of future research can feasibly integrate cannibalistic foraging and dynamics; and which types of models are most applicable for an understanding of the effect of cannibalistic foraging on population dynamics?

This chapter is divided into three sections. The first analyses the major factors correlated with, and vital to understanding, cannibalism at the individual and population levels. The second interrelates foraging theory and cannibalism to make specific predictions about the conditions that promote cannibalistic foraging. The final section outlines how to model the population level effects of cannibalism. Individual-based models are suggested as an appropriate and robust approach of cannibalistic foraging and its dynamical consequences.

The focus here is only on cannibalistic foraging and excludes other types of cannibalism such as filial or sexual cannibalism, or when cannibalism represents an extreme form of interference (see Polis 1981). Nevertheless, these types of cannibalism can be explained by foraging theory. For example, Newman and Elgar (1992) and Sargent (this volume) model mate cannibalism and filial cannibalism, respectively, as a form of foraging decisions, using stochastic dynamic programming.

MAJOR FACTORS AFFECTING CANNIBALISM

Three factors are often associated with cannibalism: prey 'vulnerability' (an individual feature), density (a population property), and the quantity and quality of food items (environmental conditions). The effects of these factors are delineated, and their analysis discussed in the following section.

Vulnerability of prey

Whether an encounter between conspecifics leads to cannibalism depends on the decision of a predator to attack, and if the attack is successful. Decision and success are each functions of the vulnerability of the potential prey individual. By definition, a more vulnerable individual is one that is more likely to be a victim. Thus, the probability and frequency of cannibalism may be strongly related to relative vulnerability.

Prey vulnerability is defined relative to the predator's characteristics, and might be represented by the size ratio of predator to prey. Vulnerability is a function of age, developmental stage, size, morphology, and even hunger. For example, cannibalism is often restricted to predation on certain relatively defenceless life stages (Polis 1981; Elgar and Crepsi, this volume). A predator estimates the vulnerability of a potential prey, compares this estimate to set criteria and decides whether to attack or not. Increases in vulnerability and lowering of the criteria each increase the tendency to cannibalize. Attack criteria may be determined by internal drive, such as hunger.

A basic rule of conventional foraging models shows how cannibalism is affected by changes in vulnerability. Consumers are thought to select a food item if the benefit (*B*) to cost (*C*) ratio (B/C_p; where *p* denotes the foraging cost for prey) is higher than a threshold value. The threshold is usually the average optimal *B*/*C* for the environment (B/C_e, *e* denotes environment; see below for details). As the vulnerability of a prey increases, B/C_p increases if gross costs decrease while gross energetic benefit remains the same. Costs decrease because the predator spends less time and energy in pursuit and handling. Therefore, prey vulnerability increases the propensity of the predator to cannibalize when B/C_c (where *c* denotes the foraging cost for conspecifics) surpasses B/C_e, thus admitting conspecifics into the selected diet.

Size

Size difference is the most frequently cited factor determining vulnerability. One feature of cannibalism, nearly universal among all taxa, is that larger individuals eat smaller ones (Polis 1981, 1988). A common assumption of foraging models and studies is that prey size is a good index of food value (Werner and Hall 1974; Goss-Custard 1977; Bailey and Polis 1987). Moreover, size is even thought to be a direct index used by predators to make decisions to include a potential prey in the diet (e.g. O'Brien *et al.* 1976).

Cannibalism occurs when the relative size of the potential prey falls within the acceptable range of the predator. The acceptable size range is established by the B/C_p relative to alternative foods; an acceptable size is such that B/C_c is greater than B/C_e. Conspecifics may be absent from the diet because they are too small or too large. Presumably, very small prey are not eaten because energetic benefits are insufficient relative to the costs of pursuit and handling, and net energetic benefits (*B*) are low relative to the average net benefit from optimal foraging. Large items may not be suitable because, as prey size increases, the relative increase in foraging costs (*C*) becomes too great relative to the increase in benefits. The ability of a prey to avoid predators increases with size (Folkvord and Hunter 1986). In each case, *B*/*C* is less than B/C_e. Large items also may be excluded because the cannibal cannot physically handle them or the threat of retaliation is too great (e.g. Polis 1980; Crowley 1984).

Developmental stage

Individuals become vulnerable to cannibalism during particular stages or behaviours. Eggs, embryos, neonates, and nestlings are often defenceless and frequently cannibalized (Polis 1981; Polis 1984; Hausfater and Hrdy 1984; Elgar and Crespi, this volume). Transition stages are particularly vulnerable, for example in moulting or callow insects and crustacea. Metamorphs become vulnerable to conspecifics of all sizes. In fact, size reversals occur when larger metamorphs fall prey to smaller cannibals, for example smaller *Tribolium* larvae eat quiescent pupae (Stevens, this volume) and *Scaphiopus* tadpoles eat emerging toadlets (Crump, this volume).

Hunger

Hunger is often reported as a primary factor affecting cannibalism (Fox 1975;

Polis 1981). However, it is rarely considered explicitly in conventional foraging models (Stephens and Krebs 1986; but see Schoener 1971; Rapport and Turner 1977; Richards 1983), but is included in dynamic programming models (Mangel and Clark 1988). Hunger is an internal state that affects behaviour, vulnerability or voracity in many ways, thus promoting cannibalism at several time scales. In short periods, food stress generally increases foraging activity and movement (Hassell and Southwood 1978; Polis 1980, 1981; Richards 1983) and may be the basic motor drive in foraging behaviour (McFarland and Houston 1981). Foraging activity increases because hunger triggers search behaviour, lowers attack threshold, and increases reaction distance and foraging time. Increased foraging activity may promote cannibalism by increasing risk through greater exposure, higher encounter and interaction rates among conspecifics, and by causing animals to forage in places (or times) with higher probability of predation and cannibalism (Gilliam and Fraser 1987; Sih 1987).

Over longer periods, failure to find food after the initial burst of energetically expensive foraging activity, further weakens individuals and makes them more vulnerable to cannibalism (Folkvord and Hunter 1986). Hunger over even longer periods can stunt growth and possibly increase the variance in body size within a population, a feature that promotes cannibalism.

Hunger also affects the propensity to become cannibalistic. One of the well substantiated predictions of foraging theory is that consumers expand their diets beyond the normal limit of acceptable prey during periods of hunger or low food availability (Bailey and Polis 1987; Stephens and Krebs 1986). This means attack thresholds (the decision to attack or not) are lowered, thus admitting previously ignored items (including conspecifics) into the diet. Lowering thresholds may also increase the probability of retaliation: if a consumer takes too big a risk, putative prey can retaliate and become the cannibal.

How is hunger operationally evaluated? In the short-term, gut content is a direct measure; in the long-term, an index of body fat can be used. Hunger is often equated with low availability of food. However, hunger is an internal state whereas food availability is an environmental state. The relationship between food availability and hunger is explicitly discussed below.

In summary, size, age, developmental state, and hunger are key determinates of relative vulnerability to cannibalism, and can be used as an operational index of vulnerability. Any variation within a population in the determinates of relative vulnerability will introduce asymmetries and consequently increase the potential for cannibalism. The magnitude of variation among individuals can change rapidly and so change the frequency of cannibalism. For example, an individual's level of satiation and hunger can change during the course of an hour–time scales that are smaller than those most frequently used in ecological studies. Nevertheless, such variation and ephemeral events can potentially affect population dynamics. Thus, constant coefficients representing the average state and models without individual variation, may be of relatively little value or even misleading in studies of cannibalism.

Density

Over 100 studies show that the rate of cannibalism increases with density (reviewed in Fox 1975; Polis 1981). Since individuals within a population differ in vulnerability, relationships are asymmetric. Thus gross (overall) density may be a poor parameter to model or to determine rates of cannibalism. Two factors must be considered when analyzing the relationship of density to cannibalism: the measure of density and the types of density dependence present.

Measures of density

Three measures of density exist: prey density (potential victims), predator density (potential cannibals), and gross density (prey density plus predator density plus density of others). Although each measure is reported to affect cannibalism (it usually increases with increases in each), each relates differently to the frequency and dynamics of cannibalism.

A few studies report that cannibalism increases with prey density, predator density, or both (e.g. Landry 1978; Hunter and Kimbrell 1980; Polis 1980; Meffe 1984; Crowley *et al.* 1987; Orr *et al.* 1990). Usually in such cases, prey and predators are easily distinguished and interactions can be described by classic predator–prey theory, especially if cannibalism primarily occurs among members of different age classes or cohorts (inter-cohort cannibalism). In such cases, the rate that a particular prey is eaten is a positive function of its density, given the assumptions of foraging models that conspecifics are highly preferred prey and encounter rates are random. Here, mortality may be described in terms of the functional response of the predator to the density of conspecific prey (see below).

The frequency of cannibalism is also a function of predator density. However, this function is not always linear. Group effects among predators can be both negative (mutual interference) and positive (mutual enhancement), and may alter both predator success (Hassell and Southwood 1978; Krebs and Davies 1987) and cannibalism rates (see Stevens, this volume). Polis (1981) presents several examples of cannibalism that are possible only because groups of conspecifics act in concert.

Gross density is often the only measure recorded, especially when it is impossible or difficult to separate densities of predator and prey. Distinguishing between predator and prey densities is particularly difficult for intra-cohort cannibalism, where the distinctions may be subtle, depending on the individual's size or vigour, and the probability of being a cannibal or victim may change rapidly. Similarly, gross density may be the only possible measure for inter-cohort cannibalism if either the time to reach other life stages is highly variable, or development rate is rapid during the observation period. Such changes increase the probability and variability of transition from potential prey to predators and vice versa. Consequently, as with traditional population models, gross density may be the only available parameter.

Clearly, it is important to differentiate between the components of a population into predators, prey, and non-interactants. Such a division may correlate with other characteristics of the population, such as age and size structure. Cannibalism, like other forms of predator-prey relationships, may be influenced by population structure. For example, Polis (1981) and Crowley (1984) suggested that age structure is a key feature affecting cannibalism and subsequent numerical dominance by particular age classes in a diversity of taxa. Both age and size structure effects of cannibalism have been examined empirically in scorpions (Polis 1988), odonates (Crowley 1984; Crowley *et al.* 1987), *Notonecta* (Orr *et al.* 1990), *Tribolium* (Mertz 1969, 1972; Higgins 1990; Stevens, this volume), amphibians (Wilbur 1988; Crump, this volume), and fish (DeAngelis *et al.* 1979; FitzGerald and Whoriskey, this volume).

Patterns of density dependence

Which types of density dependence influence the dynamics of cannibalism? Cannibalistic predators show density dependent responses to increases in potential prey by eating more individual prey, a functional response, or by increasing in number, a numerical response. These responses are discussed below.

The functional response of a predator describes the relationship between the number of prey consumed per predator and prey density. Cannibalism has been analysed as a functional response (e.g. Valdes *et al.* 1987; van den Bosch *et al.* 1988; Crowley and Martin 1989), and the numerous observations that density is frequently correlated with the rate of cannibalism suggest that a cannibalistic functional response is operating.

Three points are worth noting when applying functional response analyses to cannibalism. First, the portions of the population which are potential victims must be determined, since this is not always clear. Second, potential conspecific prey should be more profitable than the threshold for food selection. The relationship between the functional response and conventional foraging models is such that attack rates and handling times may vary, if the profitability is low (Abrams 1990). Thus, the number of prey eaten per unit time may vary with density and the type of alternative prey, in complicated patterns. The caveat is that foraging parameters cannot always be assumed constant or represented as an average. Finally, the functional response can change after predators eat conspecific prey so that more conspecifics are eaten. This occurs because cannibals receive nutritional benefits allowing growth, increasing the predator to prey size ratio (a developmental response, Murdoch 1971; also see DeAngelis *et al.* 1979); or cannibals decrease handling time (e.g. by learning; Braid 1981) thus converting foraging into a Type III (sigmoid) functional response.

The numerical response of the predator describes the relationship between the number of predators and prey density. Numerical response occurs from recruitment via birth and migration. However, a cannibalistic numerical response is quite different from the classical numerical response of a predator to heterospecific prey. The number of predators does not increase via recruitment. Instead, energy gained by cannibalism may act to increase the number of possible prey (newborn recruits) for the current class of conspecific prey. In general, the number of conspecific predators can increase only from migration; an exception might be slime moulds (Waddell, this volume). A numerical response via migration by cannibals has not been reported, although this probably reflects the operational problems of detecting it. Perhaps the appearance of large numbers of vulnerable conspecifics (eggs or neonates) would, in mobile predators, stimulate aggregations of cannibals.

Density-dependence as a function of type of density

The exact manner in which the rate of cannibalism changes with density depends on several factors. If conspecific predators and prey are clearly distinct, such as in inter-cohort cannibalism, then density-dependence can generally be similar to this classical form of interspecific predation (e.g. Holling 1966; Murdoch 1979; May 1981; Hassell 1981). If conspecific predators and prey are not distinct and gross density is used, response to density and the subsequent patterns of mortality from cannibalism can be complicated. This is especially true for functional and numerical responses; the rate of cannibalism is a function of the number of vulnerable and voracious individuals (potential prey and predators), not the total number of conspecifics. Discrimination of predators and prey is important in modelling cannibalism (e.g. Bobisud 1976; DeAngelis *et al.* 1979) and understanding fully the dynamics of many age and size structured populations (e.g. copepods, scorpions, insects, amphibians; Polis 1981, 1988; Crowley *et al.* 1987; Orr *et al.* 1990; Stephens, this volume). The analysis of density dependence using gross density has not been widely discussed, but a number of speculations are provided below.

Food

The quantity and quality of food is the most frequent environmental factor correlated with the rate of cannibalism. There are numerous examples showing that cannibalism increases as either the abundance or nutritional quality of alternative food decreases (Fox 1975; Polis 1981). Food is likely to affect cannibalism via its effects on hunger or malnutrition of individuals. The mechanisms that increase cannibalism by hunger were described above. An analysis of food selection, a central theme of foraging studies, sheds light on these mechanisms. If the quantity or quality of food is the single sufficient condition favouring cannibalism, then selection of conspecifics as prey can be regarded as a simple foraging problem.

The relationship between food and hunger is probabilistic, not deterministic. Local food availability can vary greatly among individuals in the same population, especially when food is scarce. Small physiological differences, or even chance events, can generate large differences in food acquisition which, in turn, affect hunger, vulnerability, and the variability in physiological state among individuals (e.g. size, strength). For example, variable food availability caused large variations in feeding and growth rate in individual fish (Rubenstein 1981). Increased variability in physiological states often promotes cannibalism. The general effects that food availability exert on variability in physiological states (e.g. hunger) pose problems for conventional models of foraging and cannibalism.

Interaction of density, food, and cannibalism

Food and density interact to change the relationship of each to the rate of cannibalism. Three factors potentially complicate this interaction:

(1) the presence or absence of density-dependent exploitation competition for food;

(2) the degree that prey and predator components of the population overlap in food use; and

(3) the predator–prey structure in the cannibalistic population.

The abundance of suitable food may or may not be density dependent (i.e. food levels drop when density rises). Density dependent processes, such as exploitation competition and cannibalism, only occur when increasing density depletes food to the point that per capita food level is below individual needs (Polis and McCormick 1987). How frequently such density dependence and subsequent scramble competition occur in nature is subject to much debate (Strong *et al.* 1984). An emerging consensus is that exploitation competition occurs in many populations, but only sporadically (Wiens 1977; Schoener 1983).

By definition, the interaction between food and density is unimportant if food availability is density-independent. If food is density-dependent, an increasing density of consumers decreases food supply and so increases the rate of density-dependent behaviours such as migration, interference and cannibalism. In other words, the interaction changes the relationship of both food and density to the rate of cannibalism. For example, two populations at equal densities may experience different rates of cannibalism as a function of food abundance; populations facing food shortages will exhibit higher rates. Conversely, populations with the same food supply but different densities may differ in cannibalism rates.

The second factor adding complexity to the system is the degree that the prey and predator components overlap in food use. Conspecific predators and prey can differ, sometimes substantially, in their diet, usually because they are different sizes (Polis 1984; Crowley *et al.* 1987; Polis 1988; Orr *et al.* 1990). If diet overlap is low, higher densities of prey conspecifics increase the food available for the predator component of the population, either directly when conspecifics are food resources, or indirectly, when predators eat conspecifics that feed on resources inaccessible to or not utilized by the predators. In the latter, cannibalistic predators indirectly incorporate previously unavailable resources. Polis (1981) shows that such phenomena are widespread, especially in age and size structured populations.

If prey are limiting, then this process may actually allow a higher density of predators in the cannibalistic population than would be supported by heterospecific food alone. Thus, carrying capacities in terms of food can be quite different for predators and prey.

If diet overlaps more completely, three feedback loops may exist between cannibalism, density, and food:

(1) as conspecifics are eaten, the population of intraspecific competitors declines and per capita food level increases;

(2) if density increases and food is limiting and depletable, then both per capita food and growth decrease, producing smaller and weaker individuals that are more vulnerable to cannibalism; and

(3) as cannibals become satiated, a decrease occurs in both feeding (effectively increasing per capita food) and those activities related to food stress that predispose individuals to cannibalism.

The third factor complicating the interaction between food and density is the density distribution among predators and prey. Populations with the same gross density but different predator and prey densities are likely to differ in the quantity and types of food used and exhibit different patterns of density-dependence for each food type. Again, gross density may be of little value, since it conveys little information about per capita food availability.

In summary, an understanding of the interactions between density, food, and cannibalism, requires that the densities of predators and prey, patterns of density-dependence for each food resource, and the degree to which conspecific prey use different resources compared to their predators, are specified.

FORAGING MODELS

Foraging studies focus on how, where, and what animals eat. A principal axiom of foraging theory is that natural selection has optimized foraging behaviour, and the goal of the theory is to predict the foraging decisions of individuals. Such predictions can be made using optimization models, which Stephens and Krebs (1986) argue is the most explicit and powerful approach. Foraging theory is likely to provide insights into cannibalism because it is inevitably a special case of foraging, and consequently relates to food, density, and vulnerability. The following, outlines two general models that are relevant to cannibalism, and discusses the limitations and applications of these models to cannibalism.

Foraging models consist of three components:

(1) currency;

(2) choices for decisions; and

(3) constraints (Schoener 1971; Stephens and Krebs 1986).

Currency describes design functions in an explicit way, while constraints reflect mechanisms controlling behaviour. Exact currencies are diverse but often limited to energy, nutrition, and time. Researchers usually formulate and test alternate hypotheses by asking which currencies or constraints best explain behaviour. Here, for convenience, we use energy as the currency. Nevertheless, energy is not the only or even the most important currency; others can be used (e.g. see Houston *et al.* 1988).

The most common questions in foraging theory involve search–encounter decisions. The theory assumes that consumers must decide whether to search, remain in a habitat, or attack a particular prey. Many models predict optimal rules for foragers facing a search–encounter decision, and they have been the most frequently tested (Stephens and Krebs 1986). These models are applicable to cannibalism, especially because a frequent question is, why are some individuals cannibalistic rather than feeding on other food items.

The following questions are addressed for each model that is applied to cannibalism. First is the model logically consistent with what is known about cannibalism? Second are the predictions of the model consistent with the available empirical data? Two kinds of models are examined: diet selection models (Schoener 1971; Charnov 1976*a*), which exhibit a simple structure with explicitly stated and testable predictions (Pyke 1984; Schoener 1987), and patch selection models (Charnov 1976*b*) that incorporate the conflicting demands of energy intake and risk of predation (e.g. Werner and Gilliam 1984; Gilliam 1990).

Diet selection and cannibalism

Diet selection models assume that foragers rank food items according to profitability, the net energetic value of an item is a function of the time needed to find, capture, and process that item. This is denoted E_i/T_i, where E_i is energy and T_i is time for food item *i*. A theorem is derived such that the optimal diet set of food types which maximize the net rate of gain should be those in descending rank, such that

$$E_i / T_i > E/T \qquad (2.1)$$

where E/T is the expected net gain from the environment from the optimal diet during a feeding period ($= B/C_e$); E/T is the threshold value in the decision to select a food item during a search-encounter-attack sequence. Two important predictions follow: (1) the selection of an item for the optimal diet is independent of its abundance and depends only on its E_i/Ti value and the absolute abundances of higher ranking food types, and (2) as the abundance of relatively high ranking food types increases (i.e. B/C_e increases), lower ranked types should be dropped from the optimal diet in order of their E_i/T_i rank. These predictions can be regarded as decision rules: if good food items are sufficient, do not eat poor food items, regardless of their abundance. Empirical studies reveal that these predictions are robust and generally describe diet selection by real consumers (Pyke 1984; Stephens and Krebs 1986). Cannibalistic foraging may also conform to the above predictions. The widespread observation that the rate of cannibalism is an inverse function of the availability of alternative foods (Polis 1981) is consistent with them, provided that conspecifics normally rank as items of low profitability. We now discuss four reasons why conspecifics may be of relatively low rank as food items.

An assumption of diet selection models is that the fitness (F) increases linearly with the rate of net gain from food intake (E) (Pyke 1984). This assumption is formulated as:

$$F = a + bE \qquad (2.2)$$

where a is a constant that sets a reference value for F when E equals zero, and b is a coefficient which converts each unit change in E into a change in F. Both b and E are greater than zero, because energy or nutritional values are always positive. Thus, fitness always increases, at least with respect to the food items being eaten. However, in cannibalistic foraging, a potential decrease in fitness always exists because a cannibal may eat its offspring, siblings, or other relatives. Filial and sibling cannibalism occur, sometimes frequently (Elgar and Crepsi, this volume; see also Polis 1981, 1984; Hausfater and Hrdy 1984). The implication is that cannibalism may be less frequent in some populations because there is a high probability that the victim is a relative.

Under some circumstances, the above fitness function may not apply to cannibalism. The following formulation of equation (2.2) includes the potential loss of fitness:

$$F = a + bE - rC \qquad (2.3)$$

where C is the probability of eating a relative and r is a coefficient representing the weight of C on fitness loss; r is a function of the relatedness and is always greater than or equal to zero. For heterospecific prey, $rC = 0$.

A theorem for cannibalism can now be deduced using equation (2.3). Since the net gain from food (E) can be converted into fitness (F) by coefficient b, fitness may be reconverted to E by the coefficient $1/b$ (assuming $a = 0$). Thus, we convert a fitness loss in cannibalistic foraging into an equivalent cost in terms of food gain from a victim. To adjust equation (2.1), the equivalent fitness loss (rC/b) needs to be deducted from the net gain; thus, a conspecific can only be included in the optimal diet when:

$$(E_c - rC/b)/T_c > E/T \qquad (2.4)$$

where c designates that the food item is a conspecific. Thus, a conspecific can be selected and preferred over a heterospecific food item, j, only when:

$$(E_c - rC/b)/T_c > E_j/T_j \qquad (2.5)$$

This theorm is true even if F is a nonlinear function of E; because rC is a separate component of fitness, the adjustment is simple, $F = f(E) - rC$.

Potential loss of fitness through eating relatives is one possible explanation why conspecifics are apparently ranked as low-profitability prey. There are three others.

First, the possibility of injury and even death from retaliation can be a significant cost not normally considered by foraging models. Retaliatory costs (*R*) are possibly significant where potential predator and victim are close in size and reversals may occur, such as in intra-cohort cannibalism (Polis 1980, 1981; Crowley 1984). Retaliation can also occur with other forms of predation, and its absence is a general weakness in foraging models particularly where the predator can be injured or killed by prey.

Second, conspecifics may rank low because they carry transmittable pathogens and parasites (Polis 1981). Thus, equation (2.2) must be modified further to include *R* and the mortality due to contagious pathogens and/or parasites (*P*):

$$F = a + bE - k_1R - k_2P \qquad (2.6)$$

where k_1 and k_2 are coefficients, converting the frequencies of retaliation and transmission into loss of fitness. Equations (2.3) and (2.4) then become:

$$F = a + bE - rC - k_1R - k_2P, \qquad (2.7)$$

and

$$[E_c - (rC + k_1R + k_2P)/b]/T_c > E/T. \qquad (2.8)$$

Here, the term $rC + k_1R + k_2P$ represents the potential loss of fitness during foraging, especially cannibalism, that does not exist in conventional models.

Third, conspecifics may rank low simply because they yield relatively little net energy either if they are too small, evasive, or if predators are inefficient at handling and digesting conspecifics when switching from hetero- to conspecific prey. In general, the association of cannibalism with hunger and low prey levels suggests that added costs (C, R, and P) or low net energy gain are normally important phenomena (see also Newman and Elgar 1992).

In some cases, conspecifics may be more preferred as prey than their energetic benefits might suggest, such as when cannibalism is an extreme form of interference. In these circumstances, cannibalism confers competitive benefits to the cannibal and conspecifics may still be eaten even if $E_i/T_i < E/T$ (Polis 1988). Theoretically, an examination of the relationship of E_i/T_i to E/T allows one to distinguish the selective basis of cannibalism: pure energetic benefit (if $E_i/T_i >> E/T$) or extreme competition (if $[E_c - (rC + k_1R + k_2P)/b]/T_c < E/T$, or especially, $E_iT_i < E/T$). If the E_i/T_i of conspecifics is consistently ranked low relative to the average E/T of the environment, then cannibalism may have evolved primarily as a mechanism that reduces competition. Likewise, interference is probably the mechanism in cases where conspecifics are killed but not eaten (Polis 1988). Polis *et al.* (1989) make analogous arguments for the selective basis of intraguild predation among potential interspecific competitors.

Patch selection and risk of cannibalism

The last section focused on the decision processes of potential predators. The following section investigates the decision processes of potential prey by using models of foraging and patch selection under the risk of predation.

Patch selection models initially address a central problem for foragers: when to stay and when to leave a patch containing food. These models are based on the marginal value theorem (Charnov 1976*b*) and are similar to the contingency model of diet selection detailed above. The theorem states that a rate-maximizing forager ensures that the marginal rate of energy gain in a patch is always equal to or greater than the average rate of optimal energy intake in the whole environment. Foragers choose to forage or remain within patch i only if $E_i/T_i > E/T$. Here E_i/T_i is the net gain from patch *i*; E/T is the expected net gain from the average set of patches in the environment. For cannibalistic foraging, a reduction in food may stimulate departure from patches by both potential cannibals and victims, thus increasing encounter and consequent cannibalism rates. These early patch selection models did not explicitly include the risk of predation.

Recent models recognize that foragers concurrently attempt to maximize energetic intake and avoid being eaten by a predator. A simple contingency model incorporates these conflicting demands (Werner and Gilliam 1984; Gilliam 1990; also see Gilliam and Fraser 1987). This model examines how an animal should distribute its time among different foraging sites with some risks of being killed, subject to the constraints of achieving at least some specified rate of energetic gain. The model yielded a simple result, the 'minimize μ/f' rule, which predicts that a feeding animal will adopt tactics so as to achieve the lowest ratio of mortality rate μ, the probability of death per unit time, to gross foraging rate f, the energy assimilated per unit time. This rule is more often expressed in terms of μ/g, where g is the growth rate per unit time. Foragers choose a particular patch i if $\mu_i/g_i < \mu/g$, where μ/g is the average for all patches in the habitat. This patch selection model can be expanded conceptually to include selection of time period or even prey. It was used successfully to predict habitat preferences by consumers choosing among sites differing in resource level and predation risk (e.g. Gilliam and Fraser 1987; Gilliam 1990). This model is directly applicable to cannibalism. In fact, the predators and prey in Gilliam and Fraser's (1987) study were conspecific adult and juvenile chubs (see Fraser and Cerri 1982, Sih 1982 for similar studies using cannibalistic minnows and backswimmers, respectively).

The model can be used to interpret two well known characteristics of cannibalistic populations. First, many reports (e.g. Polis 1980, 1981; Sih 1982, 1987; Schlosser 1987; Foster *et al.* 1988) show that animals in vulnerable stages (neonates, juveniles, some developmental stages) are often segregated spatially (and less often, temporally) from animals in voracious stages (usually adults). Such behaviour may be initiated to avoid cannibalistic predators and is thought to decrease cannibalism (Fox 1975; Polis 1981). Second, cannibalism increases with decreasing food levels. Low food availability in one patch may result in movement to a patch with both more food and a higher risk of predation. Potential victims may also subject themselves to a higher risk of (intra- and interspecific) predation by foraging in more dangerous times when food availability is low.

The μ/g model also predicts that juveniles of cannibalistic species should sometimes exhibit low foraging success and growth rates because they live in habitats and time periods which are sub-optimal in terms of energetic gain but exhibit a low risk of cannibalism. The studies described above show that juveniles in the presence of conspecific adults will decrease foraging activity, segregate to inferior locations, and thus lower feeding rates, which presumably reduce growth rates (Gilliam and Fraser 1987; Gilliam 1990; Sih 1982, 1987).

In summary, the rate of cannibalism appears to be consistent with simple optimal foraging models of predators encountering prey in environments of variable food quality. Encounter models (e.g. DeVita *et al.* 1982; Pyke 1984; Stephens and Krebs 1986) may be important for understanding cannibalistic foraging, since cannibalistic attack rates are proportional to the probability of encountering vulnerable individuals (Fox 1975; Polis 1981). Encounter rates may increase with density, search area, speed, and hunger. Unfortunately, our knowledge of factors affecting encounter behaviour is generally inadequate (see Persson and Diehl 1990).

MODELS OF CANNIBALISM: FROM INDIVIDUAL BEHAVIOUR TO POPULATION DYNAMICS

In many species, cannibalism is observed infrequently, especially compared with foraging on heterospecific food. Nevertheless, even rare cannibalism can greatly influence the dynamics of populations (Polis 1981). The study of low-frequency phenomena can be facilitated by either controlled experiments in which the sufficient, or at least necessary, conditions for the phenomenon are provided, or by theoretical modelling. Although models require a good understanding of mechanistic relationships, they can estimate the possible effects of important factors, even those for which it is impossible to control experimentally. The following discusses how to model cannibalism in order to reveal its effects on population dynamics.

Individual behaviours can determine the properties and dynamics of populations (Koehl 1989, Persson and Diehl 1990). This is particularly true for cannibalism and foraging. However, the population consequences of behaviour have, until recently, attracted relatively little attention (Hassell and May 1985). Recent population models that incorporate behaviour fall into two categories, according to the level of their state variables: population-state oriented models and individual-state based models. Population-state models emphasize population level processes (e.g. birth, migration, mortality) and use densities as state variables. These powerful and common forms of models often produce rather simple analytic solutions, and allow logical generalizations.

Individual-state based models retain and use individual features to describe population dynamics. The state variables in these models are characteristics of individuals, such as size, age, hunger, and life history stage. Population-state variables can also be incorporated into these models by focusing on the individual and working towards population level processes. These population-state variables sum individual-state variables and form important components of the environment within which each individual must interact. Thus, interactions between physiological and behavioural processes of individuals and population states can be explicitly simulated and manipulated.

Few models describe the population consequences of cannibalism, and most of these are population state models using age or size structures (e.g. Ricker 1962; Mertz and Davies 1968; Mertz 1969; Gabriel 1985; Gabriel and Lampert 1985; Crowley *et al.* 1987; Hastings and Constantino 1987; Valdes *et al.* 1987; van den Bosch *et al.* 1988; Higgins 1990). Individual-based models of cannibalism are rare by comparison. DeAngelis *et al.* (1979) constructed such a model to simulate the dynamics of cannibalism among bass fry. Although each type of model is designed to predict dynamics, each differs in its tools, questions, and solutions. If cannibalism can be reasonably viewed as a form of foraging behaviour, and if foraging models produce good predictions about cannibalism, individual-based models incorporating foraging rules of individuals are surely the best way to describe the population level consequences of cannibalism. The basic structure of individual-based models and their advantages and disadvantages are discussed below.

DeAngelis *et al.* (1991) developed an individual-based model to study population regulation in bass. This model delineates the basic structure of individual-based models, although it does not include cannibalism. For demonstration, a submodel is presented that simulates the dynamics from fry to one-year-olds on a daily basis (Fig. 2.1). Components in the sub-model include foraging, growth, and mortality.

Two foraging processes exist in the sub-model: search and encounter of prey, and prey selection. Prey are assumed to be randomly distributed and each individual searches a volume of space per day. Volume is determined by length of the search period, swimming speed, and distance that prey are perceived, all of which are functions of body size. Encounter rate is determined by density, the spatial distribution of prey and the predator's effective search volume. Prey are selected and consumed if:

(1) they are smaller than a set fraction of the predator's length;

(2) they are the largest edible prey available on that day; and

(3) if space exists in the predator's stomach.

Assuming size indicates the rank of net energy gain of each prey, criterion (2) follows the rules for selection of the optimal diet discussed earlier. DeAngelis *et al.* (1991) argue that other selection strategies could be modelled, including cannibalistic foraging.

A flow chart of a simple and general individual-based model for cannibalism, which is currently under construction, is shown in Fig. 2.2. In this model, individuals search a volume of space during a certain time period. Since hunger affects cannibalism, this period is often short (e.g. hours). An individual exhibits

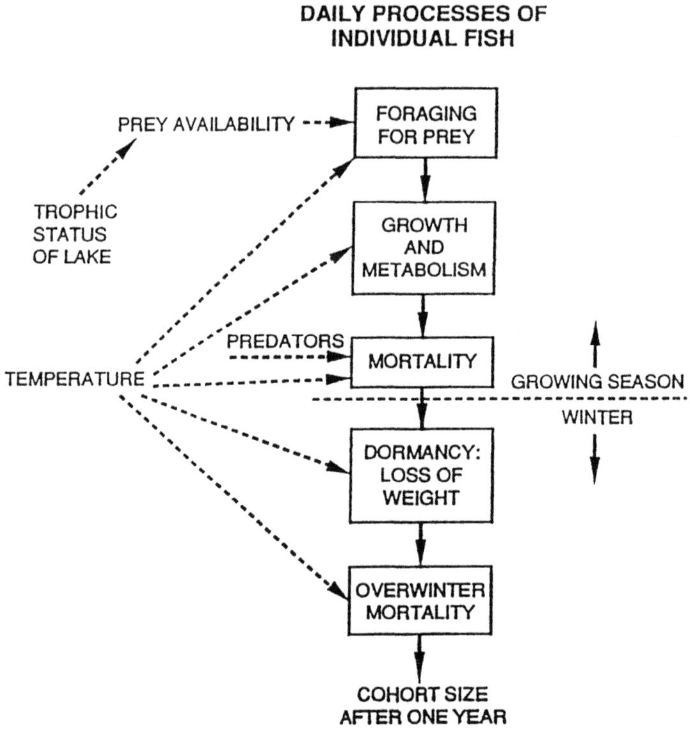

Fig. 2.1 Major daily processes in an individual-based model for a smallmouth bass population (from DeAngelis *et al.* 1991). The important processes that each individual experiences, the important environmental effects and the interactions among individuals are combined in this model. Each individual is described uniquely by a set of characters and each is simulated daily; the aggregate of individual fates is used to determine the population dynamics.

probabilities of encounter with other entities, either heterospecific or conspecific. The result of encounters are predicted by the individual's decision rules and what is encountered. For example, if the conspecific is below a certain threshold size, it will be selected as food. Encounters can result in energy and nutritional gain, escape to another place, or death. Encounter rates are functions of the densities of each entity; for potential conspecific prey under a certain size, rates are determined by the size/density distribution. Different scenarios are simulated to produce predictions for empirical testing. Thus, the model may vary the initial densities, size distributions, availability of alternative prey, or other factors such as habitat complexity.

The model has the advantage of flexibility, and can be modified with additions or deletions of states or processes, such as interference, in order to fit the specifics of a particular cannibalistic population. Consequently, the model can simulate the interactions between cannibalism, food, and size- or age-specific densities. These

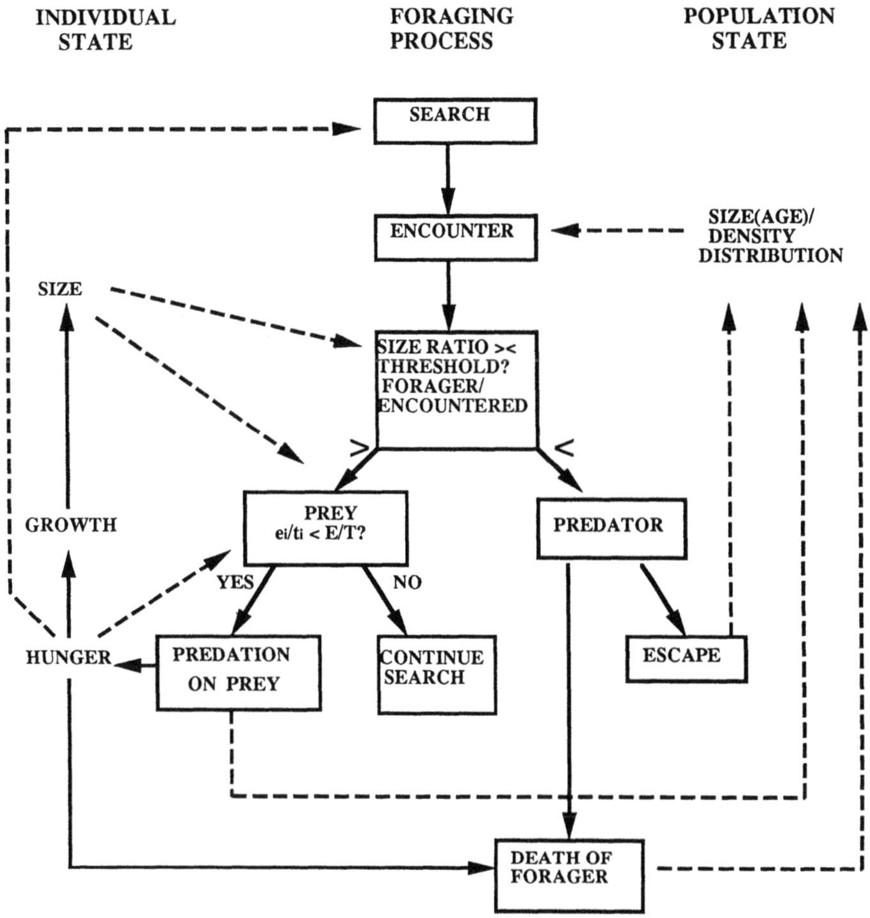

Fig. 2.2 Flow chart of processes leading to cannibalistic foraging that would be incorporated into an individual-based model. Here, we exclude encounters with heterospecific individuals and consider only encounters among conspecifics. Bifurcations indicate alternative outcomes of an encounter. For example, foraging decision rules based on characteristics of the prey and predator determine if the individual feeds or leaves. Cannibalism can occur in two places, the forager on a smaller conspecific or a larger conspecific on the predator. (See the text for details.)

simulations can be used to address several specific questions about cannibalism. For example, can cannibalism cause varying initial densities to converge to a certain level; what ranges of initial densities can cannibalism compensate; how can differences in the food regime affect cannibalism, population dynamics, and final densities?

Advantages of individual-based models of cannibalism

There are two reasons why individual-based models are likely to be a better approach to examining cannibalism. First, individual-based models do not make three assumptions, implicitly made by population-state models, and are particularly inappropriate for cannibalism. These assumptions are that all individuals are approximately identical; interactions among individuals are approximately symmetric; and stable emergent properties exist independently of individual variation. The population properties, under these assumptions, are linear functions of density; the population structure is irrelevant to individual interactions; local, nonlinear and asymmetric interactions among individuals are ignored; and the effects of individual behaviours are represented as average coefficients. Operationally, the assumptions of these population-state models are reasonable if at least one of the following criteria is met (Metz 1992). First, that all interactions are symmetric, as a result of environmental and individual homogeneity. Second, that the population consequences of behaviour are additive of individual states, and therefore linear. Third, that the population sample size is large relative to individual variability.

However, there is inevitable individual variation in behaviour, physiological state and even morphology. Such variation is often a key factor for population dynamics, especially of cannibalistic species (Huston *et al.* 1988), and cannibalism almost always involves an asymmetric interaction (Polis 1981; Polis 1984; Polis 1988). Although cannibals and victims coexist locally, through time each can experience different resources, food availability, hunger, effects of density, and risk of death. Cannibalism is a function of hunger and relative vulnerability, states that can vary on the scale of hours. This scale is much smaller than that of other important processes such as reproduction. Furthermore, large variation in cannibalism on small temporal scales often significantly affect dynamics (Fox 1975). These considerations suggest that the assumptions and criteria of population state models are unsatisfied in many cases, and particularly for intra-cohort cannibalism.

The second reason for advocating individual-based models is the difficulty for population state models to discard the above assumptions and incorporate explicitly variation and interaction among individuals. Including this variation usually requires the addition of complex functions, and the parameters representing individual processes may vary nonlinearly and may even be functions of density. Indeed, constant parameters may not be the proper way to represent active multiple behavioural choices. For instance, Abrams (1990) found the constant parameter version of the Type II functional response model inappropriate to predict functional response in the presence of alternative prey. If alternative prey are preferable to conspecifics, then using the constant parameter version of the Type II response in population state models of cannibalism is incorrect. Finally, large differences in temporal scale among different processes can make it difficult and complex for population state models to find proper analytic solutions to population dynamics. By attempting to include these considerations, population-state models are likely to become intractable, or even as complicated as the system under investigation.

In conclusion, individual-based models of cannibalism focus on foraging processes. Since data from actual processes and states of individuals are used, these models do not make the unreasonable assumptions made by population-state models and thus yield more general predictions. Further, measurements of individuals and the population are independent, thus making predictions more testable and refutable than those of population-state models (Murdoch *et al.* 1992). However, individual-based models are not problem free (Murdoch *et al.* 1992). They frequently demand considerable data sets and computer time, and can be so complex that they often lose tractability and even biological realism. Nevertheless, it seems highly likely that their application will considerably advance our understanding of population biology in general and the dynamics of cannibalism in particular.

DISCUSSION AND CONCLUSIONS

Density, changes in food levels, size, age, and hunger are the most important factors that affect the frequency of cannibalism in natural populations. These factors must be explicit in studies of the population consequences of cannibalism; the most useful empirical studies and models should include both the behaviour of individuals in response to these factors and variation in response and physiological state among individuals in the same population. This approach is a potentially powerful way to analyze cannibalism and the dynamics of cannibalistic populations. In fact, an individual-based approach is frequently advocated as a general way to understand and analyze the processes and dynamics of populations and communities (Schoener 1986; Persson and Diehl 1990; Murdoch *et al.* 1992).

Specifically, foraging theory facilitates this approach (Schoener 1987) and also provides a particularly appropriate way to analyze cannibalism (Polis 1981). Cannibalism can be viewed as a series of foraging decisions by both the predator and the prey. Decisions include: which foods to select, where foraging will occur, and the frequency and type of movement by foragers. These decisions, based on the physiological state of the forager, can affect the rate of cannibalism and subsequent population dynamics. The major variables in foraging decisions are the same that influence cannibalism: hunger, relative size, the densities, and age or size distributions of con- and heterospecific prey and predators. Unfortunately, neither the patterns nor effects of these factors are well understood. For example, how these factors influence changes in the threshold to cannibalize is largely unknown. Measurement and manipulation of these factors will enhance our understanding of cannibalism.

Density is a key factor for the population dynamics of cannibalism. Although cannibalism generally increases with density, its effects are not always a clear function of population density. Not all measures of density are useful in the analysis of cannibalism. For example, the use of gross (overall) density is inappropriate for inter-cohort cannibalism where cannibals and their conspecific prey are clearly different entities. If possible, it is better to separate the densities of the predator and prey components of one population. However, this is often difficult or impossible, especially if cannibalism occurs within a cohort of more or less equal individuals. Further, an individual may alternate (often during short periods) between being a potential prey or a predator as a function of its relative size or developmental stage. Cannibalism and foraging decisions are a function of the densities of the different components of the population. For potential prey, the density of potential predators is correlated with the risk of death and is important in patch selection.

The interaction among density, food, and vulnerability also affects rates of cannibalism. There is a general expectation that when density increases, food availability decreases, and vulnerability increases, resulting in a higher rate of cannibalism. However, this expectation is approximate, because internal structures of density and food availability are complicated and exert significant effects on this interaction.

Two contingency models of optimal foraging theory are applied to cannibalistic foraging in this chapter. The first modified Schoener's (1971) and Charnov's (1976*a*) diet selection models by more fully delineating the costs and benefits specific to cannibalism. As a result, a criterion was derived to assess the underlying evolutionary mechanism of cannibalism: net energy gain or competition. The second model, of patch selection under risk of predation (Werner and Gilliam 1984), is applicable to cannibalism with little or no change. Both models are heuristic about the foraging conditions under which cannibalism occurs. Each provides a partial explanation for the well documented observations that cannibalism increases with decreasing food levels and increasing density. These models show that decreases in food availability initiate changes in prey selection, patch occupancy and other foraging activities, and behaviours in ways that increase cannibalism.

Recently, foraging theory has been improved by incorporating a common currency that includes non-energetic costs and benefits (e.g. Mangel and Clark 1988; Houston *et al.* 1988). This is particularly appropriate for cannibalistic foraging models, which must include the potential loss of fitness when relatives are eaten. Other costs, such as the probabilities of retaliation, parasite/pathogen transmission, and benefits, including competitive advantages must also be incorporated into all optimal foraging models of cannibalism. Other aspects of foraging theory can also be applied to cannibalism. For example, it is interesting to ask how cannibalism varies as a function of normal variations in food, since food availability and quality vary considerably both temporally and spatially. The risk sensitivity models (e.g. Caraco 1980) of optimal foraging theory investigates foraging strategies of consumers that experience changing prey availability. Such models could be applied profitably to cannibalistic foraging.

Cannibalism is a strongly asymmetric interaction, usually occuring when there are differences between individuals in size, age, life history stage, or hunger. This variation makes some (e.g. smaller) individuals vulnerable to cannibalism relative to others. It is clearly essential to focus on the individual and include variation among individuals when analyzing the dynamics of cannibalistic populations.

Population models incorporating individual behaviours can be placed into two categories, according to the level of their state variables: population-state models and individual-based models. Although the traditional, holistic population-state models are most common, they cannot incorporate individual variation, a key predictor of cannibalism, without great complexity and difficulty. In contrast, individual-based models explicitly represent variations and interactions, and are sensitive to the physiological structure of populations. Thus, these models require fewer biologically unrealistic assumptions, and behavioural parameters of individuals can usually be estimated more accurately. The use of individual-based models should incorporate both individual variation in those traits important for cannibalism, and the foraging rules of individuals. Such an approach will provide a more detailed and richer analysis of the population consequences of cannibalism.

ACKNOWLEDGEMENTS

We thank C. Catterall, P. Crowley, B. Crespi, D. DeAngelis, M. Elgar, J. Gilliam, A. Hastings, D. McCauley, P. Raimondi, and T. Yamashita for their comments and suggestions for improving this chapter. We acknowledge support from the Fellowship in population dynamics of the Electric Power Research Institute (QD) and from Vanderbilt University's Natural Science Committee and University Research Council (GAP).

REFERENCES

Abrams, P. A. (1990). The effects of adaptive behavior on the type-2 functional response. *Ecology*, **71**, 877–85.

Bailey, K. H. and Polis, G. A. (1987). An experimental analysis of optimal and central place foraging by the harvester ant, *Pogonomyrmex californicus*. *Oecologia*, **72**, 440–8.

Barber, F. (1979*a*). Disparity of pink salmon runs. *Fisheries and Marine Services, MS Report*, 1504.

Barber, F. G. (1979*b*). On ocean migration, speciation, cycle dominance, and density dependence in Pacific Salmon. *Fisheries and Marine Services Technical Report*, 872.

Bobisud, L. (1976). Cannibalism as an evolutionary strategy. *Bulletin of Mathematical Biology*, **38**, 359–68.

Braid, J. D. (1981). Incidence of cannibalism among striped bass fry in an intensive culture system. *Progressive Fish Culturist*, **43**, 210–12.

Caraco, T. (1980). On foraging time allocation in a stochastic environment. *Ecology*, **61**, 119–28.

Charnov, E. L. (1976*a*). Optimal foraging: attack strategy of a mantid. *American Naturalist*, **110**, 141–51.

Charnov, E. L. (1976*b*). Optimal foraging: the marginal value theorem. *Theoretical Population Biology*, **9**, 129–36.

Crowley, P. (1984). Evolutionarily stable strategies for larval dragonflies. In *Lecture notes in Biomathematics*, **54**, (ed. S. Levin and T. Hallam), pp. 55–74. Springer Verlag, Berlin.

Crowley, P., Nisbet, R, Gurney, W., and Lawton, J. (1987). Population regulation in animals with complex life histories: formulation and analysis of a damselfly model. *Advances in Ecological Research*, **17**, 1–59.

Crowley, P. and Martin, E. (1989). Functional responses and interference within and between year classes of a dragonfly population. *Journal of the North American Benthological Society*, **8**, 211–21.

DeAngelis, D., Cox, D., and Coutant, C. (1979). Cannibalism and size dispersal in young-of-the-year-largemouth bass: experiment and model. *Ecological Modeling*, **8**, 133–48.

DeAngelis, D., Godbout, L., and Shuter, B. J. (1991). An individual-based approach to predicting density-dependent dynamics in smallmouth bass populations. *Ecological Modeling*. (In press.)

De Vita, J., Kelly, D., and Payne, S. (1982). Arthropod encounter rate: a null model based on random motion. *American Naturalist*, **119**, 499–510.

Foster, S. , Garcia, V., and Town, M. (1988). Cannibalism as the cause of an ontogenetic shift in habitat use by fry of the threespine stickleback. *Oecologia*, **74**, 577–85.

Fox, L. R. (1975). Cannibalism in natural populations. *Annual Review of Ecology and Systematics*, **6**, 87–106.

Folkvord, A. and Hunter, J. R. (1986). Size-specific vulnerability of northern anchovy, *Engraulis mordax*, larvae to predation by fishes. *Fishery Bulletin* (US), **84**, 859–69.

Fraser, D. F. and Cerri, R. D. (1982). Experimental evaluation of predator-prey relationships in a patchy environment: consequences for habitat use patterns in minnows. *Ecology*, **63**, 307–13.

Gabriel, W. (1985). Overcoming food limitation by cannibalism: a model study of cyclopoids. *Archiv für Hydrobiologie*, **21**, 373–81.

Gabriel, W. and Lampert, W. (1985). Can cannibalism be advantageous in cyclopods? A mathematical model. *Verhandlungen der Internationalen Vereinigung för Theoretische und Angewandte Limnologie*, **22**, 3164–8.

Gilliam, J. (1990). Hunting by the hunted: optimal prey selection by foragers under predation hazard. In *Behavioral mechanisms of food selection*, (ed. R. Hughes), pp.797–820. NATO ASI Series, Series A: Life Sciences, Springer-Verlag, New York.

Gilliam, J. and Fraser, D. (1987). Habitat selection under predation hazard: test of a model with foraging minnows. *Ecology*, **68**, 1856–62.

Goss-Custard, J. D. (1977). The energetics of prey selection redshank, *Tringa totanus* (L.), in relation to prey density. *Journal of Animal Ecology*, **46**, 1–9.

Hassell, M. (1981). Arthropod predator–prey systems. In *Theoretical ecology: principles and applications*, (ed. R. M. May), pp. 105–31. Sinauer Associates, Sunderland, Mass.

Hassell, M. and May, R. (1985). From individual behaviour to population dynamics. In *Behavioural ecology*, (ed. R. M. Sibly and R. H. Smith), pp. 3–32. Blackwell Scientific Publications, Oxford.

Hassell, M. and Southwood, T. R. E. (1978). Foraging strategies of insects. *Annual Review of Entomology*, **9**, 75–98.

Hastings, A. and Costantino, R. F. (1987). Cannibalistic egg-larva interactions in *Tribolium*: an explanation for the oscillations in population numbers. *American Naturalist*, **130**, 36–52.

Hausfater, G. and Hrdy, S. (1984). *Infanticide: comparative and evolutionary perspectives.* Aldine Publishing Company, New York.

Higgins, J. (1990). Age structure and population dynamics of flour beetles. Dissertation, University of Illinois at Chicago Circle.

Holling, C. S. (1966). The strategy of building models of complex ecological systems. In *Systems analysis in ecology,* (ed. K. Watt), pp. 195–214. Academic Press, New York.

Houston, A., Clark, C., McNamara, J., and Mangel, M. (1988). Dynamic models in behavioural and evolutionary ecology. *Nature*, **332**, 29–34.

Huston, M., DeAngelis, D., and Post, W. (1988). New computer models unify ecological theory. *BioScience*, **38**, 682–91.

Hunter, J. R. and Kimbrell, C. A. (1980). Egg cannibalism in the northern anchovy, *Engraulis mordax. Fishery Bulletin* (US), **78**, 811–16.

Jude, D., Mansfield, P., Schneeberger, P., and Wojcik J. (1987). Compensation in fish populations subject to catastrophic impact. In *Compensatory mechanisms in fish populations: Literature reviews.* Vol. 2. Research Project Report 1633–6, Electric Power Research Institute, Palo Alto, California.

Koehl, M. A. R. (1989). Discussion: from individuals to populations. In *Perspectives in ecological theory*, (ed. J. Roughgarden, R. M. May and S. A. Levin), pp. 39–52. Princeton University Press.

Krebs, J. R. and Davies, N. B. (1987). *An introduction to behavioral ecology,* 2nd edition. Sinauer Associates, Sunderland, Mass.

Landry, M. (1978). Population dynamics and population of a planktonic marine copepod *Acarcia clausi* in a small temperate lagoon on San Juan Island, Washington. *Internationale Revue der Gesamten Hydrobiologie*, **63**: 77–119.

McFarland, D. and Houston, A. (1981). *Quantitative ethology: the state space approach.* Pitman Advanced Publishing Program, Boston, Mass.

Mangel, M. and Clark, C. W. (1988). *Dynamic modeling in behavioral ecology.* Princeton University Press.

May, R. (1981). Models for two interacting populations. In *Theoretical ecology: principles and applications,* (ed. R. May), pp. 78–104. Sinauer Associates, Sunderland, Mass.

Meffe, G. K. (1984). Density-dependent cannibalism in the endangered Sonoran topminnow (*Poeciliopsis occidentalis*). *Southwestern Naturalist*, **29**, 500–3.

Mertz, D. B. (1969). Age-distribution and abundance in populations of flour beetles. I. experimental studies. *Ecological Monographs*, **39**, 1–31.

Mertz, D. B. (1972). The *Tribolium* model and the mathematics of population growth. *Annual Review of Ecology and Systematics,* **3**, 51–78.

Mertz, D. B. and Davies, R. B. (1968). Cannibalism of the pupal stage by adult flour beetles: an experiment and a stochastic model. *Biometrics*, **24**, 247–275.

Metz, J. A. J. (1992). The role of physiologically structured models within a general individual-based modelling perspective. In *Individual-based models and approaches in ecology*, (ed. D. DeAngelis and L. J. Gross). Routledge, Chapman, and Hall, New York. (In press.)

Murdoch, W. (1971). The developmental response of predators to changes in prey density. *Ecology*, **52**, 132–7.

Murdoch, W. (1979). Predation and the dynamics of prey populations. *Forschrift für Zoologie*, **52**, 295–310.

Murdoch, W. W., McCauley, E., Nisbet, R. M., Gurney, W. S. C., and de Ross, A. M. (1992). Individual-based models: combining testability and generality. In *Individual-based models and approaches in ecology*, (ed. D. DeAngelis and L. J. Gross). Routledge, Chapman, and Hall, New York. (In press.)

Newman, J. A. and Elgar, M. A. (1992). Sexual cannibalism in orb-weaving spiders: an economic model. *American Naturalist*. (In press.)

O'Brien, W., Slade, N., and Vinyard, G. (1976). Apparent size as the determinant of prey selection by bluegill sunfish (*Lepomis macrochirus*). *Ecology*, **57**, 1304–10.

Orr, B. K, Murdoch, W. W. ,and Bence, J. R. (1990). Population regulation, convergence, and cannibalism in *Notonecta* (Hemiptera). *Ecology*, **71**, 68–82.

Persson, L. and Diehl, S. (1990). Mechanistic individual-based approaches in the population/community ecology of fish. *Annals Zoology Fennici*, **27**, 165–82.

Polis, G. (1980). The significance of cannibalism on the population dynamics and surface activity of a natural population of desert scorpions. *Behavioral Ecology and Sociobiology*, **7**, 25–35.

Polis, G. (1981). The evolution and dynamics of intraspecific predation. *Annual Review of Ecology and Systematics*, **12**, 125–51.

Polis, G. (1984). Age structure component of niche width and intraspecific resource partitioning by predators: can age groups function as ecological species? *American Naturalist*, **123**, 541–64.

Polis, G. (1988). Exploitation competition and the evolution of interference, cannibalism, and intraguild predation in age/size-structured populations. In *Size-structured populations*, (ed. B. Ebenman and L. Persson), pp. 185–202. Springer-Verlag, Berlin.

Polis, G. and McCormick, S. (1987). Competition and predation among species of desert scorpions. *Ecology*, **68**, 332–43.

Polis, G., Myers, C., and Holt, R. (1989). The ecology and evolution of intraguild predation: potential competitors that eat each other. *Annual Review Ecology and Systematics*, **20**, 297–330.

Pyke, G. (1984). Optimal foraging theory: a critical review. *Annual Review of Ecology and Systematics*, **15**, 523–75.

Radovich, J. (1962). Effects of sardine spawning stock size and environment on year-class production. *California Fish and Game Bulletin*, **48**, 123–40.

Rapport, D. and Turner, J. (1977). Economic models in ecology. *Science*, **195**, 367–73.

Richards, L. (1983). Hunger and optimal diet. *American Naturalist*, **122**, 326–34.

Ricker, W. E. (1962). Regulation of the abundance of pink salmon populations. *Symposium on Pink Salmon*, (ed. N. J. Wilimovsky), pp. 155–201. University of British Columbia, Vancouver.

Rubenstein, D. I. (1981). Individual variation and competition in the Everglades pygmy sunfish. *Journal of Animal Ecology*, 50, 337–50.

Saila, S. B., Chen, X., Erzini, K., and Martin, B. (1987). Critical evaluation of case histories of fish populations experiencing chronic exploitation or impact. In *Compensatory mechanisms in fish populations: Literature reviews*. Vol. 1. Research Project Report 1633–6, Electric Power Research Institute, Palo Alto.

Schlosser, I. (1987). The role of predation in age- and size-related habitat use by stream fishes. *Ecology*, **68**, 651–9.

Schoener, T. W. (1971). Theory of feeding strategy. *Annual Review of Ecology and Systematics*, **11**, 369–404.

Schoener, T. W. (1983). Field experiments on interspecific competion. *American Naturalist*, **122**, 240–85.

Schoener, T. W. (1986). Mechanistic approaches to community ecology: a new reductionism. *American Zoologist*, **26**, 81–106.
Schoener, T. W. (1987). A brief history of optimal foraging ecology. In *Foraging behavior* (ed. A. C. Kamil, J. R. Krebs, and H. R. Pulliam), pp. 5–67. Plenum, New York.
Sih, A. (1982). Foraging strategies and the avoidance of predation by an aquatic insect, *Notonecta hoffmanni*. *Ecology*, **63**, 786–96.
Sih, A. (1987). Predator and prey lifestyles: an evolutionary and ecological overview. In *Predation: direct and indirect impacts on aquatic communities*, (ed. W. C. Kerfoot and A. Sih), pp. 203–24. University Press of New England, Hanover, NH.
Stephens, D. W. and Krebs, J. R. (1986). *Foraging theory*. Princeton University Press.
Strong, D., Simberloff, D., Abele, L., and Thistle, A. (1984). *Ecological communities: conceptual issues and the evidence*. Princeton University Press.
Valdes, E., Shelton, P., Armstrong, M., and Field J. (1987). Cannibalism in South African anchovy: egg mortality and egg consumption rates. *South African Journal of Marine Science*, **5**, 613–22.
van den Bosch, F., de Roos, A. M., and Gabriel, W. (1988). Cannibalism as a life boat mechanism. *Journal of Mathematical Biology*, **26**, 619–33.
Werner, E. and Gilliam, J. (1984). The ontogenetic niche and species interactions in size-structured populations. *Annual Review of Ecology and Systematics*, **15**, 393–425.
Werner, E. and Hall, D. (1974). Optimal foraging and the size selection of prey by the Bluegill Sunfish (*Leponis macrochirus*). *Ecology*, **55**, 1042–52.
Wiens, J. A. (1977). On competition and variable environments. *American Scientist*, **65**, 590–7.
Wilbur, H. (1988). Interactions between growing predators and growing prey. In *Size-structured populations*, (ed. B. Ebenman and L. Persson), pp. 157–72. Springer-Verlag, Berlin.

3

Ecology of filial cannibalism in fish: theoretical perspectives

Robert Craig Sargent

Rohwer (1978) proposed a theory of filial cannibalism for fishes with paternal care, where filial cannibalism is defined as a parent eating its own offspring. Since that time, filial cannibalism has been widely documented, not only in fishes with paternal care, but also in species with maternal and biparental care (see Dominey and Blumer 1984; FitzGerald and Whorisky this volume, for review). Rohwer (1978) begins with two assumptions: (1) nesting male fishes are food limited and forced to fast; and, (2) a male's own offspring may be an alternative source of food. Rohwer (1978) proposes that a filial cannibal trades off current offspring against an energetic investment in future reproductive success. From these simple assumptions, Rohwer (1978) provides a verbal model that potentially explains the diversity of filial cannibalism, and associated behaviours, in natural populations.

Despite numerous studies that document filial cannibalism, there are relatively few that examine Rohwer's (1978) two basic assumptions of energy limitation, and offspring as an alternate source of energy (e.g. FitzGerald and Whorisky this volume). One study by Hoelzer (1989) provides convincing evidence that filial cannibalism is energy-based in the Cortez damselfish, *Stegastes rectifraenum*. Hoelzer (1989) found that parental male Cortez damselfish in field populations obtain food from two sources: (1) the algal mats on their territories, and (2) the eggs in their nests. Hoelzer (1989) experimentally supplemented the diets of some males with eggs from the nests of other males. He found that those males with diets that were supplemented with foreign eggs were less likely to eat their own eggs, and were less likely to feed on their algal mats. These results suggest an energetic basis for filial cannibalism in this species.

On the other hand, two studies with the threespine stickleback *Gasterosteus aculeatus*, are less conclusive. Field experiments by Whorisky and FitzGerald (1985) indicate that in the guts of parental males and foraging females there is the same diversity of food items. This finding challenges Rohwer's (1978) assumption that parental males are forced to 'fast', at least in some populations of this species. Perhaps 'fasting' by a parental fish, when it does occur, represents a decision, rather than a constraint imposed by the environment. In a recent laboratory experiment, Belles-Isles and FitzGerald (1991) experimentally manipulated the food ration for parental males, including one treatment of unfed males. They found that males under all food levels ate small numbers of their own eggs, but that the level of filial cannibalism was independent of the level of food. This finding challenges Rohwer's (1978) assumption of a parent's offspring as an alternate food source. Possibly, male sticklebacks eat some of their own eggs for reasons other than energy. For example, males will remove and eat diseased eggs from their clutches (personal observation); the consumption of healthy eggs may be an incidental biproduct of removing diseased eggs. Alternatively, eggs may be rich in certain nutrients important to male sticklebacks (e.g. carotenoids).

Despite Belles-Isles and FitzGerald's (1991) failure to demonstrate an energetic basis for filial cannibalism in threespine sticklebacks, two other studies with this species suggest energetic limitations for territory defence and parental care. For example, Sargent (1985) found that breeding males lose weight over the course of the breeding season, and that males who routinely defend their nests against rivals lose weight faster than solitary males. Moreover, FitzGerald *et al.* (1989) found that males with nests and eggs lost more weight than control males without nests and eggs. Because breeding male sticklebacks apparently do deplete their energy reserves (Sargent 1985; FitzGerald *et al.* 1989), it seems likely that energy limitation may still be important in determining at least some instances of filial cannibalism in this species.

Thus far, virtually all filial cannibalism theory is based on verbal models (e.g. Rohwer 1978), which have predictions that can be difficult to evaluate (e.g. Dominey and Blumer 1984; FitzGerald and Whorisky this volume). Therefore, as a first approximation to quantifying the phenomenon of filial cannibalism, I have constructed a dynamic-programming model based on a single energy resource. Like Rohwer (1978), the model makes the following assumptions: (1) parental males are food limited, and (2) offspring are an alternative source of food (i.e. energy obtained from offspring and energy obtained from more traditional food are measured in the same units). However, it is not assumed that a parent is forced to fast. Any fasting generated by this model will be the parent's decision, rather than a constraint imposed by the environment.

Following Rohwer (1978), an energy-based, adaptive basis for filial cannibalism may best be understood within the framework of parental-investment and life history theory (e.g. Williams 1966; Trivers 1972). A parent defending a brood of offspring has two components to its future reproductive success: (1) survival of the brood at stake, and (2) future broods (Williams 1966). Assuming that natural selection favours behaviour that maximizes remaining lifetime reproductive success, parents should base their parental investment on the value of the current brood relative to the value of the parent's own expected future reproduction (Pressley 1976; Carlisle 1982; Sargent and Gross 1985, 1986). Conditions that favour high levels of parental investment are likely to favour low levels of filial cannibalism, and vice versa.

For example, older clutches are more valuable to a parent than younger clutches, simply because they are closer to hatching; moreover, older clutches typically receive higher levels of parental care than younger clutches (e.g. Pressley 1976;

Carlisle 1982; Taborsky 1985; Sargent and Gross 1986; Sargent 1990). Thus, one would expect younger clutches to be more vulnerable to filial cannibalism than older clutches. Empirical studies support this prediction (DeMartini 1987; Mrowka 1987; Sargent 1988; Peterson and Marchetti 1989).

In addition, Rohwer (1978) suggests that filial cannibalism would be most likely when a parent was low on energy reserves. Recently, Sargent (1990) constructed simple dynamic-programming models for parental care, and found that behaviours associated with reproduction (i.e. mating and parental care) increase as energy reserve increases, whereas feeding increases as energy reserve decreases. These results suggest that if filial cannibalism had been included in these models, then its likelihood would also increase as parental energy reserve decreases. However, there is apparently no direct experimental evidence for this prediction.

Finally, the optimal level of parental care is predicted to be an an increasing function of offspring number, and fishes are known to give more care to larger clutches (e.g. Pressley 1976; Carlisle 1982; Sargent and Gross 1986; Sargent 1990). It is interesting to ask whether filial cannibalism also depends on offspring number. To answer this question, it is necessary to recognize that there are two potential benefits to a filial cannibal: (1) increased reproduction in future brood cycles, and (2) increased offspring survival within the present brood cycle.

Rohwer (1978) mainly emphasized filial cannibalism as an investment by a parent in future brood cycles. In this case, one would expect filial cannibalism to be more likely in small clutches. If a parent's clutch size is too small, then the net benefits of care may be less than the net benefits of abandoning the offspring and reproducing again in the future. If the clutch were also edible, then the parent would gain more from eating the clutch than it would from caring for it (e.g. Dominey and Blumer 1984; Taborsky 1985). There is considerable empirical evidence for such total filial cannibalism of small clutches (e.g. DeMartini 1987; Mrowka 1987; Hoelzer 1988; Sargent 1988; Peterson and Marchetti 1989). Thus, with respect to future brood cycles, total filial cannibalism should increase, and parental care should decrease, as clutch size decreases.

Rohwer (1978) also argued that in species with prolonged brood cycles, filial cannibalism may represent an investment within the present brood cycle. If food is limiting, then partial consumption of a clutch within a brood cycle may enable a parent to remain alive and capable of caring for the remainder of its current clutch. DeMartini (1987), Mrowka (1987), Sargent (1988), Hoelzer (1989), and Peterson and Marchetti (1989) all provide evidence for this partial filial cannibalism. The negative effects of partial filial cannibalism should be more dilute in larger clutches. Thus, with respect to survival of the current clutch, partial filial cannibalism and parental care may both increase as clutch size increases.

Thus, based on Rohwer (1978) and the preceding arguments, four predictions can be addressed: (1) the likelihood of filial cannibalism increases as clutch age decreases; (2) the likelihood of filial cannibalism increases as parental energy reserve decreases; (3) if filial cannibalism represents investment in future brood cycles, then the likelihood of total filial cannibalism increases as clutch size decreases; and (4) if filial cannibalism represents investment in current offspring, then the likelihood of partial filial cannibalism increases as clutch size increases. This chapter presents a simple dynamic-programming model that was constructed to examine these predictions.

Dynamic programming

Stochastic dynamic programming falls within the realm of dynamic optimization. Dynamic optimization in behavioural ecology uses a state–space approach to model the dynamics of behaviour over time (e.g. McFarland and Houston 1981; McFarland 1982). With this approach, several kinds of behaviour can be modelled simultaneously, and their consequences examined in terms of a common currency such as lifetime reproductive success. Dynamic optimization allows decision variables (e.g. the particular behaviour chosen from the range of available options, or strategy set) to depend on state variables (e.g. hunger, energy reserve, body size, offspring number, offspring age), which in turn depend on past behavioural decisions. Thus, dynamic optimization finds behavioural trajectories through an individual's lifetime (e.g. Stephens and Krebs 1986). The dependence of the decision variables on the state variables, and the state dynamics may be stochastic. In this case, dynamic optimization finds expected behavioural trajectories through time, and all alternative trajectories.

Oster and Wilson (1978) identify five essential components of dynamic optimization models (for a lucid discussion, see Mangel and Clark 1988, chapter 8). These are:

(1) *the optimization criterion:* the currency being maximized;

(2) *the strategy set:* the set of behavioural options available to an animal;

(3) *the state space:* all combinations of magnitudes of variables that depend cumulatively on past decisions, and that determine the present optimal policy;

(4) *the constraints:* the limitations on the state space;

(5) *the state dynamics:* the rules for moving through the state space.

To this list, Sargent (1990) added a sixth component:

(6) *the trade-offs:* the biotic or abiotic factors that prevent an animal from maximizing all of its options simultaneously.

In order to construct and solve a dynamic optimization model, one needs to explicitly specify each of these components.

Dynamic programming is a technique in dynamic optimization that finds numerical solutions, usually with the aid of a computer (Bellman 1957; Mangel and Clark 1986, 1988; McNamara and Houston 1982, 1986; Houston and McNamara 1988). Dynamic-programming models are solved by iterating backward through time, following Bellman's (1957) principle of optimality: 'An optimal policy has the property that, whatever the initial state and initial decision are, the remaining decisions must constitute an optimal policy with regard to the state resulting from the first decision.' Put another way, the optimal trajectory from some initial time, t, to some final time, T (also known as the time horizon), contains the optimal trajectory from time T-1 to T. Thus, to find the optimal trajectory from t to T, one first finds the optimal trajectory from T-1 to T, then the optimal trajectory from T-2 to T-1, and so on back to t. The basic task of constructing a dynamic-programming model is in specifying the dynamic-programming equation. Specifying the dynamic-programming equation first requires specifying the state dynamics. In the model that follows, two state variables are used: (i) parental state (i.e. energy reserve), and (ii) nest state (i.e. the number of clutches within a parent's nest).

At each time t, for each level of parental state $X(t)$ and nest state $Y(t)$, one can find a behaviour $B^*(t, X(t), Y(t))$ that maximizes the expectation of terminal fitness, $F(t, X(t), Y(t))$, where the maximum is written as F^*. This requires that we already know $F^*(t+1, X(t+1), Y(t+1))$, for all possible $X(t+1)$ and $Y(t+1)$. $B^*(t, X(t), Y(t))$, $X(t)$, and $Y(t)$ jointly determine $X(t+1)$ and $Y(t+1)$. The recursions for the state variables and for fitness in general form are:

$$X(t+1) = f_1(t, B(t, X(t), Y(t)), X(t), Y(t)) \tag{3.1}$$

$$Y(t+1) = f_2(t, B(t, X(t), Y(t)), X(t), Y(t)) \tag{3.2}$$

$$F(t) = f_3(t, B(t, X(t), Y(t)), X(t), Y(t), F^*(t+1, X(t+1), Y(t+1))) \tag{3.3}$$

where f_1, f_2, and f_3 represent functions. Equations (3.1) and (3.2) specify the state dynamics, and equation (3.3) is the dynamic-programming equation for two state variables.

To solve the dynamic-programming equation, one begins by setting the initial values for the terminal fitnesses at the time horizon, T. Terminal fitness is zero if the animal is dead at the time horizon; otherwise it may be positive, and represents future reproduction after time T. Then starting at T-1, one iterates backward over time over all combinations of the state variables, and finds the optimal behaviour for each state variable combination at each time step. We begin by finding the optimal behaviours and the associated fitnesses for all combinations of the state variables for T-1. Once this is achieved, we now have a new set of terminal fitnesses. We then repeat the process for T-2, and so on back to $t = 1$.

Because dynamic programming is a numerical technique, the solution obtained depends on the values chosen for the parameters, and on the assumed state dynamics. Generally, the stability of the solution is examined with a sensitivity analysis, in which the parameters are manipulated independently to explore each one's effect on the solution. Ideally, the parameter values and state dynamics are chosen from detailed knowledge of the system being modelled; however, such knowledge is often lacking. Thus, the following approach is adopted. First, a model is constructed with assumed parameters and state dynamics that are plausible, and with a default solution that qualitatively mimics known fish behaviour. Second, each parameter is varied plus or minus a small amount (e.g. 10 per cent) to determine whether or not the qualitative solution is an artifact of the parameter values. (To save space, the sensitivity analysis for this model is not presented.) Third, one or more of the model's parameters is manipulated to examine whether the resulting change in the optimal policy qualitatively mimics what happens in a natural system under similar circumstances. If so, then insights into the regulation of behaviour are gained, but if not, then one or more of the model's assumptions is false.

The following model examines how filial cannibalism is affected by the following:

(1) the parent's expectation of future brood cycles;

(2) the cost of parental feeding to offspring survival; and

(3) the relative value of feeding versus filial cannibalism in augmenting the parental energy reserve.

THE MODEL: FEED, EAT EGGS OR CARE?

A brooding fish may be confronted with the conflicting demands of predator avoidance, feeding, and parental care. These demands represent very different kinds of behaviour, and incorporating all of them into one model will require several simplifying assumptions to keep the analysis manageable. Thus, the following model and its analysis are intended to provide qualitative predictions on the phenomenon of filial cannibalism, and not quantitative predictions.

In the past, the conflicting demands of avoiding predators, feeding, and parental care have been modelled with static resource-allocation models (e.g. Pressley 1976; Carlisle 1982; Sargent and Gross 1986). In these models, an animal is assumed to be able to vary continuously the proportion of resources it invests in each of its behavioural options, subject to the constraint that any resource invested in one option cannot be invested in another. There have been relatively few dynamic resource-allocation models (but see Oster and Wilson 1978; Mangel and Clark 1986, 1988 chapter 6), probably because the state dynamics of such models can be prohibitively complex. Therefore, as a first approximation to dynamic resource allocation, the model is based on Mangel and Clark's (1988, chapter 2) simple patch-selection model, which assumes that an animal must choose from a discrete menu of behavioural options, rather than have continuous decision variables. In spite of this assumption, dynamics were found that resemble those of resource-allocation models.

Trade-offs are incorporated in two ways. These are behaviour-dependent increments or decrements to parental energy reserve, and behaviour-dependent multiplicative probability reducing coefficients, which are decimal fractions. These coefficients represent costs to parental survival and offspring survival, depending on the behaviour chosen.

This model is an extension of the model, 'Feed or Care?', of Sargent (1990). Imagine that a parental male fish has just finished mating and has 0, 1, 2, 3, or 4 female clutches in his nest. Assume that further mating is impossible, and that this hypothetical fish is confronted with the conflicting demands of feeding and parental care. An example of such a species is the bluegill sunfish, *Lepomis macrochirus*; females are only available for spawning with males for the first one or two days of the nesting cycle (Coleman *et al.* 1985). This model can also be applied to species with maternal care, such as maternal mouthbrooding cichlids (e.g. Mrowka 1987), in which case the number of clutches would represent fractions of the average total brood. The time horizon, T, is defined as the time at which all offspring are hatched and independent of parental care.

Following Oster and Wilson (1978) the essential components of this model are the following (Table 3.1 lists the model's parameters).

1. *Optimization criterion.* The currency being maximized is terminal fitness at time T: i.e. the number of clutches, plus the parent's expectation of future brood cycles.

2. *Strategy Set B(t, X(t), Y(t)).* To keep the analysis simple, I assume that a parent's strategy set consists of eight discrete options: the individual behaviours, *nothing, feed, eat eggs* (i.e. filial cannibalism), and *care;* and the combined behaviours, *feed & eat eggs, feed & care, eat eggs & care,* and *feed, eat eggs & care. Nothing* represents an inactive fish whose energy expenditure is basal metabolism. *Feed* represents a parent who briefly leaves its nest to eat available food items (e.g. food that drifts by the territory). *Eat eggs* represents a parent who commits filial cannibalism. *Care* represents a parent who defends its nest against egg predation. To simplify the analysis, it is assumed that the individual behaviours *feed, eat eggs,* and *care* are not mutually exclusive in terms of the energy gains from feeding and filial cannibalism. This assumption may seem unrealistic, but it assures that a parental male is not forced to fast. A male who *feeds & cares* gains the same energy from feeding as a male that *feeds* (see State Dynamics, below).

3. *The State Space,* (X(t), Y(t)). The state variables are parental energy reserve, $X(t)$, and the number of clutches in a male's nest $Y(t)$.

4. *Constraints.* Parental energy reserve is constrained as follows: $0 < X(t) <$ Capacity, where Capacity represents the upper limit to a parent's

Table 3.1 The default parameters

	Feed, Eat Eggs or Care?
$T = 6$	time horizon, or end of the nesting cycle; there are five discrete time steps.
Capacity = 12	maximum level of parental energy reserve.
$u = 0.3$	probability that a clutch survives a time step without care.
$w = 0.95$	probability that a clutch survives a time step with care
$P = 0.99$	probability that the parent survives a time step
$p_f = 0.9$	parental survival reducing coefficient due to feeding
$p_c = 0.9$	parental survival reducing coefficient due to caring
$q_f = 0.7$	clutch survival reducing coefficient due to feeding
$\phi(T) = Y(T) + 0.15\,X(T)$	terminal fitness equals number of clutches hatched at time T, plus 0.15 times the amount of parental energy reserve at time T

energy reserve. Clutch number is constrained as follows: $Y(t) = 0, 1, 2, 3$, or 4 clutches.

5. *State Dynamics*. (i) Parental state. Parental energy reserve at time $t + 1$ depends deterministically on the behaviour chosen at time t; i.e., $X(t+1) = X(t) + \Delta X(B(t), X(t), Y(t))$. The state dynamics are based on known qualitative trends between fish energetics and behaviour. These are: (a) a fish that only feeds, gains weight; (b) a fish that is starved, loses weight; (c) a fish that defends a breeding territory, loses weight (e.g. Unger 1983; Sargent 1985); (d) a fish that cares for eggs, loses more weight than a fish that does not care for eggs (e.g. FitzGerald *et al.* 1989); and (e) a parental male fish that has a diet supplemented with eggs exhibits less filial cannibalism, and less feeding on other foods (Hoelzer 1989). The units of energy reserve are arbitrary, and the following state dynamics are based on qualitative trends in fish energetics; thus, the analysis of the model is intended to be qualitative.

   A fish that does *nothing* pays a metabolic cost of one unit of energy. This metabolic cost is subtracted from the energy reserve of a fish, no matter what behaviour it chooses. *Feeding* results in a gain of two units of energy, which when added to the metabolic cost yields a net gain of one. To keep the analysis simple, the energy gain from feeding is unconstrained by combining *feed* with other behaviours.

   The model assumes that a fish can eat only one clutch per time step. This implies that a fish in this model which commits total filial cannibalism eats all of its clutches among time steps, but within one brood cycle; whereas, a fish that commits partial filial cannibalism eats some but not all of his clutches. Without any information to the contrary, a fish is assumed to gain the same amount of energy from *eating eggs* as it would from *feeding*. Thus, a fish that *eats eggs* gains two units of energy, which when added to the metabolic cost yields a gain of one. A fish that *cares* for his clutches loses one unit of energy, which when added to the metabolic cost yields a net loss of two.

The energy gains or losses for the combined behaviours are simply the sum of gains and losses from each individual behaviour, plus the loss due to basal metabolism. Qualitative support for these behaviour-dependent energy dynamics is found in empirical studies on fish with paternal care (e.g. Unger 1983; Sargent 1985; FitzGerald *et al.* 1989).

(ii) Nest State. Clutch number at time t+1 depends stochastically on the behaviour chosen at time t, the energy reserve at time t, and the number of clutches present at time t. The survival of each clutch is independent of the survival of the other clutches. Because further mating within this brood cycle is assumed to be impossible, $Y(t+1) \leq Y(t)$. Although these nest-state dynamics may appear somewhat restrictive, they nevertheless are broad enough to yield predictions on much of the range of filial cannibalism that is observed in nature.

6. *Trade-offs.* The trade-offs are produced by multiplicative probability-reducing coefficients, and by the behaviour-dependent state dynamics for parental energy reserve. Parental survival is reduced by p_f for *feed*, by p_c for care, and by the product of the two for *feed & care*. Offspring survival is increased from u to w if the parent *cares*; offspring survival with care is reduced by q_f if the parent *feeds*.

Table 3.1 lists assumed values (i.e. default values) for the model's parameters. Table 3.2 lists each behaviour in the strategy set, and its consequences for parental energy reserve, parental survival, and clutch survival. The assumed sequence of events within a time step are: (1) choose behaviour; (2) evaluate its consequences for clutch survival; and (3) evaluate its consequences for parental survival. The expected fitness that a fish would enjoy from choosing any of the behaviours from the strategy set can now be written as:

$$F_B(t, X(t), Y(t)) = Pr(\text{parent dies in } t)\, Pr(\text{clutch survival until } T)\, Y(t) +$$
$$Pr(\text{parent surviving } t) \{\sum_{i=0}^{Y(t)} [Pr(Y(t+1)= i)\, F^*(t+1, X+\Delta X(B\,(t, X(t), Y(t))), i)]\} \quad (3.4)$$

Equation (3.4) is evaluated for each behaviour in the strategy set, and a parent is assumed to choose the behaviour that maximizes expected fitness (F_B). The first term on the right-hand side of (3.4) tallies clutch survival if the parent dies at the end of t. If any behaviour results in a parent's energy reserve falling to zero, then the parent's probability of death is set to one. Behaviours that would reduce a

Table 3.2 The tradeoffs: behaviour-dependent consequences for parental energy reserve, parental survival, and clutch survival

Behaviour	Change in parental state	Parental survival	Clutch survival
Nothing	-1	P	u
Feed	+1	Pp_f	u
Eat eggs	+1	P	$u/0$
Care	-2	Pp_c	w
Feed and eat eggs	+3	Pp_f	$u/0$
Feed and care	0	Pp_fp_c	wq_f
Eat eggs and care	0	Pp_c	$w/0$
Feed, eat eggs and care	+2	Pp_fp_c	$wq_f/0$

Note: for those behaviours that include filial cannibalism, I give the survival for the uneaten clutches, followed by the survival for the eaten clutch.

parent's energy reserve to a negative number are not permitted. Clutch survival during t depends on $B(t, X(t), Y(t))$; clutch survival from t+1 to T equals $u^{(T-t-1)}$. The second term on the right hand side of equation (3.4) tallies future reproductive success if the parent survives to t+1. This term sums over all independent parental state by nest state transition probabilities between t and t+1, which depend on $X(t)$, $Y(t)$, and $B(t, X(t), Y(t))$ (Table 3.2). These transition probabilities are then multiplied by their respective F^*s evaluated at $X+\Delta X(B(t, X(t), Y(t)))$ and $Y(t+1) = i$. Equation (3.4) is the dynamic-programming equation.

To solve the dynamic-programming equation, the terminal fitnesses are first set at T. A common feature in dynamic-programming models of behaviour is that expected cumulative fitness is an increasing function of energy reserve (e.g. Mangel and Clark 1988; Sargent 1990). Therefore, the terminal fitness (i.e. the prospect of future brood cycles) in the following model is assumed to be an increasing function of energy reserve. In the default run, the terminal fitness is set at $\phi(T, X(T), Y(T)) = Y(T) + 0.15 X(T)$. Because parental care costs energy, whereas feeding and eating eggs increase energy, this form of terminal fitness function sets up an energetic trade-off between present and future brood cycles. Equation (3.4) is then iterated backward through time over all levels of parental state and nest state.

RESULTS

The default run

The default parameters are rough estimates from research with fathead minnows *Pimephales promelas* (e.g. Unger and Sargent 1988; Sargent 1988, 1989), and threespine sticklebacks *G. aculeatus* (e.g. Sargent and Gebler 1980; Sargent 1985). The energy-state dynamics, and assumed maximum level of energy reserve (i.e. Capacity = 12) were chosen under the assumption that most reproductively successful parents enjoy only one brood cycle, or finish the brood cycle with a low energy reserve and a low probability of breeding again. Although solitary male *G. aculeatus* may undertake several brood cycles in the laboratory, most males in nature have only one (see FitzGerald *et al.* 1989, for review). By assuming that *care* costs a parent two units of energy, only parents with 11 or more units of energy at $t = 1$ can survive the five time steps until T, without some form of feeding or filial cannibalism. Without any data either way, the model assumes that feeding and filial cannibalism both have the same effect on parental energy reserve. This assumption is changed later. The results of the default run are presented in Fig. 3.1. As was found in similar models (Sargent 1990), expected fitness increases with increasing clutch number, and with increasing energy reserve.

One can evaluate the dependence of the eight behaviours on the state variables and time by examining their distribution over these variables. One can also track behavioural trajectories through time, remembering that energy-state dynamics are deterministic, but that nest-state dynamics are stochastic. As would be expected from parental-investment theory, the likelihood of caring increases as clutch number increases, and as clutch age increases. Note that a parent with zero clutches feeds in an attempt to balance the trade-off between maximizing survival to breed again versus maximizing energy reserve at the next breeding.

As expected, filial cannibalism increases as clutch age decreases, and as parental energy reserve decreases. Note that filial cannibalism appears singly, and in combination with other behaviours: *eat eggs*, *feed & eat eggs*, *eat eggs & care*, and *feed, eat eggs & care*. All behaviours that include filial cannibalism increase as clutch age and parental energy reserve decrease. However, both *eat eggs* and *feed & eat eggs* increase as clutch number decreases, whereas *eat eggs & care* increases as clutch number increases. *Feed, eat eggs & care* occurs too infrequently to make any generalizations.

A parent that *eats eggs* or *feeds & eats eggs* increases its energy reserve at the expense of one clutch, and any uneaten clutches have low survival for that time step. For parents with one or two clutches at $t = 1$, all behavioural trajectories result in all clutches being eaten, or *total* filial cannibalism. A parent who starts with three or four clutches at $t = 1$, however, may later *care* for any uneaten clutches that survive. An example is the transition between $t = 1$, $X(1) = 5$, $Y(1) = 4$, where the parent *feeds & eats eggs*, and $t = 2$, $X(2) = 8$, $Y(2) = 3$, where the parent *cares*. Because clutch survival without *care* is low, this transition is unlikely.

The transition probabilities for a parent that starts at time $t = 1$, $X(1) = 5$, $Y(1) = 4$ and adopts the behaviour *feed & eat eggs* are easily calculated from the binomial theorem, or by the expansion of $(u + (1\text{-}u))^3$. Any particular term in the expansion is given by:

$$C(v, r)\, u^r \ (1\text{-}u)^{v\text{-}r},$$

where $C(v, r)$ stands for the number of combinations that can be formed from v

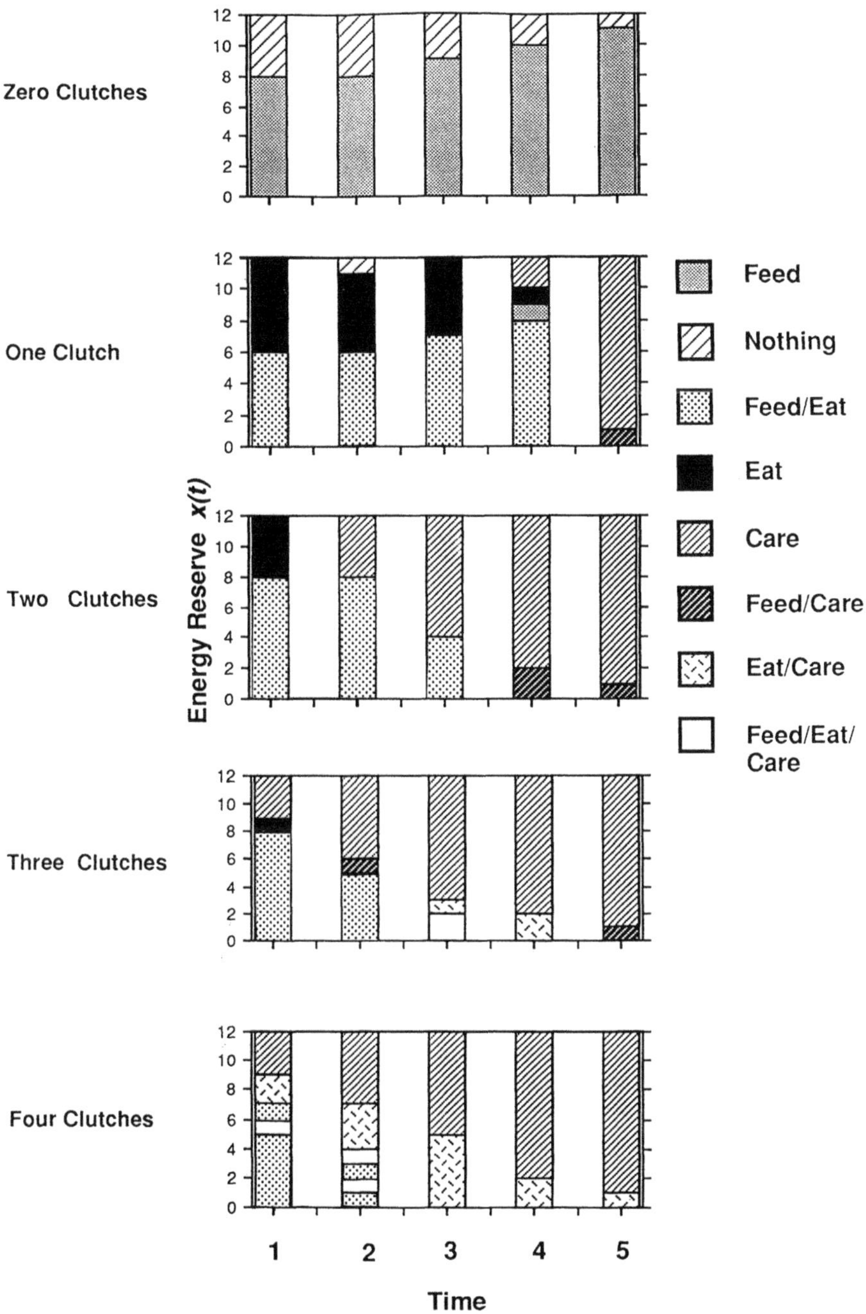

Fig. 3.1 The default run. A graphical representation of the default run depicted in Table 3.1. Along the horizontal axis are the five time steps, and along the vertical axis is the parental energy reserves. The five nest states (i.e. 0, 1, 2, 3, or 4 clutches) are stacked in descending order.

items taken r at a time; $C(v, r)=v!/[r!(v-r)!]$ (Sokol and Rohlf 1981). Clutch survival without parental care is $u = 0.30$. The probability that the three remaining clutches (after one is eaten) survive to time $t+1$ is given by:

$$C(3,3)\, u^3\ (1-u)^0\ =\ 0.027,$$

and the optimal policy in this state ($t = 2, x(2) = 8, y(2) = 3$) is to *care*. Similarly, the probability that two of the three remaining clutches survive to time $t+1$ is given by:

$$C(3,2)\, u^2\ (1-u)^1\ =\ 0.189,$$

and the optimal policy in this state ($t = 2, x(2) = 8, y(2) = 2$) is to *feed & eat eggs*. The probability that only one of the three clutches survives to time $t+1$ is given by:

$$C(3,1)\, u^1\ (1-u)^2\ =\ 0.441,$$

and that optimal policy in this state ($t = 2, x(2) = 8, y(2) = 1$) is to *eat eggs*. Finally the probability that no clutches survive to time $t+1$ is given by:

$$C(3,0)\, u^0\ (1-u)^3\ =\ 0.343,$$

and the optimal policy in this state ($t = 2, x(2) = 8, y(2) = 0$) is to *feed*. Thus, only three per cent of the trajectories from $t = 1, X(1) = 5, Y(1) = 4$ result in the survival of three clutches followed by *care* at time $t = 2$. All other trajectories result in low clutch survival and some combination of *feed* and *eat eggs*. Similar calculations for all other starting conditions that begin with *feed & eat eggs* or *eat eggs* indicate that these behaviours are more likely to benefit the parent's future reproduction than the survival of the remaining clutches.

In contrast, a parent that *eats eggs & cares* appears to be investing in the survival of its remaining clutches. First note that this behaviour does not occur unless the parent has three or more clutches. Second, the behavioural trajectories that include *eat eggs & care* are likely to be followed by high offspring survival. For example, consider the starting point, $t = 1, X(1) = 9, Y(1) = 4$; the parent *eats eggs & cares*. The nest-state transition probabilities can be calculated from the binomial theorem, or by expansion of $(w+(1-w))^3$. With clutch survival with care equal to 0.95, these transition probabilities are: 0.857 to remain at three clutches ($Y(2) = 3$), where the parent *cares*; 0.135 to be reduced to two clutches ($Y(2) = 2$), where the parent *cares*; 0.00713 to be reduced to one clutch ($Y(2) = 1$), where the parent *eats eggs*; and, 0.000125 to be reduced to zero clutches ($Y(2) = 0$), where the parent does *nothing*. Thus, more than 99 per cent of the trajectories from $t = 1, X(1) = 9, Y(1) = 4$ result in the survival of two or three clutches to time $t = 2$, whereupon the parent *cares*. Similar calculations for all other starting conditions that begin with *eat eggs & care* indicate that this behaviour is more likely to benefit the survival of the remaining clutches than the parent's future reproduction.

Thus, it appears that *eat eggs* and *feed & eat eggs* lead to *total* filial cannibalism and represent an investment by the parent into future brood cycles. On the other hand, *eat eggs & care* represents *partial* filial cannibalism, which is an investment by the parent into the survival of the remaining clutches in its current brood. The model was run again, but this time with no future reproduction for the parent, in order to investigate these possibilities.

No future brood cycles

The default run of the model suggested that *eat eggs* and *feed & eat eggs* represent investment in future brood cycles; whereas *eat eggs & care* represents investment in the present brood cycle. To test this hypothesis further, the model was run again with a zero probability for future brood cycles, by setting the terminal fitness function as follows: $\phi(T,Y(T)) = Y(T)$. The parent dies at the end of the brood cycle; therefore, any filial cannibalism that occurs must be an investment in the current brood. The results of this run are presented in Fig. 3.2. Note that the only filial cannibalism that remains is *eat eggs & care*, and that this behaviour increases as clutch number increases. A parent that *eats eggs & cares* increases its energy reserve and provides maximal survival for its remaining clutches, all at the expense of one clutch. Thus, this behaviour represents investment in survival of the current brood. The observation that *eat eggs* and *feed & eat eggs* both disappeared from this run further supports the hypothesis that these behaviours represent investment in future brood cycles.

It now becomes interesting to ask under what conditions would a parent not commit filial cannibalism as an investment in survival of its current brood. Under the first two runs of the model, a large cost to clutch survival was assumed when a parent left its nest to feed; that is $q_f = 0.7$, which represents a 30 per cent reduction in clutch survival with *care* if the parent *feeds & cares*. The effect of reducing this cost is described in the next section.

Reducing the cost of parental feeding to offspring survival

In the default run, the costs of feeding were a reduction in parental survival (e.g. by exposing itself to predators), and a reduction in offspring survival (e.g. by exposing the nest to predators). Because *eat eggs & care* represents an investment in the survival of a parent's current brood, at the expense of one clutch, this behaviour is expected to decrease as the trade-off between feeding and clutch survival is reduced. Therefore, the dependence of *eat eggs & care* on q_f is examined by varying this parameter +0.2 with respect to its default value, while all other parameters and the terminal fitness function were at their default values.

The tendency was for *eat eggs & care* to increase as q_f decreases, or as the cost of feeding to clutch survival increases. The results when $q_f = 0.9$ are presented in Fig. 3.3. Note that *eat eggs & care* no longer appears, but that *eat eggs* and *feed & eat eggs* both do. Thus, partial filial cannibalism appears to be more sensitive than total filial cannibalism to the trade-off between feeding and clutch survival.

The relative value of feeding versus filial cannibalism

In all of the preceding runs, *feed* and *eat eggs* both yielded one unit of energy. Here this assumption is investigated by allowing the energy gains of *feed* and *eat eggs* to each equal two, in separate runs, while holding all other parameters at their default values.

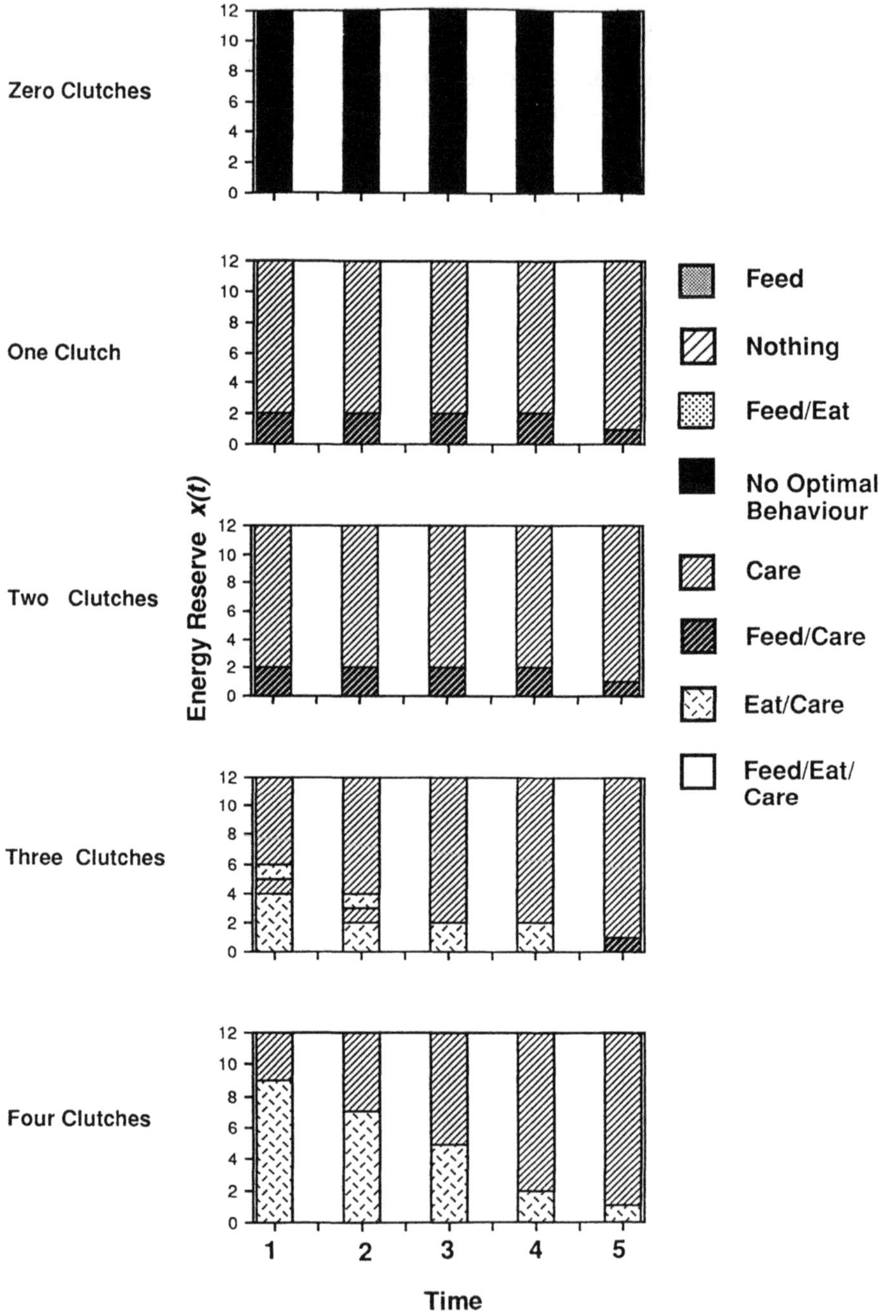

Fig. 3.2 No future brood cycles. Terminal fitness is given by $\phi(T, Y(T)) = Y(T)$; all other parameters are at their default values. The axes are as in Fig. 3.1. Note that when the parent has no future brood cycles, the behaviours, *eat eggs* and *feed & eat eggs*, both disappear; however, the behaviour, *eat eggs & care*, remains.

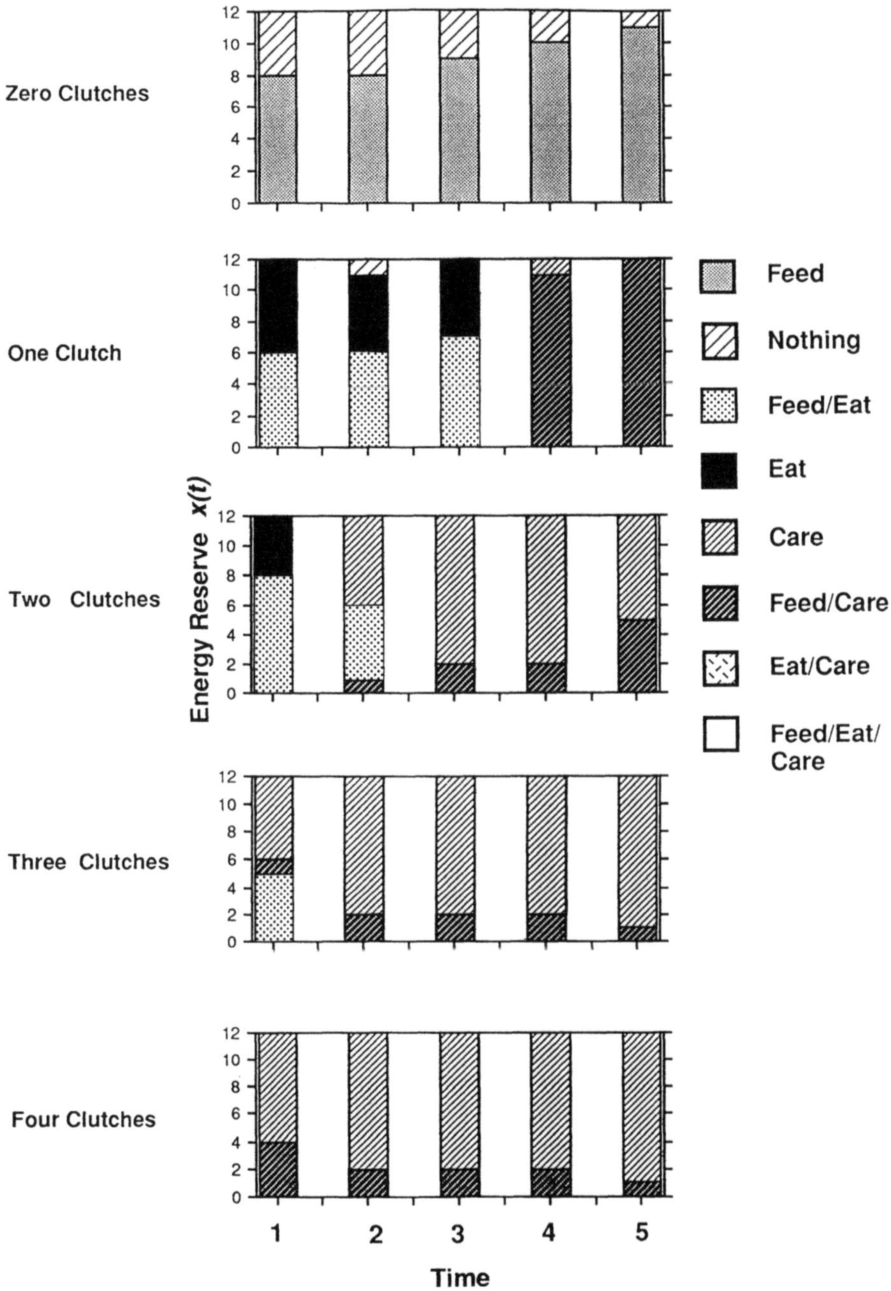

Fig. 3.3 Reducing the cost of parental feeding to offspring survival. In this run, $q_f = 0.90$; all other parameters and terminal fitness are at their default values. The axes are as in Fig. 3.1. Note that relative to the default run, *eat eggs & care* is no longer observed.

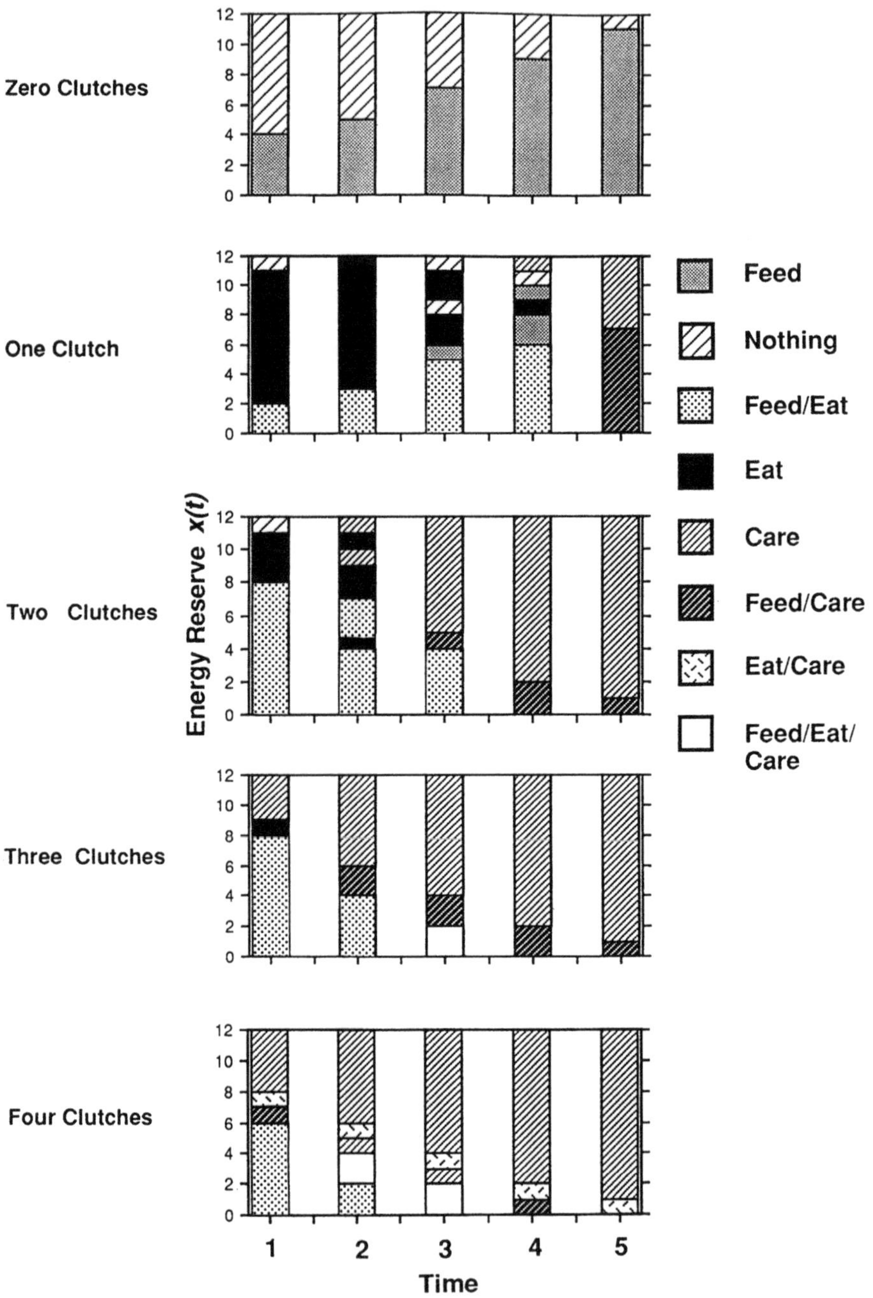

Fig. 3.4 Feeding is more profitable than filial cannibalism. In this run, $\Delta X(F) = 2$. All other parameters and terminal fitness are at their default runs; axes are as in Fig. 3.1. Note that relative to the default run, *eat eggs & care*, is now confined to parents with four clutches, and to a narrower range of energy reserve.

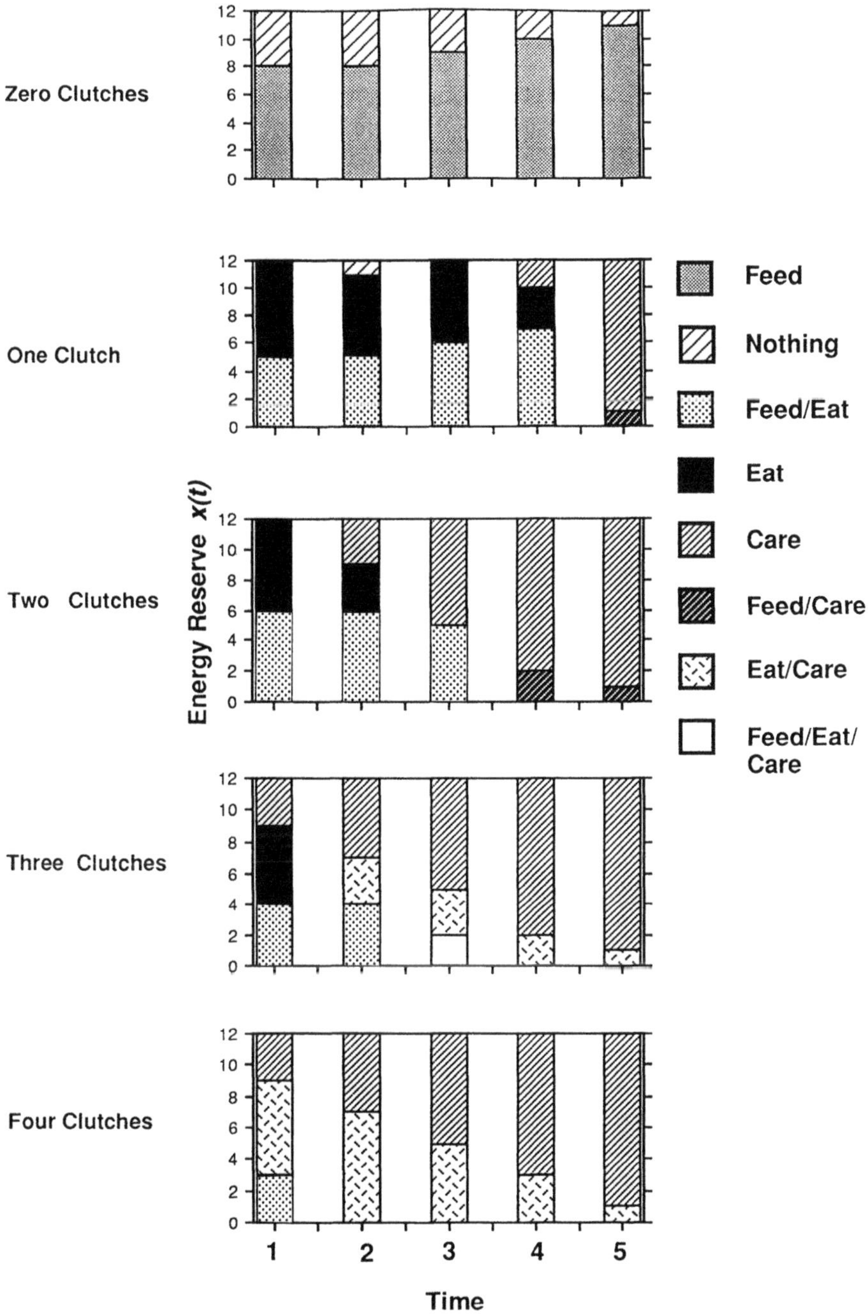

Fig. 3.5 Filial cannibalism is more profitable than feeding. In this run, $\Delta X(E) = 2$. All other parameters and terminal fitness are at their default values; axes are as in Fig. 3.1. Note that relative to the default run, *eat eggs & care* is observed over a wider range of parental energy reserve and time.

When the energy gain from *feeding* is greater than that from *eating eggs*, then the incidence of *eat eggs & care* is reduced; however, *eat eggs* and *feed & eat eggs* are largely unaffected (Fig.3.4). Conversely, when the energy gain from *eating eggs* is greater than that from *feeding*, then the incidence of *eat eggs & care* is increased; again *eat eggs* and *feed & eat eggs* are largely unaffected (Fig. 3.5). Once again, partial filial cannibalism appears to be more sensitive than total filial cannibalism to the relative profitability of *feeding* versus *eating eggs*.

DISCUSSION

A simple dynamic-programming model is presented which investigates the adaptive bases for filial cannibalism. The state variables were: parental energy reserve and clutch number; clutch age was congruent with time. As in previous dynamic-programming models for parental care (Sargent 1990), parental care increases, and feeding decreases, as clutch number increases, as clutch age increases, and as parental energy reserve increases. The model then examined the four predictions outlined in the introduction:

(1) filial cannibalism increases as clutch age decreases;

(2) filial cannibalism increases as energy reserve decreases;

(3) total filial cannibalism increases as clutch size decreases; and

(4) partial filial cannibalism increases as clutch size increases.

Filial cannibalism paralleled feeding with respect to clutch age and parental energy reserve; thus, predictions (1) and (2) were corroborated by the model. Rohwer (1978) postulated two adaptive bases for filial cannibalism: (1) investment in the parent's future brood cycles; and (2) investment in the survival of the current brood. If filial cannibalism represents an investment in future brood cycles, then it should parallel feeding with respect to clutch number. The model showed that filial cannibalism, in the form of *eat eggs* and *feed & eat eggs*, increases as clutch number decreases. These two behaviours are usually followed by more filial cannibalism, and ultimately result in *total* filial cannibalism. Thus, prediction (3) was corroborated by the model. If on the other hand, filial cannibalism represents an investment in the current brood, then it should parallel parental care with respect to clutch number. The model predicted that filial cannibalism, in the form of *eat eggs & care*, increases as clutch number increases; this behaviour usually represents partial filial cannibalism, and is followed by parental care. Thus, prediction (4) was corroborated by the model.

Each of the predictions also has empirical support. Although there are apparently no direct empirical data relating filial cannibalism to parental energy reserve to address prediction (2), Hoelzer (1989) found that supplementing the diet of parental male Cortez damselfish reduces their rate of filial cannibalism, which is indirect support for this prediction. On the other hand, Belles-Isles and FitzGerald (1991) found no relationship between food ration and filial cannibalism in the threespine stickleback, which is indirect evidence against prediction (2).

Three empirical studies directly support the other three predictions. In two paternal egg-guarding species, *Oxylebius pictus* (DeMartini 1987) and Cortez damselfish (Peterson and Marchetti 1989), and in one maternal mouthbrooding species *Pseudocrenilabrus multicolor* (Mrowka 1987) there is empirical evidence that supports predictions (1), (3), and (4). Moreover, circumstantial evidence for all four predictions exists for the paternal egg-guarding species *Pimephales promelas* (Sargent 1988, 1989). In all four species, filial cannibalism decreases as clutch age increases (prediction (1)); parents completely cannibalize small clutches (prediction (3)); and parents partially cannibalize large clutches (prediction (4)). There are apparently no data that have teased apart predictions (3) and (4).

Although it is encouraging that this dynamic-programming model corroborates predictions generated from verbal models (e.g. Rohwer 1978) and static models (e.g. Pressley 1976, Carlisle 1982; Taborsky 1985; Sargent and Gross 1986), the utility of this model will be in experimental manipulations of its parameters. Examples of three parameter manipulations might include: (1) the parent's expectation of future brood cycles; (2) the cost of feeding to clutch survival; and, (3) the relative value of feeding versus filial cannibalism in augmenting the parental energy reserve. The effects of these manipulations of the model may be tested experimentally in the following ways.

1. *Future brood cycles.* As the expectation of future brood cycles decreases toward zero, filial cannibalism as an investment in future brood cycles decreases toward zero; however, filial cannibalism as an investment in the survival of the current brood is largely unaffected. Thus, this manipulation of the model further clarifies the distinction between *total* filial cannibalism as an investment in future brood cycles, and *partial* filial cannibalism as an investment in the current brood cycle.

   One could easily expand this model to study the ontogeny of filial cannibalism over an animal's lifetime, by connecting several brood cycles or breeding seasons together with the technique of sequential coupling (e.g. Mangel and Clark 1988). However, before constructing such a model, one would want to know any seasonal patterns in parent and offspring survival. For example, it may be tempting to use the *no future brood cycles* results above, to predict a decrease in the incidence of filial cannibalism at the end of the breeding season, because of the presumed low residual reproductive value (*sensu* Williams 1966) of the parent at this time. Often, however, one sees an increase in filial cannibalism at the end of the breeding season (e.g. Belles-Isles and FitzGerald 1991), which seems to contradict the model. One missing ingredient is seasonal variation in offspring survival. If end-of-season offspring have substantially lower survival than the parent itself, then a high incidence of filial cannibalism may be favoured at this time. Thus, experimental corroboration of the effects of end-of-brood-cycle terminal fitness may be difficult without detailed demographic information about the study population. An alternative approach may be comparisons among populations that differ in their number of breeding seasons.

2. *Feeding versus clutch survival.* As the cost of feeding to clutch survival decreases, filial cannibalism as an investment in the survival of the current brood decreases. It may be possible to test this prediction by examining how filial cannibalism depends on the density of egg predators, and perhaps on the distance between a parent's nest and its food source.

3. *Feeding versus filial cannibalism.* As the relative profitability of feeding versus filial cannibalism increases, partial filial cannibalism decreases. It may be possible to test this prediction by selectively feeding or starving parents on their nests, and examining the resulting filial cannibalism. Hoelzer (1989), and Belles-Isles and FitzGerald (1991) performed similar manipulations. The results of Hoelzer (1989) support the model, whereas those of Belles-Isles and FitzGerald do not.

It is now interesting to ask, 'What determines filial cannibalism?'. Rohwer (1978) suggested that the conflicting demands of feeding and parental care are so severe as to make these two behaviours virtually mutually exclusive. Although this assumption might be true for many species, such as mouthbrooding cichlids (e.g. Mrowka 1987), it is not necessary for filial cannibalism in this model. In this model, food may be abundant; however, filial cannibalism persists (Fig. 3.1–3.5). There appear to be two separate trade-offs that determine filial cannibalism: (1) the present brood cycle versus future brood cycles; and (2) within the present brood cycle, offspring number versus offspring survival. These two trade-offs correspond with the distinction between *total* and *partial* filial cannibalism respectively. *Total* filial cannibalism is favoured if the benefits of care are outweighed by the expectation of future brood cycles, and is very sensitive to the terminal fitness function (Fig. 3.2). *Total* filial cannibalism increases as offspring number decreases, and thus represents investment in future brood cycles. *Partial* filial cannibalism is favoured if the parent must build its energy reserve to survive to the time horizon, and if there are feeding costs to offspring or parent survival. *Partial* filial cannibalism is very sensitive to the cost of feeding to offspring survival (Fig. 3.3), and to the relative profitability of feeding versus filial cannibalism (Fig. 3.4–3.5). *Partial* filial cannibalism increases as offspring number increases, and thus represents investment in the survival of the present brood.

The model developed here provides insights into the phenomena of parental care and filial cannibalism as reflecting dynamic shifts in the trade-off between present and future reproduction. It also provides insights into when filial cannibalism represents an investment in future brood cycles versus investment in offspring survival during the current brood cycle. However, there are five potentially important phenomena that the model does not address: (1) biparental care; (2) asynchronous clutches; (3) filial cannibalism versus heterocannibalism; (4) resource allocation; (5) more realistic energetics.

1. *Biparental care.* Filial cannibalism is known to occur in fishes with biparental care (see FitzGerald and Whorisky, this volume for review). In such species, there is a conflict of interest between the sexes in the value of offspring as units of reproductive success, and probably in their value as food. Thus, biparental care requires a game-theoretical approach. The analysis of dynamic games with state variables in behavioural ecology is still in its infancy, and beyond the scope of this chapter. However, filial cannibalism in species with biparental care deserves serious theoretical investigation.

2. *Asynchronous clutches.* In this model, all clutches within a nest are assumed to be the same age, and brood cycling, or clutch cycling (*sensu* Peterson and Marchetti 1989) is a constraint. In many species of fish, clutches overlap in age. Sargent (1990) constructed a dynamic-programming model in which a parent faces the conflicting demands of feeding, mating, and providing parental care. In this model, nest-state dynamics are constrained such that no two clutches may be the same age. Moreover, brood cycling is not imposed, because the time horizon is the end of the breeding season, rather than the end of a brood cycle. Nevertheless, brood cycling was found (Sargent 1990). The original version of this model did not explore filial cannibalism; however, in subsequent analyses, we found basically the same patterns of filial cannibalism as is done here (Sargent *et al.* in prep.).

3. *Filial cannibalism versus heterocannibalism.* Thus far, the subject of filial cannibalism versus heterocannibalism (*sensu* Rohwer 1978) has been avoided. Although it might be tempting to explore this subject by manipulating the payoffs to *feeding* versus *eating eggs* (assuming that *feeding* represents heterocannibalism), such an analysis may give inaccurate predictions. Heterocannibalism is best modelled with game theory, because a parent that cannibalizes its neighbour's clutch may have its own clutch cannibalized by the neighbour. Dynamic games with state variables are beyond the scope of this chapter. Nevertheless, the subject of filial versus heterocannibalism is worthy of further investigation.

4. *Resource allocation.* Animals, in life, do not choose their behaviour from a discrete menu; rather, they appear to vary their level of behaviour continuously. Although this model does display similar dynamics to what might be

expected from a resource-allocation model, it is nevertheless inadequate for addressing questions of resource allocation. To convert this model into one of resource allocation would require far more complex state dynamics, which could easily exceed the limitations of a microcomputer. However, dynamic resource-allocation models are likely to yield additional insights into the phenomenon of filial cannibalism, and these models will become more common in the future.

5. *More realistic energetics*. The studies of Whorisky and FitzGerald (1985), Belles-Isles and FitzGerald (1991), and Hoelzer (1989) all suggest that male fish defending eggs may be limited by more than one food-related resource. Belles-Isles and FitzGerald (1991) found that filial cannibalism in the threespine stickleback is independent of ration. Hoelzer (1989) found that filial cannibalism is energy-based in the Cortez damselfish, and that males typically eat algae on their territories and their own eggs. In both species, well-fed males with experimentally supplemented diets still ate small numbers of their own eggs, which cannot be explained by the current model. Thus, it may be necessary to use a multiple-resource model to explain this phenomenon.

   This model also assumes a very simple basis for energy reserve. A more realistic model might track two or more distinct components of a parent's overall energy reserve, such as gut contents and fat reserve (e.g. Mangel and Clark 1988). The inclusion of more biologically realistic energetics in future models may explain a broader set of phenomena.

In summary, in spite of its simplifying assumptions, this model may give broad insights into the diversity of filial cannibalism in nature. This model clearly specifies the energetic trade-offs that can determine filial cannibalism, as originally argued by Rohwer (1978), and it provides insight into the distinction between *partial* and *total* filial cannibalism. Although filial cannibalism in nature is undoubtedly more complex than any model, simple models such as this one may help clarify our thinking, and stimulate new experimental research.

ACKNOWLEDGEMENTS

I thank T. Bakker, A. Berglund, B. Crespi, P. Crowley, M. Elgar, A. Fales, G. FitzGerald, H. Guderley, G. Hoelzer, A. Houston, C. LaCharité, K. Ledford, K. Lindström, J. McNamara, D. Neergaard, J. Newman, B. Rogers, A. Sih, and R. Warner for advice and discussion on the theory and biology of filial cannibalism. Any errors, however, are my own. Some of the ideas expressed in this paper were developed during my participation in two Nordic Council for Ecology minicourses: (1) The Behaviour of Fishes: Ecological Consequences, Tvärminne Zoological Station, Finland, 2–6 October 1989; (2) Alternative Reproductive Strategies, Erken

Biological Station, Sweden, 5–9 March 1990. I thank the organisers and participants in these minicourses for many stimulating discussions. This research was partially supported by National Science Foundation grants BSR-86-14640 and BSR-89-18871.

REFERENCES

Belles-Isles, J.-C. and FitzGerald, G.J. (1991). Filial cannibalism in sticklebacks: a reproductive management strategy? *Ethology Ecology and Evolution*, **3**, 49–62.

Bellman, R.E. (1957). *Dynamic programming*. Princeton University Press.

Carlisle, T.R. (1982). Brood success in variable environments: implication for parental care allocation. *Animal Behaviour*, **30**, 824–36.

Coleman, R.M., Gross, M.R., and Sargent, R.C. (1985). Parental investment decision rules: a test with bluegill sunfish. *Behavioral Ecology and Sociobiology*, **18**, 59–66.

DeMartini, E.E. (1987). Paternal defence, cannibalism and polygamy: factors influencing the reproductive success of painted greenling (Pisces, Hexigrammidae). *Animal Behaviour*, **35**, 1145–58.

Dominey, W. and Blumer, L.S. (1984). Cannibalism of early life stages in fishes. In *Infanticide: comparative and evolutionary perspectives*, (ed. G. Hausfater and S.B. Hrdy), pp. 43–64. Aldine, New York.

FitzGerald, G.J., Guderley, H., and Picard, P. (1989). Hidden reproductive costs in the three-spined stickleback (*Gasterosteus aculeatus*). *Experimental Biology*, **48**, 295–300.

Hoelzer, G. (1988). Filial cannibalism in a non-brood cycling fish. *Environmental Biology of Fishes*, **21**, 309–13.

Hoelzer, G. (1989). Sexual selection and reproductive behavior in the Cortez damselfish (*Stegastes rectifraenum*). Ph.D. dissertation, University of Arizona.

Houston, A.I. and McNamara, J.M. (1988). A framework for the functional analysis of behaviour. *Behavioral and Brain Sciences*, **11**, 117–63.

Mangel, M. and Clark, C.W. (1986). Toward a unified foraging theory. *Ecology*, **67**, 1127–38.

Mangel, M. and Clark, C.W. (1988). *Dynamic modeling in behavioral ecology*. Princeton University Press.

McFarland, D. (ed.) (1982). *Functional ontogeny*. Pitman, London.

McFarland, D. and Houston, A.I. (1981). *Quantitative ethology: the state space approach*. Pitman, London.

McNamara, J.M. and Houston, A.I. (1982). Short term behaviour and lifetime fitness. In *Functional ontogeny*, (ed. D. McFarland), pp. 60–87, Pitman, London.

McNamara, J.M. and Houston, A.I. (1986). The common currency for behavioural decisions. *American Naturalist*, **127**, 358–78.

Mrowka, W. (1987). Filial cannibalism and reproductive success in the maternal mouthbrooding cichlid fish *Pseudocrenilabrus multicolor*. *Behavioral Ecology and Sociobiology*, **21**, 257–65.

Oster, G.F. and Wilson, E.O. (1978). *Caste and ecology in the social insects*. Princeton University Press.

Peterson, C.W. and Marchetti, K. (1989). Filial cannibalism in the Cortez damselfish *Stegastes rectifraenum*. *Evolution*, **43**, 158–68.

Pressley, P.H. (1976). Parental investment in the threespine stickleback, *Gasterosteus aculeatus*. M.Sc. thesis, University of British Columbia.

Rohwer, S. (1978). Parent cannibalism of offspring and egg raiding as a courtship strategy. *American Naturalist*, **112**, 429–40.

Sargent, R.C. (1985). Territoriality and reproductive trade-offs in the threespine stickleback, *Gasterosteus aculeatus*. *Behaviour*, **93**, 217–26.

Sargent, R.C. (1988). Paternal care and egg survival both increase with clutch size in the fathead minnow, *Pimephales promelas*. *Behavioral Ecology and Sociobiology*, **23**, 33–8.

Sargent, R.C. (1989). Allopaternal care in the fathead minnow, *Pimephales promelas*: stepfathers discriminate against their adopted eggs. *Behavioral Ecology and Sociobiology*, 25, 379–85.

Sargent, R.C. (1990). Behavioural and evolutionary ecology of fishes: conflicting demands during the breeding season. *Annales Zoologici Fennici*, **27**, 108–18.

Sargent, R.C. and Gebler, J.B. (1980). Effects of nest site concealment on hatching success, reproductive success, and paternal behavior of the threespine stickleback, *Gasterosteus aculeatus*. *Behavioral Ecology and Sociobiology*, **7**, 137–42.

Sargent, R.C. and Gross, M.R. (1985). Parental investment decision rules and the Concorde fallacy. *Behavioral Ecology and Sociobiology*, 17, 43–5.

Sargent, R.C. and Gross, M.R. (1986). Williams' principle: an explanation of parental care in teleost fishes. In *The behaviour of teleost fishes*, (ed. T.J. Pitcher), pp. 275–93, Croom Helm, Ltd, Beckenham.

Sokol, R.R. and Rohlf, F.J. (1981). *Biometry*. W.H. Freeman, San Francisco.

Stephens, D.W. and Krebs, J.R. (1986). *Foraging theory*. Princeton University Press.

Taborsky, M. (1985). On optimal parental care. *Zeitschrift für Tierpsychologie*, **70**, 331–6.

Trivers, R.L. (1972). Parental investment and sexual selection. In *Sexual selection and the descent of man*, (ed. B. Campbell), pp. 136–79, Aldine, Chicago.

Unger, L.M. (1983). Nest defense by deceit in the fathead minnow, *Pimephales promelas*. *Behavioral Ecology and Sociobiology*, **13**, 125–30.

Unger, L.M. and Sargent, R.C. (1988). Allopaternal care in the fathead minnow, *Pimephales promelas:* females prefer males with eggs. *Behavioral Ecology and Sociobiology*, **23**, 27–32.

Williams, G.C. (1966). Natural selection, costs of reproduction and a refinement of Lack's principle. *American Naturalist*, **100**, 687–90.

Whorisky, F.G. and FitzGerald G.J. (1985). Sex, cannibalism and sticklebacks. *Behavioral Ecology and Sociobiology*, **18**, 15–8.

4

Brood reduction in gymnosperms

David Haig

Cannibalism in animals is often associated with the killing of an offspring by a parent or of one sibling by another. Similar phenomena occur in plants. Mothers abort seeds and ovules, and sibling embryos compete in a life-or-death struggle within seeds. For convenience, I refer to both processes as brood reduction. This chapter addresses the mechanisms and functions of brood reduction in gymnosperms, with a brief comparison to analogous processes in angiosperms. I emphasize gymnosperms because seed and ovule abortion in angiosperms have been extensively reviewed elsewhere (Stephenson 1981; Lee 1988). There is no conclusive evidence that gymnosperms utilize aborted ovules and defeated embryos as food, though this would clearly be adaptive in terms of the efficient use of resources. I discuss brood reduction in gymnosperms because these processes possibly have similar evolutionary causes to filicide and siblicide (associated with cannibalism) in animals.

A BRIEF OVERVIEW OF GYMNOSPERM REPRODUCTION

The life cycle of gymnosperms alternates between haploid and diploid generations (Fig. 4.1). The individual that germinates from a seed and develops roots, stems, and leaves is the diploid *sporophyte*. A sporophyte produces haploid spores by meiosis. A spore does not function directly as a gamete but develops into a multicellular *gametophyte* that produces gametes by mitosis. As a result all gametes produced by a gametophyte are genetically identical. A zygote is produced when an egg is fertilized by a sperm and divides mitotically to form an *embryo*, which is a young sporophyte of the next generation.

Spores and gametophytes are of two kinds. Microspores develop into male gametophytes which produce sperm. Male gametophytes are dispersed from paternal sporophytes as pollen grains. After pollination, pollen grains germinate to form *pollen tubes*. Megaspores develop into female gametophytes which produce egg nuclei, and each egg is situated at the base of an *archegonium*. The female gametophyte is enclosed within the *nucellus*, which is in turn surrounded by one or more *integuments*. The integuments completely enclose the nucellus except for a narrow channel to the outside called the *micropyle*. The female gametophyte

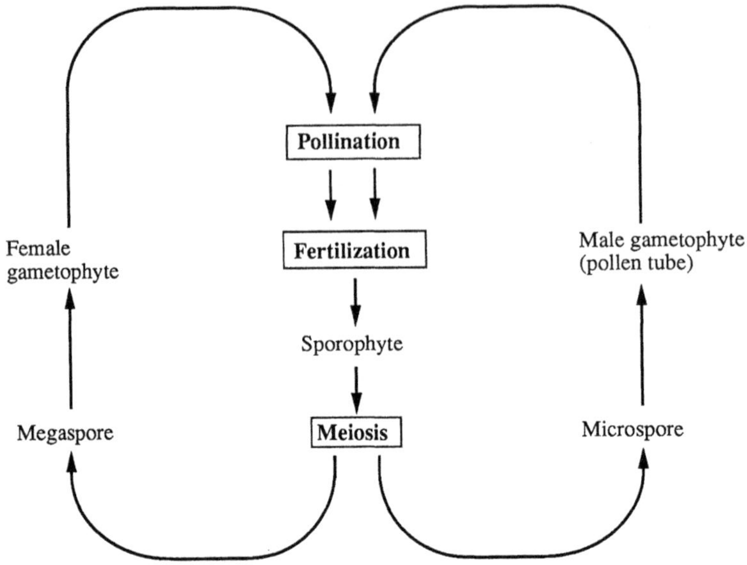

Fig. 4.1 Simplified life cycle of seed plants. The diagram does not represent the telescoping of generations whereby the maternal sporophyte, female gametophyte, and embryo sporophyte all contribute tissues to a seed, nor does it represent variation in the time of female meiosis relative to pollination of the ovule.

(haploid), plus the integuments and nucellus (diploid), form an *ovule* (Fig. 4.2). Once an ovule is fertilized and contains an embryo, it is called a *seed.* Seeds are dispersed from the maternal sporophyte.

Pollination refers to the capture of a pollen grain by an ovule, whereas fertilization refers to the union between sperm and egg nuclei to form a zygote. Archegonia are fertilized by free-swimming sperm in *Ginkgo* and cycads (*zooidogamy).* The male gametes of other gymnosperms are delivered directly to archegonia by pollen tubes (*siphonogamy).* The interval between pollination and fertilization varies greatly among gymnosperms. The shortest interval is less than ten hours in *Ephedra* (Land 1907), but a more typical delay is three to four months. Fertilization occurs one year after pollination in *Agathis* (Kaur and Bhatnagar 1986) and most *Pinus.* These species take two years to complete their reproductive cycle. A few species of *Pinus* have three-year cycles, with a two-year delay between pollination and fertilization (see Singh, 1978 for an overview of time relations). By comparison, the delay is 24 to 48 hours in most angiosperms, though it can be as brief as 15 minutes (*Taraxacum*) or as long as 14 months (*Quercus*) (Maheshwari 1950).

Female gametophytes (and ovules) of modern gymnosperms are poorly developed at the time of pollination, but most grow substantially before fertilization. The ovules of many gymnosperms are close to their final size before fertilization (e.g. *Ginkgo*, cycads: Favre-Duchartre 1958; *Pinus*, *Picea*: Sarvas 1962, 1968). Podocarps and araucarians undergo substantial growth of ovules and female

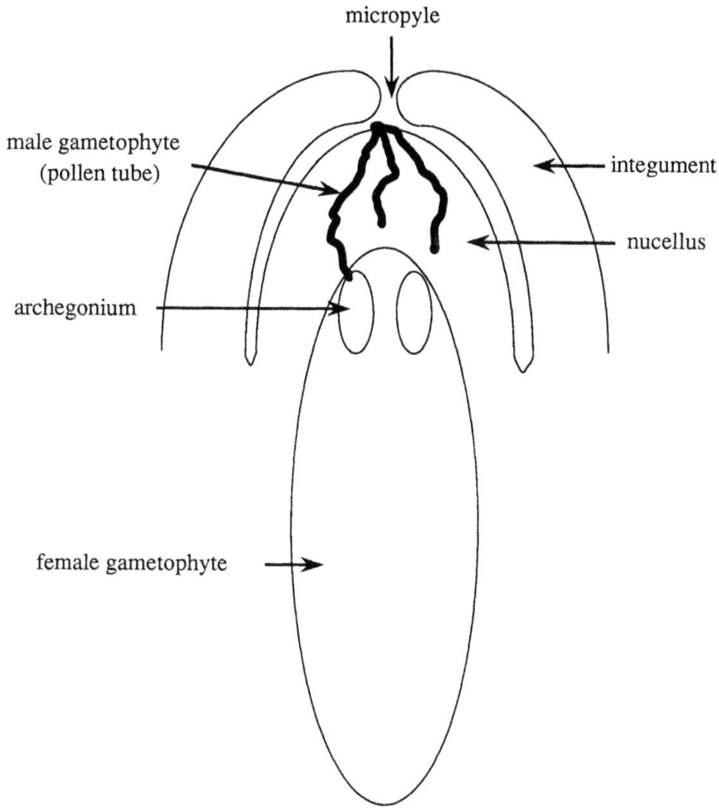

Fig. 4.2 Schematic diagram of a pine ovule showing structures discussed in the text.

gametophytes after fertilization (Burlingame 1915; Osborn 1960; Konar and Oberoi 1969; Haines 1983; Kaur and Bhatnagar 1986) and, in a few other taxa, the greatest part of the female gametophyte's growth occurs after fertilization (*Torreya*: Coulter and Land 1905; *Taxus*: Dupler 1917; *Ephedra*: Land 1907; *Gnetum*: Maheshwari and Vasil 1961; *Welwitschia*: Martens 1971).

Almost all gymnosperms produce female gametophytes with more than one archegonium (see Figure 20 of Favre-Duchartre 1970 and the Appendix to Willson and Burley 1983 for useful reviews). Exceptions are rare; female gametophytes of *Torreya taxifolia* always produce a single archegonium (Land 1907), and some species produce a proportion of female gametophytes with only one archegonium (e.g. Sarvas 1962, 1968). *Simple polyembryony* refers to the development of embryos from more than one zygote within a seed, whereas *cleavage polyembryony* refers to the formation of multiple embryos from the embryonal mass derived from a single zygote. Cleavage polyembryony has a sporadic distribution among gymnosperms (Doyle and Brennan 1971, 1972; Singh 1978; Willson and Burley 1983) whereas simple polyembryony is almost universal.

This brief overview of gymnosperm life cycles has identified interactions in which one genetic individual is known to consume the tissues of a conspecific: male gametophytes form pollen tubes that feed on maternal tissues (Friedman 1987; Choi and Friedman 1991); female gametophytes grow at the expense of the nucellus; and embryos digest the female gametophyte. Such interactions are not conventionally described as cannibalism, but it is far from clear that the interests of the eater and the eaten are identical. For example, male gametophytes may originally have evolved pollen tubes as feeding structures to gain a nutritional advantage over other pollen grains in the pollen chamber. Parasitism on maternal tissues by male gametophytes would then have allowed paternal sporophytes to produce smaller pollen grains, because the costs of maintaining male gametophytes were transferred to maternal sporophytes. Pollen tubes could then have acquired their secondary function (in siphonogamous gymnosperms) of delivering sperm to archegonia before their competitors (Haig and Westoby 1989*a*). These forms of cannibalism will not be addressed further in this chapter.

SEED AND OVULE ABORTION

Brood reduction in gymnosperms occurs at a number of levels. Seed cones may abort, as may individual ovules or seeds within cones, and individual embryos within seeds. The abortion of ovules, seeds, and cones reduces a sporophyte's potential seed crop. These processes mainly involve interactions between gametophytes and the maternal sporophyte. On the other hand, polyembryony primarily involves interactions between embryos and gametophytes, and the death of individual embryos within seeds does not necessarily reduce seed production. This section discusses the causes and functions of ovule and seed abortion, and a subsequent section discusses polyembryony. Ovule abortion occurs before fertilization, and seed abortion occurs after fertilization (by definition of a seed). I define ovule abortion as a form of brood reduction because the female gametophyte that is aborted in an ovule is the sporophyte's 'offspring'.

Ovule, seed, and cone abortion can have a major impact on seed production. The best quantitative data come from long-term studies of *Pinus sylvestris* and *Picea abies* growing in monospecific stands in Finland (Sarvas 1962, 1968). Stands of *P. sylvestris* abscised 19–77 per cent of conelets. In the remaining cones, 13–54 per cent of ovules aborted. As a result, actual seed production was about 20 per cent of potential seed production at poor sites, but up to 80 per cent of potential seed production at good sites. Moreover, roughly 14 per cent of 'actual' seeds were empty (Sarvas 1962). The average proportion of empty seeds was 66 per cent for *P. abies*, though in this species the 'empty seed' category included aborted ovules (Sarvas 1968). Bartram and Miller (1988) have reported seed production from four seed orchards of *Pseudotsuga menziesii* over three successive harvests. From 10 per cent to 87 per cent of conelets aborted, and in the surviving cones, from 44 per cent to 94 per cent of ovules failed to produce a filled seed.

Ovule abortion

Resources committed to an aborted ovule are a cost to the sporophyte. The magnitude of the cost depends on how much has been committed to the ovule at the time it is aborted. A clear distinction must be made between the size of an ovule and its energy content, because the bulk of nutrients accumulate after growth is complete.

Ginkgo and cycads differ from other gymnosperms in that fertilization occurs after nutrient accumulation is complete and sterile ovules contain the same food reserves as seeds with an embryo (Favre-Duchartre 1958). Ovules are small at the time of pollination, and unpollinated ovules abscise (Baird 1939; Lee 1955; Favre-Duchartre 1958; Norstog *et al.* 1986; Friedman 1987; Tang 1987). These species do not provision unpollinated ovules but do provision pollinated ovules which remain unfertilized and fertilized ovules in which all embryos have died. Seeds of other gymnosperms accumulate most of their food reserves after fertilization (Schnarf 1933; Favre-Duchartre 1958), and never provision unfertilized ovules.

Pinus aborts unpollinated ovules shortly after pollination would normally have occurred, when ovules are little-developed (Sarvas 1962; Owens *et al.* 1981*a*). Starch begins to accumulate in the female gametophyte during the earliest stages of embryo development, but it is not until the seed is almost mature that the female gametophyte becomes packed with lipids, protein and starch (Lill 1976; Owens and Molder 1977*a*; Owens *et al.* 1982). *Juniperus communis* also has a two-year cycle and, like *Pinus*, aborts unpollinated ovules in the first growing season (Sarvas 1962).

A different pattern is found in members of the Pinaceae with a one-year reproductive cycle. The integuments develop into a hard seed-coat whether or not ovules are pollinated. Unpollinated ovules form full-sized empty 'seeds' which contain the degenerate remains of the nucellus and female gametophyte (Owens and Molder 1979*a*, 1979*b*, 1980; Owens *et al.* 1981*b*). Degeneration of unpollinated *Picea* ovules begins before pollinated ovules are fertilized (Sarvas 1968; Mikkola 1969), although occasional ovules will develop to the archegonial stage without pollen (Mikkola 1969; Koski 1973).

Female gametophytes of *Araucaria* (Haines 1983) and *Agathis* (Kaur and Bhatnagar 1986) develop archegonia independently of pollination, but do not accumulate nutrients unless fertilized. Unpollinated and pollinated (but unfertilized) ovules abort after other ovules are fertilized, and female gametophytes in fertilized ovules undergo substantial further growth. Thus, the abortion of unpollinated ovules at the archegonial stage is probably less expensive than would be the case in species that complete growth before fertilization. *Metasequoia* appears to be similar to *Agathis* and *Araucaria* (Engels and Gianordoli 1983).

Some ovules abort for reasons other than non-pollination. *Juniperus monosperma* preferentially abscises 'fruits' that are parasitized by insects (Fernandez and Whitham 1989), and unseasonal frosts can cause increased proportions of empty pine seed (Sarvas 1962). In crosses between *Pinus* species, foreign pollen sometimes initiates normal development but ovules abort before fertilization (McWilliam 1959; Mikkola 1969).

Seed abortion

Not all fertilized ovules develop into filled seeds. In years of abundant pollination, most empty *Picea* seeds result from the abortion of fertilized ovules during early embryo stages (Sarvas 1968; Owens and Molder 1979*a*). In two stands of *P. abies*, 8 and 6 per cent of seeds were empty because of abortion before fertilization (including unpollinated ovules), but 30 and 16 per cent of seeds were empty because of abortion after fertilization (Sarvas 1968).

Typically, many more empty seeds are formed after self-pollination than after cross-pollination (Sarvas 1962, 1968; Koski 1973; Franklin 1970; Plym Forshell 1974; Griffin and Lindgren 1985; Smith *et al.* 1988). Self-pollen fertilizes archegonia, but many of the resulting embryos abort during early development. Within a given population, some sporophytes are more self-sterile than others (Plym Forshell 1974; Johnsson 1976) as would be expected if some sporophytes carried more deleterious recessives than others.

Monoecious sporophytes usually receive a mixture of self- and cross-pollen. Koski (1973) concluded that roughly 10 per cent of seedlings are selfed in monospecific conifer stands. This was considerably less than the proportion of own-pollen in all pollen received. Many selfed embryos were probably eliminated in empty seeds or perished in competition with outcrossed embryos in filled seeds. Rudin *et al.* (1986) found 12 per cent selfed embryos in mature seeds from a *Pinus sylvestris* stand. The proportion of self-pollination is likely to be greater for isolated trees and trees in mixed-species stands.

Selfing cannot explain all empty seeds. Sarvas (1962) and Johnsson (1976) recorded 10 and 19 per cent empty seeds following cross-pollination of *P. sylvestris*. There was either a very high frequency of recessive lethal homozygotes among outcrossed progeny, or some empty seeds were produced despite containing viable embryos. Burdon and Low (1973) reported a higher proportion of empty pine seed on a phosphorus-deficient site, which suggested resource-limitation as an explanation of some empty seeds.

Conelet abscission

Pines regularly abscise cones with many aborted ovules, after pollination but before cones enlarge, during the first year of the reproductive cycle. *Pinus sylvestris* abscises conelets if more than 60 per cent of ovules are aborted. In this species, conelet abscission approaches 100 per cent when pollination is poor, but never falls much below 20 per cent, even when pollen is superabundant (Sarvas 1962). Hagman (reported in Sweet 1973) observed 30 per cent conelet abortion in three pine species, despite adequate pollination. Conelet abortion is a problem in pine-seed orchards even when there is no evidence that pollen is limiting (Sweet and Thulin 1967; Sweet 1973; White *et al.* 1977). *Juniperus communis* also abscises poorly-pollinated conelets in the first year of its reproductive cycle (Sarvas 1962), whereas *Picea abies* (which matures seeds in the same year as pollination) does not abort poorly-pollinated cones, even if the cones have 100 per cent empty seeds (Sarvas 1968). Thus, conelet abortion appears to be associated with two-year reproductive cycles (Sarvas 1962).

Functions of seed and ovule abortion

Brood reduction appears paradoxical. Why should a parent initiate an ovule, only to abort it at a later stage? The paradox is easily resolved. An ovule's fate is uncertain at the time when it is initiated. Some ovules will outcross, some will self, and others will remain unpollinated. Moreover, each ovule is subject to the vagaries of parasitism, frost-damage, drought, etc. Thus, sporophytes can gain higher fitness returns from some ovules than from others. Brood reduction can be adaptive if resources are redirected from low-yielding to high-yielding ovules. A sporophyte will obviously benefit from aborting unpollinated ovules before they are fully provisioned, because these ovules will give no return on investment. If so, a sporophyte should also benefit from aborting ovules that give a very low return on investment. For example, some viable ovules may be aborted when poorly-pollinated conelets abscise, but these ovules would give a low return if matured, because the cost per seed should be greater for few-seeded cones.

There are two senses in which a sporophyte's fitness returns can be 'uncertain'. These will be illustrated using the example of pollinated and unpollinated ovules. (1) The proportion of pollinated ovules could be predictable when ovules are initiated, but which ovules will be pollinated could be uncertain. Such uncertainty is necessary for brood reduction to be adaptive because a sporophyte should initiate only those ovules that would be pollinated, if the fate of individual ovules were known. (2) The proportion of ovules to be pollinated could be uncertain at the time of ovule initiation. Such uncertainty could explain why a sporophyte might occasionally abort ovules with high expected fitness.

If ovules can be produced for less than the cost of a seed, a sporophyte might initiate more ovules than it is able to provision, abort unpollinated ovules, and provision the remainder. If the proportion of pollinated ovules is predictable in advance, then the number of ovules should be adjusted so that there are sufficient resources remaining to provision all pollinated ovules. However, if the proportion of pollinated ovules is uncertain, some pollinated ovules may abort in years of abundant pollination because the sporophyte has insufficient resources. In such years, the sporophyte could have matured more seeds if it had produced fewer ovules. On the other hand, if pollination is poor, resources may remain after all pollinated ovules have been provisioned. In such years, the sporophyte could have matured more seeds if it had produced more ovules. The abortion of 'excess' ovules when pollination is good is the price that is paid for a buffer against low seed production when pollination is poor (Haig and Westoby 1988*a*; Cohen and Dukas 1990).

There is ample evidence that pollination frequency varies from year to year. This can be illustrated by two Finnish examples. At the Tuusula XXIII stand of *Pinus sylvestris*, four per cent of ovules were unpollinated in 1958 but 32 per cent were unpollinated in 1959 (Sarvas 1962). Over a period of 13 years at the Tuusula XXX stand of *Picea abies*, the proportion of empty seeds ranged from 22 to 100 per cent, mainly because of differences in pollination (Sarvas 1968). There is also circumstantial evidence that resources may be limiting when pollen is abundant. Conelet drop in *Pinus* is reduced but never eliminated in years of high pollen density (Sarvas 1962; Sweet 1973), and repeated hand-pollination of *Picea* does not reduce the proportion of empty seeds below 50 per cent (Ho 1985). The evidence for resource limitation is circumstantial because it has not been shown that aborted ovules would develop if other ovules were removed.

The argument developed here for pollinated and unpollinated ovules applies generally to other unpredictable causes of ovule loss. If more ovules are initiated than can be provisioned, then substantial variation in levels of pollination or ovule predation can occur without corresponding variation in seed production. The number of ovules initiated should be such that some viable ovules are aborted because of resource limitation in the best years but reproductive resources are not fully utilized in the worst years. Such resources could, of course, be redeployed for vegetative growth and future reproduction. Other things being equal, the cheaper an aborted ovule is relative to a seed, the greater the number of ovules that should be initiated and the greater the number of years in which seed production is predicted to be resource-limited. The argument also works in reverse, if the quantity of resources that will be available for seed maturation is uncertain when ovules are initiated. In this case, some viable ovules should be aborted in the worst years (when resources are scarce) but some resources should remain unutilized in the best years (when resources are abundant) (Kozlowski and Stearns 1989; Cohen and Dukas 1990).

This section has identified two possible functions of brood reduction that are not mutually exclusive. The first is the increased reproductive efficiency achieved by aborting low quality offspring. The second is the *buffer* against environmental uncertainty that is provided by large numbers of cheap ovules.

POLYEMBRYONY

Embryos are the progeny of gametophytes, and only indirectly the progeny of sporophytes. Since gametophytes are genetic individuals in their own right, it is interesting to consider the functions of polyembryony from their perspective. A gametophyte passes on its genes when it contributes a gamete to the zygote that becomes the successful embryo in a seed. Each gametophyte can at best contribute one gamete to the next (sporophyte) generation because only one embryo germinates from each seed, yet most female gametophytes and many male gametophytes contribute gametes to more than one zygote. This section discusses the factors that determine how many embryos are produced within a seed and discusses why gametophytes should produce more than one gamete.

Number of archegonia

Female gametophytes of almost all gymnosperms produce two or more archegonia (see Figure 20 of Favre-Duchartre 1970 and the Appendix to Willson and Burley 1983 for useful reviews). Exceptions are rare. Female gametophytes of *Torreya taxifolia* only ever produce a single archegonium (Land 1907), and some other species produce a proportion of female gametophytes with only one archegonium (e.g. *Abies balsamea* (15 per cent): Favre-Duchartre 1970). The number of zygotes produced within an ovule cannot exceed the number of archegonia. Therefore, counts of archegonia give an upper estimate as to the size of a female gametophyte's brood.

Most zooidogamous gymnosperms have low numbers of archegonia per female gametophyte. Female gametophytes of *Ginkgo* usually produce two archegonia (Lee 1955; Favre-Duchartre 1958), and those of *Macrozamia* spp. usually four to seven archegonia (Chamberlain 1913; Baird 1939; Brough and Taylor 1940). *Microcycas* is an exception. Female gametophytes of this species are reported to produce more than 64 archegonia (Caldwell 1907; Reynolds 1924).

Archegonia of most siphonogamous gymnosperms occur singly, rather than being grouped into complexes. In these species, female gametophytes typically produce four to seven archegonia (Singh 1978). The Cupressaceae (including Taxodiaceae: Price and Lowenstein 1989) are the exception. Members of this family have archegonia grouped into complexes and typically produce higher numbers of archegonia (Favre-Duchartre 1970; Singh 1978; Willson and Burley 1983). *Biota*, for example, produces 15–28 archegonia (Singh and Oberoi 1962). Within the Cupressaceae, female gametophytes produce either a single apical complex (as in *Biota*), or one or more lateral complexes each associated with a pollen tube. Very large numbers of archegonia have been reported for species with lateral complexes (Saxton 1934; Moseley 1943) though these appear to be counts of archegonial initials rather than functional archegonia.

Other factors influencing polyembryony

The number of embryos within a seed, and hence the severity of brood reduction, are influenced by factors other than the number of archegonia. These include the quantity of pollen received by an ovule, the capacity of the pollen chamber, and how many pollen tubes are able to fertilize an egg. Three examples from the Pinaceae will illustrate the interplay of factors.

Sarvas' (1962) study of *Pinus sylvestris* is once again invaluable. Simple polyembryony was common but some ovules produced only one zygote because the female gametophyte had only one archegonium or because the pollen chamber contained only one pollen grain. When pollinated ovules were examined, about 30 per cent of pollen chambers contained one pollen grain, roughly 50 per cent contained two grains, and the rest contained three or more grains. Pollen chambers contained a single pollen grain either because this was the maximum capacity of the chamber or because the ovule received insufficient pollen. Most pollen chambers had a maximum capacity of two to three pollen grains, but about 18 per cent of pollen chambers had room for only one grain. When pollen was abundant, most ovules received more pollen at the micropyle than could be accommodated within the pollen chamber. Almost all pollen grains that reached the pollen chamber fertilized an archegonium if there was one available, and most archegonia in pollinated ovules were fertilized. The most frequent number of archegonia was two (about 70 per cent of ovules), but some trees produced up to 30 per cent of ovules with only one archegonium.

Sarvas (1968) reported similar observations for *Picea abies*. About 30 per cent of ovules with embryos contained only one embryo, but this figure overestimated the frequency of single fertilizations because many embryos aborted before the third zygotic division and were not recorded. Pollen chambers had an average capacity of five pollen grains, and female gametophytes produced an average of three archegonia. Therefore, some pollen tubes were unable to fertilize an egg. In a year of abundant pollen, 90 per cent of archegonia were penetrated by a pollen tube. About six per cent of pollen chambers had room for only one grain, though less than one per cent of ovules contained a single archegonium.

Pinus monticola revealed a different pattern (Owens and Molder 1977*a*). Female gametophytes produced three to five archegonia, but it was unusual for more than one archegonium to be fertilized, even if the nucellus contained abundant pollen tubes. Fertilization of one egg appeared to inhibit the fertilization of adjacent eggs.

Functions of polyembryony

Buchholz (1922) proposed that polyembryony is a mechanism of 'developmental selection', such that competition among embryos within seeds results in the elimination of less vigorous embryos (cf. Stearns 1987). Developmental selection remains the most plausible function of simple polyembryony. By the time of fertilization, most gymnosperms have already made a substantial investment in each ovule. Therefore, any seed that lacks a viable embryo constitutes a significant loss of reproductive resources. However, if more than one egg is fertilized, an ovule has an increased probability of producing at least one viable embryo, and some seeds will contain more than one viable embryo. Buchholz suggested that competition among viable embryos would be adaptive because the surviving embryos have better than average viability. Clearly, the benefit to be gained from each additional zygote decreases as the number of zygotes per seed increases.

Buchholz emphasized the importance of fair competition among embryos, but fertilizations are not always synchronous within an ovule (Kozinski 1986). Mikkola

(1969) has questioned the evidence for effective competition among *Picea* embryos, because one embryo always seemed to have a favoured position within the corrosion cavity. Such observations are not fatal to Buchholz's hypothesis. Developmental selection is analagous to choosing a candidate for a job. Some species could select embryos by a 'competitive audition' in which all genotypes compete on even terms, whereas other species could give 'probationary appointments' in which the first embryo always wins unless it is inviable. The second strategy could be more efficient if competition is costly and an embryo's competitive ability is poorly correlated with seedling fitness.

For developmental selection to be effective, some pollen grains must be more likely than others to produce successful embryos. Fowler (1964) reported a shortage of mutant homozygotes among the selfed progeny of a heterozygous sporophyte. Matheson (1980) and Smith *et al.* (1988) found that mixtures of self-and outcross-pollen produced fewer than expected inbred seedlings. Apsit *et al.* (1989) showed that some sporophytes were more successful than others in mixtures of outcross-pollen. These studies did not identify whether differential male success was determined before or after fertilization.

Self-pollination of gymnosperms usually results in a greater proportion of empty seeds and poor growth of seedlings from filled seeds (Franklin 1970; Koski 1973; Matheson 1980). If inbreeding depression is caused by deleterious recessives and ovules receive a mixture of self- and cross-pollen, simple polyembryony would result in fewer empty seeds and a disproportionate number of outcrossed progeny (Sorensen 1982). Simple polyembryony could also be adaptive if ovules only received self-pollen, because each ovule would have more than one chance to produce a viable embryo. Dioecious gymnosperms also possess simple polyembryony, which suggests that developmental selection can be adaptive when there is no possibility of self-fertilization.

The percentage of empty seeds produced after selfing has been used to estimate the average number of recessive lethals per sporophyte in several gymnosperm populations (Sorensen 1969; Franklin 1972; Koski 1973; Park and Fowler 1984). Griffin and Lindgren (1985) obtained a better fit to data when they assumed that embryo death was caused by homozygosity at two or more loci that were individually non-lethal. Such studies have revealed higher genetic loads than similar estimates from animals. Klekowski (1982) proposed that simple polyembryony reduces the fitness costs of each genetic death because seeds containing embryos homozygous for a recessive lethal are also likely to contain viable embryos that are heterozygous for the allele. As a result, natural selection against recessive lethals is relaxed, and the alleles are maintained at higher equilibrium frequencies.

A tacit assumption of all these models is that the abortion of a fertilized ovule indicates that all embryos within the seed have lethal genotypes. In effect, the female gametophyte and maternal sporophyte are assumed to be passive participants in an embryo's death. As argued previously, some viable seeds may be aborted because resources are limiting or because the expected fitness return from the seed is low. For example, seeds could be actively aborted if the vigour of the 'best' embryo falls below some threshold. In this case, vigour (and embryo death) would be a quantitative character determined by the segregation of alleles at many loci (cf. Haig and Westoby 1988*b*). Consistent with this view, Bishir and Namkoong (1987) concluded that five to 30 per cent of empty seeds were the result of 'maternal effects' rather than embryonic lethals. However, I am generally sceptical about all genetic models based on the frequency of empty seeds. If a model allows sufficient flexibility in its choice of parameters and number of loci, agreement with observation is hardly strong confirmation of the model, particularly as embryo deaths are unlikely to have a single cause.

Developmental selection appears to be an unsatisfactory explanation of cleavage polyembryony, because cleavage produces embryos with identical genotypes. Sarvas (1962) and Willson and Burley (1983) have suggested that cleavage might increase a genotype's competitiveness in its struggle with other zygotic genotypes within the seed. Sorensen (1982) suggested that cleavage could reduce the effects of climatic disturbance, provided that cleavage embryos develop asynchronously and developmental stages vary in their sensitivity to environmental stress. Hurst (1990) suggested that competition among cleavage products might be adaptive if the embryos have different cytoplasmic parasites.

Reproductive strategies of male gametophytes

Ovules often contain two or more male gametophytes (pollen tubes), but only one contributes its genome to a successful embryo. Therefore, the male gametophytes within an ovule are strict competitors for reproductive success. In theory, male gametophytes could compete by fertilizing as many archegonia as possible, thus pre-empting fertilization by other gametophytes. In this case, the interests of male and female gametophytes conflict, because all zygotes fathered by a pollen tube are genetically identical, thus decreasing the efficiency of developmental selection.

Male gametophytes of most zooidogamous gymnosperms produce two functional sperm (Chamberlain 1912, 1916; Lawson 1926; Baird 1939; Brough and Taylor 1940; Friedman 1987). *Microcycas* is an exception and produces up to 16 sperm per pollen tube (Caldwell 1907). The motile sperm of *Ginkgo* and cycads are enormous. *Zamia* sperm can reach 400 μm in diameter and have over 10,000 flagella (Norstog 1975). Frequently, several sperm force their way into an archegonium though only one can fertilize the archegonium's egg (Chamberlain 1912; Brough and Taylor 1940; Lee 1955). Male gametophytes of cycads produce fewer sperm than there are archegonia to fertilize, suggesting that there is a trade-off between sperm size and number with smaller sperm being less competitive in the race for fertilizations (Haig and Westoby 1989*a*).

The pollen tubes of siphonogamous gymnosperms deliver their gametes directly to archegonia. A pollen tube can deliver sperm only to one archegonium when archegonia occur singly (Lawson 1907). Curiously, these pollen tubes produce two sperm though only one can fertilize an egg. When archegonia are grouped into complexes, a pollen tube can fertilize multiple archegonia. Both sperm released by a pollen tube have been observed to fertilize an egg in several members of the Cupressaceae (Lawson 1907; Saxton 1909; Buchholz 1939*a*, 1939*b*; Baird 1953), and individual male gametophytes of *Cupressus* produce up to 14 sperm (Doak 1932, 1937; Mehra and Malhotra 1947).

Reproductive strategies of female gametophytes

All zygotes within an ovule carry a copy of the female gametophyte's genome. From the perspective of a gene in the female gametophyte, the important issue is not whether it will be present in the successful zygote, but with what other genes it will be associated. The female gametophyte's genome is often combined with different sets of paternal genes because of simple polyembryony. Natural selection will favour any process in the female gametophyte that ensures the most compatible combination becomes the successful embryo. Selection among embryos within a seed acts on the paternal genetic contribution because the maternal genetic contributions are identical.

The more paternal genotypes are sampled, the greater the expected fitness of the best genotype. The greatest increment in expected fitness occurs when the number of genotypes is increased from one to two. Thereafter, the increase in expected fitness decreases for each additional genotype. Moreover, developmental selection has costs as well as benefits. A female gametophyte receives a limited amount of resources from its mother and allocates these resources between extra embryos or greater food reserves. There must be some finite number of embryos which maximizes the gametophyte's expected fitness, because the increase in expected fitness obtained from each additional embryo decreases but the cost of an embryo does not.

The benefits of simple polyembryony are influenced by the variation in quality among embryos within an ovule. Monoecious sporophytes can receive a mixture of self and cross pollen, but dioecious sporophytes never receive self pollen. One might predict that, as a result, dioecious species would produce fewer archegonia than related monoecious species. It is not possible to quantitatively test this idea because monoecy versus dioecy is usually a constant character within a genus, often constant within families (see Fig. 1 of Favre-Duchartre 1970), and comparisons between higher taxa are usually confounded with other factors. For example, the dioecious cycads and the monoecious Pinaceae produce similar numbers of archegonia (*Microcycas* excluded from comparison). This appears to contradict the prediction based on breeding system, but cycads are zooidogamous and have male gametophytes that can fertilize more than one archegonium, thus reducing the variation in quality among embryos.

If a male gametophyte fertilizes two or more eggs, the resulting embryos are genetically identical and the female gametophyte would have to produce additional zygotes to obtain the same benefit from developmental selection. Therefore, other things being equal, female gametophytes should produce more archegonia in those species with male gametophytes that fertilize more than one archegonium. Thus,

Microcycas has more archegonia per female gametophyte and produces more sperm per male gametophyte than other cycads. Similarly, siphonogamous species with archegonial complexes produce more archegonia than species with scattered archegonia. These comparisons are not confounded by breeding system, because all cycads are dioecious and the predominantly monoecious Pinaceae (scattered archegonia) have fewer archegonia than the predominantly monoecious Cupressaceae (archegonial complexes).

Gymnosperms are an exception to the rule that chloroplasts are usually maternally inherited (Whatley 1982; Neale *et al.* 1986; Szmidt *et al.* 1987; Wagner *et al.* 1987; Neale and Sederoff 1989). In at least some species, maternal plastids are excluded from the neighbourhood of the egg nucleus *before* fertilization (Whatley 1982; Owens and Morris 1990). This suggests that the female gametophyte's nuclear genome has an interest in paternal rather than maternal inheritance of the plastid genome. Because plastids are paternally inherited, simple polyembryony involves competition among embryos with different paternal plastids, whereas, if plastids were maternally inherited, all plastids would have the same maternal genome. Thus, paternal inheritance of organelles possibly increases the effectiveness of developmental selection by increasing the variance in quality among embryos.

An embryo's perspective on polyembryony

Within a seed, two embryos fathered by different male gametophytes are genetically identical for maternal alleles but may have different paternal alleles. Such embryos do not correspond to any of the conventional categories of full sibs, half-sibs, or selfed-sibs. The male gametophytes that are the 'fathers' of a pair of embryos may be the progeny of the same sporophyte or of different sporophytes, and either 'father' may be the progeny of the same sporophyte as the female gametophyte that is fertilized. Embryos fathered by a single male gametophyte are equivalent to monozygotic twins. Such twins are produced when a male gametophyte fertilizes two archegonia, or when the products of a single zygote cleave to form multiple embryos.

At any locus in an embryo there are two alleles, one maternally derived and the other paternally derived. Clearly, the genetic interests of the alleles are different with respect to polyembryony. The maternally derived allele is present in every other embryo within the seed. Therefore, this allele's genetic interests are the same as those of the female gametophyte, and the allele's inclusive fitness is maximized if the fittest embryo within the seed is successful in embryonic competition. On the other hand, the paternally derived allele may be absent from the other embryos within the seed. Therefore, this allele's inclusive fitness will usually be maximized by the victory of its own embryo, rather than by the success of another embryo.

A conventional assumption of evolutionary models has been that an allele has the same phenotypic effect whether it is maternally or paternally derived. Therefore, gene expression in the embryo would be expected to be a compromise between the different interests of maternal and paternal alleles. However, this assumption is not always valid. Differential expression of maternal and paternal alleles at a locus is predicted, and observed, when the alleles have different genetic interests and when gene expression at the locus can influence these interests (Haig and Westoby 1989*b*).

Loci that are involved in embryonic competition satisfy the conditions for differential gene expression. Therefore, one might predict paternal genes in embryos would be particularly active in embryonic selection and maternal genes would have a more passive role. There is only indirect evidence to support this prediction. Parthenogenesis is unknown among mammals, and its absence is explained by 'genomic imprinting' and the requirement for both a maternal and a paternal genome during development (Solter 1988). Parthenogenetic gymnosperms are also unknown, and this could be explained in a similar manner if paternal and maternal genes have distinct roles during embryogenesis. If these speculations are well-founded, there is no such thing as an embryo's perspective on polyembryony. Rather, maternal and paternal alleles should pursue their separate interests.

DEVELOPMENTAL SELECTION AND SEED ABORTION

Embryonic selection within seeds can enhance the average quality of seedlings, provided that the surviving embryos are an improved subset of all embryos. However, there will be residual variation in quality among the embryos that survive in different seeds. This section discusses whether seed abortion could have a role in further enhancing seedling quality.

Haig (1990) has described the optimal allocation of a sporophyte's resources among seeds, when the seeds differ in expected fitness. The return in seedling fitness per unit investment is maximized when the sporophyte aborts all ovules where the expected fitness falls below some threshold and then provisions the rest. The allocation pattern is an ESS (evolutionarily stable strategy) if the decision whether to abort is determined by genes expressed in the sporophyte. This is because a gene in the sporophyte has an equal likelihood of being present in the female gametophyte of each ovule, and because the distribution of resources which maximizes the number of successful seedlings also maximizes the number of seedlings which carry the gene.

At the ESS, whether or not a seed is aborted depends on the sporophyte's expected return on investment if the seed were provisioned. The expected return depends on the information available to the sporophyte. If the only information available is whether a seed contains an embryo, a sporophyte should provision seeds with embryos and abort the rest. If the sporophyte has additional information that allows the relative viability of embryos to be assessed, more sophisticated strategies are possible.

Gymnosperm sporophytes probably have very limited information about embryo quality. Embryos are surrounded by the female gametophyte which solicits nutrients from the sporophyte. Sporophytes probably must rely on the relative vigour of female gametophytes to assess the relative quality of embryos. There should be only limited variation in vigour among female gametophytes because grossly defective gametophytes would be eliminated before fertilization, and the remaining variation should be only weakly correlated with differences in embryo quality. On the other hand, the female gametophyte should have much better information about embryo vigour but should be less likely to abort the ovule if the embryo has low vigour.

Suppose that a gene expressed in the female gametophyte determines whether or not an ovule will be aborted. If all zygotes produced by the female gametophyte were inviable, the gene would benefit from its own ovule's abortion because resources could be redeployed to other ovules on the same sporophyte. These ovules would have a 50 per cent chance of carrying the gene in their female gametophyte. Since the gene benefits from the abortion of its own ovule when all embryos are inviable, the gene could also benefit from abortion if the best embryo has very low viability. The gametophytic ESS has the same form as the sporophytic ESS. That is, all ovules where the expected fitness falls below some threshold should be aborted and the rest provisioned.

Given equivalent information about embryo quality, genes expressed in female gametophytes should evolve a lower viability threshold for abortion than genes expressed in sporophytes, because genes expressed in gametophytes are subject to a 'prisoner's dilemma' (Axelrod 1984). That is, there are two alleles at each locus in a sporophyte. Half of the female gametophytes produced by the sporophyte will carry one allele and half will carry the other. Resources are optimally allocated when seeds that fall below the sporophytic threshold are aborted. However, if one allele aborts its own ovules when they fall below the threshold, the other allele can benefit by provisioning ovules with slightly lower viability because, by so doing, the second allele's ovules obtain more than 50 per cent of the sporophyte's reproductive resources. This is mathematically analogous to meiotic drive. If both alleles adopt the gametophytic ESS, both receive 50 per cent of reproductive resources on average, but each ovule receives more than the optimal amount of resources (see Queller 1984 for a more formal development of ESS models).

In summary, the female gametophyte of gymnosperms controls access to information about embryo quality. As a consequence, genes expressed in the female gametophyte are likely to determine which seeds are aborted in the course of developmental selection, and the sporophyte's resources will be suboptimally allocated among seeds.

Developmental selection in angiosperms

Brood reduction has many similarities between angiosperms and gymnosperms. Angiosperms abort unpollinated ovules and individual seeds within fruits, as well as abscising entire fruits. These processes probably have analogous functions to those discussed above for gymnosperms (see Stephenson 1981; Lee 1988).

Angiosperms differ from gymnosperms in at least one crucial respect: angiosperms produce only one egg per female gametophyte, in contrast to the multiple eggs of most gymnosperms. Therefore, angiosperms lack simple polyembryony, and developmental selection among embryos (if it occurs) would require selective seed abortion.

Two factors may have favoured angiosperms adopting selection among ovules as an alternative to selection within ovules. First, mature female gametophytes of angiosperms usually contain no more than eight nuclei, compared with hundreds to several thousand nuclei in gymnosperms (Favre-Duchartre 1970). Thus, gymnosperms make a greater investment in each ovule before fertilization, and abortion after fertilization would be correspondingly more costly. Second, an angiosperm sporophyte probably has better information about embryo quality than its gymnosperm counterpart because the sporophyte interacts directly with embryos or endosperms (see below) rather than interacting indirectly with embryos through the mediation of female gametophytes.

Endosperm is perhaps the most distinctive feature of angiosperms. At fertilization, a pollen tube releases two sperm into the female gametophyte. One fuses with the egg nucleus to form the zygote and the other fuses with two other nuclei of the female gametophyte to form a triploid nucleus which gives rise to the endosperm. The female gametophyte usually does not persist long after fertilization, and the seed's food reserves are stored in the endosperm or in the embryo itself. The endosperm is genotypically identical to its associated embryo, except for the second copy of the maternal genome. Thus, an angiosperm sporophyte may be able to assess embryo fitness directly, or from properties of the endosperm, and use this information in choosing which seeds to abort (Westoby and Rice 1982). If so, developmental selection in angiosperms may be more effective than that of gymnosperms because the sporophytic ESS can be enforced (see previous section).

CONCLUSIONS

Among gymnosperms, sporophytes regularly produce more ovules than develop into filled seeds, and female gametophytes often produce two or more embryos even though only one survives in the mature seed. Sporophytes and female gametophytes probably benefit from brood reduction because it allows resources to be redirected from low-yielding to high-yielding offspring (unpollinated ovules are classed here as low-yielding offspring), and because the 'overproduction' of offspring creates a buffer against environmental uncertainty. Seed and ovule abortion can increase reproductive efficiency because gymnosperm ovules do not accumulate food reserves until after pollination or fertilization. Similarly, the additional embryos in polyembryonic seeds are produced for a fraction of the cost of a mature seed. The costs of abortion would be further reduced if nutrients are recovered from aborting offspring.

Finally, the gymnosperm life cycle involves interactions in which one genetic individual is known to consume the tissues of a conspecific. Male gametophytes form haustoria that penetrate maternal tissues (Friedman 1987; Choi and Friedman 1991), female gametophytes grow at the expense of the nucellus, and embryos digest the female gametophyte. Such interactions are not conventionally described as cannibalism, but it is far from clear that the interests of the eater and the eaten are identical. Male gametophytes may have evolved haustoria to gain a nutritional advantage over other pollen grains in the pollen chamber. These haustoria would then be preadapted for the delivery of sperm to archegonia. Moreover, male gametophytes that were parasitic on maternal tissues would have allowed paternal sporophytes to produce smaller pollen grains (Haig and Westoby 1989*a*).

ACKNOWLEDGEMENTS

This chapter has benefited from helpful comments by B. J. Crespi, M. A. Elgar, W. D. Koenig, D. G. Lloyd, M. T. Stanback, and M. Westoby.

REFERENCES

Apsit, V. J., Nakamura, R. R., and Wheeler, N. C. (1989). Differential male reproductive success in Douglas fir. *Theoretical and Applied Genetics*, 77, 681–4.

Axelrod, R. (1984). *The evolution of cooperation*. Basic Books, New York.

Baird, A. M. (1939). A contribution to the life-history of *Macrozamia reidlei*. *Journal of the Royal Society of Western Australia*, **25**, 153–75.

Baird, A. M. (1953). The life-history of *Callitris*. *Phytomorphology*, **3**, 258–84.

Bartram, C. and Miller, G. (1988). Estimation of seed orchard efficiencies by means of multistage variable probability sampling. *Canadian Journal of Forest Research*, **18**, 1397–404.

Bishir, J. and Namkoong, G. (1987). Unsound seed in conifers, estimation of number of lethal alleles and of magnitudes of effects associated with the maternal parent. *Silvae Genetica*, **36**, 3–4.

Brough, P. and Taylor, M. H. (1940). An investigation of the life cycle of *Macrozamia spiralis* Miq. *Proceedings of the Linnean Society of New South Wales*, **65**, 494–524.

Buchholz, J. T. (1922). Developmental selection in vascular plants. *Botanical Gazette*, **73**, 249–86.

Buchholz, J. T. (1939*a*). The morphology and embryogeny of *Sequoia gigantea*. *American Journal of Botany*, **26**, 93–101.

Buchholz, J. T. (1939*b*). The embryogeny of *Sequoia sempervirens* with a comparison of the sequoias. *American Journal of Botany*, **26**, 248–57.

Burdon, R. D. and Low, C. B. (1973). Seed production in radiata pine clones on four different sites. *New Zealand Journal of Forestry Science*, **3**, 211–9.

Burlingame, L. L. (1915). The morphology of *Araucaria brasiliensis*. III. Fertilization, the embryo and the seed. *Botanical Gazette*, **59**, 1–39.

Caldwell, O. W. (1907). *Microcycas calocoma*. *Botanical Gazette*, **44**, 118–41.

Chamberlain, C. J. (1912). Morphology of *Ceratozamia*. *Botanical Gazette*, **53**, 1–19.

Chamberlain, C. J. (1913). *Macrozamia moorei*, a connecting link between living and fossil cycads. *Botanical Gazette*, **55**, 141–54.
Chamberlain, C. J. (1916). *Stangeria paradoxa*. *Botanical Gazette*, **61**, 353–72.
Choi, J.-S. and Friedman, W. E. (1991). Development of the pollen tube in *Zamia furfuracea* (Zamiaceae) and its evolutionary implications. *American Journal of Botany*, **78**, 544–60.
Cohen, D. and Dukas, R. (1990). The optimal number of female flowers and the fruits-to-flowers ratio in plants under pollination and resources limitation. *American Naturalist*, **135**, 218–41.
Coulter, J. M. and Land, W. J. G. (1905). Gametophytes and embryos of *Torreya taxifolia*. *Botanical Gazette*, **39**, 161–78.
Doak, C. C. (1932). Multiple male cells in *Cupressus arizonica*. *Botanical Gazette*, **94**, 168–82.
Doak, C. C. (1937). Morphology of *Cupressus arizonica*: gametophytes and embryogeny. *Botanical Gazette*, **98**, 808–15.
Doyle, J. and Brennan, M. (1971). Cleavage polyembryony in conifers and taxads–a survey. I. Podocarps, taxads and taxodioids. *Scientific Proceedings of the Royal Dublin Society, Series A*, **4**, 57–88.
Doyle, J. and Brennan, M. (1972). Cleavage polyembryony in conifers and taxads–a survey. II. Cupressaceae, Pinaceae and conclusions. *Scientific Proceedings of the Royal Dublin Society, Series A*, **4**, 137–58.
Dupler, A. W. (1917). The gametophytes of *Taxus canadensis* Marsh. *Botanical Gazette*, **64**, 115–36.
Engels, F. M. and Gianordoli, M. (1983). The basic autonomy of *Metasequoia* female gametophytes. *Acta Botanica Neerlandica*, **32**, 295–305.
Favre-Duchartre, M. (1958). *Ginkgo*, an oviparous plant. *Phytomorphology*, **8**, 377–90.
Favre-Duchartre, M. (1970). *Des ovules aux graines*. Masson et Cie, Paris.
Fernandez, G. W. and Whitham, T. G. (1989). Selective fruit abscission by *Juniperus monosperma* as an induced defense against predators. *American Midland Naturalist*, **121**, 389–92.
Fowler, D. P. (1964). Pre-germination selection against a deleterious mutant in red pine. *Forest Science*, **10**, 335–6.
Franklin, E. C. (1970). Survey of mutant forms and inbreeding depression in species of the family Pinaceae. *USDA Forest Service Research Paper SE-61*.
Franklin, E. C. (1972). Genetic load in loblolly pine. *American Naturalist*, **106**, 262–5.
Friedman, W. E. (1987). Growth and development of the male gametophyte of *Ginkgo biloba* within the ovule (in vivo). *American Journal of Botany*, **74**, 1797–815.
Griffin, A. R. and Lindgren, D. (1985). Effect of inbreeding on production of filled seed in *Pinus radiata*–experimental results and a model of gene action. *Theoretical and Applied Genetics*, **71**, 334–43.
Haig, D. (1990). Brood reduction and optimal parental investment when offspring differ in quality. *American Naturalist*, **136,** 550–66.
Haig, D. and Westoby, M. (1988*a*). On limits to seed production. *American Naturalist*, **131**, 757–9.
Haig, D. and Westoby, M. (1988*b*). Sex expression in homosporous ferns: an evolutionary perspective. *Evolutionary Trends in Plants*, **2**, 111–9.
Haig, D. and Westoby, M. (1989*a*). Selective forces in the emergence of the seed habit. *Biological Journal of the Linnean Society*, **38**, 215–38.

Haig, D. and Westoby, M. (1989*b*). Parent specific gene expression and the triploid endosperm. *American Naturalist*, **134**, 147–55.

Haines, R. J. (1983). Seed development in *Araucaria* Juss. *Australian Journal of Botany*, **31**, 255–67.

Ho, R. H. (1985). Effect of repeated pollination upon filled seed in white spruce. *Canadian Journal of Forest Research*, **15**, 1195–7.

Hurst, L. D. (1990). Parasite diversity and the evolution of diploidy, multicellularity and anisogamy. *Journal of Theoretical Biology*, **144**, 429–43.

Johnsson, H. (1976). Contributions to the genetics of empty grains in the seed of pine (*Pinus silvestris*). *Silvae Genetica*, **25**, 10–15.

Kaur, D. and Bhatnagar, S. P. (1986). Studies in the family Araucariaceae, female cone development in *Agathis robusta*. *Phytomorphology*, **36**, 129–43.

Klekowski, E. J., Jr. (1982). Genetic load and soft selection in ferns. *Heredity*, **49**, 191–7.

Konar, R. N. and Oberoi, Y. P. (1969). Studies on the morphology and embryology of *Podocarpus gracilior* Pilger. *Beiträge zur Biologie der Pflanzen*, **45**, 329–76.

Kosinski, G. (1986). Megagametogenesis, fertilization, and embryo development in *Larix decidua*. *Canadian Journal of Forest Research*, **16**, 1301–9.

Koski, V. (1973). On self-pollination, genetic load, and subsequent inbreeding in some conifers. *Communicationes Instituti Forestalis Fenniae*, 78.10, 1–40.

Kozlowski, J. and Stearns, S. C. (1989). Hypotheses for the production of excess zygotes: models of bet-hedging and selective abortion. *Evolution*, **43**, 1369–77.

Land, W. J. G. (1907). Fertilization and embryogeny in *Ephedra trifurca*. *Botanical Gazette*, **44**, 273–92.

Lawson, A. A. (1907). The gametophytes and embryo of the Cupressineae with special reference to *Libocedrus decurrens*. *Annals of Botany*, **21**, 281–301.

Lawson, A. A. (1926). A contribution to the life history of *Bowenia*. *Proceedings of the Royal Society of Edinburgh*, **54**, 357–94.

Lee, C. L. (1955). Fertilization in *Ginkgo biloba*. *Botanical Gazette*, **117**, 79–100.

Lee, T. D. (1988). Patterns of fruit and seed production. In *Plant reproductive ecology*, (ed. J. Lovett Doust and L. Lovett Doust), pp. 179–202. Oxford University Press.

Lill, B. S. (1976). Ovule and seed development in *Pinus radiata*, postmeiotic development, fertilization, and embryogeny. *Canadian Journal of Botany*, **54**, 2141–54.

Maheshwari, P. (1950). *An introduction to the embryology of angiosperms*. McGraw-Hill, New York.

Maheshwari, P. and Vasil, V. (1961). *Gnetum*. Council of Scientific and Industrial Research, New Delhi.

Martens, P. (1971). *Les Gnétophytes*. Borntraeger, Berlin.

Matheson, A. C. (1980). Unexpectedly high frequencies of outcrossed seedlings among offspring from mixtures of self and cross pollen in *Pinus radiata* D. Don. *Australian Forest Research*, **10**, 21–7.

McWilliam, J. R. (1959). Interspecific incompatibility in *Pinus*. *American Journal of Botany*, **46**, 425–33.

Mehra, P. N. and Malhotra, R. K. (1947). Stages in the embryology of *Cupressus sempervirens* Linn. with particular reference to the occurrence of multiple male cells in the male gametophyte. *Proceedings of the National Academy of Sciences, India B*, **17**, 129–53.

Mikkola, L. (1969). Observations on interspecific sterility in *Picea*. *Annales Botanici Fennici*, **6**, 285–339.

Moseley, M. F. (1943). Contributions to the life history, morphology and phylogeny of *Widdringtonia cupressoides*. *Lloydia*, **6**, 109–32.
Neale, D. B. and Sederoff, R. R. (1989). Paternal inheritance of chloroplast DNA and maternal inheritance of mitochondrial DNA in loblolly pine. *Theoretical and Applied Genetics*, **77**, 212–6.
Neale, D. B., Wheeler, N. C., and Allard, R. W. (1986). Paternal inheritance of chloroplast DNA in Douglas-fir. *Canadian Journal of Forest Research*, **16**, 1152–4.
Norstog, K. (1975). The motility of cycad spermatozoids in relation to structure and function. *Biological Journal of the Linnean Society*, **7** (suppl. 1), 135–42.
Norstog, K. J., Stevenson, D. W., and Niklas, K. J. (1986). The role of beetles in the pollination of *Zamia furfuracea* L. fil. (Zamiaceae). *Biotropica*, **18**, 300–6.
Osborn, T. G. B. (1960). Some observations on the life history of *Podocarpus falcatus*. *Australian Journal of Botany*, **8**, 243–55.
Owens, J. N. and Molder, M. (1977*a*). Seed-cone differentiation and sexual reproduction in western white pine (*Pinus monticola*). *Canadian Journal of Botany*, **55**, 2574–90.
Owens, J. N. and Molder, M. (1977*b*). Sexual reproduction of *Abies amabilis*. *Canadian Journal of Botany*, **55**, 2653–67.
Owens, J. N. and Molder, M. (1979*a*). Sexual reproduction of white spruce (*Picea glauca*). *Canadian Journal of Botany*, **57**, 152–69.
Owens, J. N. and Molder, M. (1979*b*). Sexual reproduction of *Larix occidentalis*. *Canadian Journal of Botany*, **57**, 2673–90.
Owens, J. N. and Molder, M. (1980). Sexual reproduction of Sitka spruce (*Picea sitchensis*). *Canadian Journal of Botany*, **58**, 886–901.
Owens, J. N. and Morris, S. J. (1990). Cytological basis for cytoplasmic inheritance in *Pseudotsuga menziesii*. I. Pollen tube and archegonial development. *American Journal of Botany*, **77**, 433–45.
Owens, J. N., Simpson, S. J., and Molder, M. (1981*a*). Sexual reproduction of *Pinus contorta*. I. Pollen development, the pollination mechanism, and early ovule development. *Canadian Journal of Botany*, **59**, 1828–43.
Owens, J. N., Simpson, S. J., and Molder, M. (1981*b*). The pollination mechanism and the optimal time of pollination in Douglas-fir (*Pseudotsuga menziesii*). *Canadian Journal of Forest Research*, **11**, 36–50.
Owens, J. N., Simpson, S. J., and Molder, M. (1982). Sexual reproduction of *Pinus contorta*. II. Postdormancy ovule, embryo, and seed development. *Canadian Journal of Botany*, **60**, 2071–83.
Park, Y. S. and Fowler, D. P. (1984). Inbreeding in black spruce (*Picea mariana* (Mill.)) B.S.P., self-fertility, genetic load, and performance. *Canadian Journal of Forest Research*, **14**, 17–21.
Plym Forshell, C. (1974). Seed development after self-pollination and cross-pollination of Scots pine, *Pinus sylvestris* L. *Studia Forestalia Suecica*, **118**, 1-24.
Price, R. A. and Lowenstein, J. M. (1989). An immunological comparison of the Sciadopityaceae, Taxodiaceae, and Cupressaceae. *Systematic Botany*, **14**, 141–9.
Queller, D. C. (1984). Models of kin selection on seed provisioning. *Heredity*, **53**, 151–65.
Reynolds, L. G. (1924). The female gametophyte of *Microcycas*. *Botanical Gazette*, **77**, 391–403.
Rudin, D., Muona, O., and Yazdani, R. (1986). Comparison of the mating system of *Pinus sylvestris* in natural stands and seed orchards. *Hereditas*, **104**, 15–19.

Sarvas, R. (1962). Investigations on the flowering and seed crop of *Pinus silvestris*. *Communicationes Instituti Forestalis Fenniae*, **53**.4, 1–198.

Sarvas, R. (1968). Investigations on the flowering and seed crop of *Picea abies*. *Communicationes Instituti Forestalis Fenniae*, **67**.5, 1–84.

Saxton, W. T. (1909). Preliminary account of the ovule, gametophytes, and embryo of *Widdringtonia cupressoides*. *Botanical Gazette*, **48**, 161–78.

Saxton, W. T. (1934). Notes on conifers. IX. The ovule and embryogeny of *Widdringtonia*. *Annals of Botany*, **48**, 429–31.

Schnarf, K. (1933). *Embryologie der Gymnospermen*. Borntraeger, Berlin.

Singh, H. (1978). *Embryology of gymnosperms*. Borntraeger, Berlin.

Singh, H. and Oberoi, Y. P. (1962). A contribution to the life history of *Biota orientalis*. *Phytomorphology*, **12**, 373–93.

Smith, C. C., Hamrick, J. L., and Kramer, C. L. (1988). The effects of stand density on frequency of filled seeds and fecundity in lodgepole pine (*Pinus contorta* Dougl.). *Canadian Journal of Forest Research*, **18**, 453–60.

Solter, D. (1988). Differential imprinting and expression of maternal and paternal genomes. *Annual Review of Genetics*, **22**, 127–46.

Sorensen, F. C. (1969). Embryonic genetic load in coastal Douglas fir, *Pseudotsuga menziesii* var. menziesii. *American Naturalist*, **103**, 389–98.

Sorensen, F. C. (1982). The roles of polyembryony and embryo viability in the genetic system of conifers. *Evolution*, **36**, 725–33.

Stearns, S. C. (1987). The selection-arena hypothesis. In *The evolution of sex and its consequences* ,(ed. S. C. Stearns), pp. 337–49. Birkhäuser Verlag, Basel.

Stephenson, A. G. (1981). Flower and fruit abortion, proximate causes and ultimate functions. *Annual Review of Ecology and Systematics*, **12**, 253–80.

Sweet, G. B. (1973). Shedding of reproductive structures in forest trees. In *Shedding of plant parts*, Vol. 2, (ed. T. T. Kozlowski), pp. 341–82. Academic Press, New York.

Sweet, G. B. and Thulin, I. J. (1967). The abortion of conelets in *Pinus radiata*. *New Zealand Journal of Forestry*, **14**, 59–67.

Szmidt, A. E., Aldén, T., and Hällgren, J.-E. (1987). Paternal inheritance of chloroplast DNA in *Larix*. *Plant Molecular Biology*, **9**, 59–64.

Tang, W. (1987). Insect pollination in the cycad *Zamia pumila* (Zamiaceae). *American Journal of Botany*, **74**, 90–9.

Wagner, D. B., Furnier, G. R., Sagai-Maroff, M. A., Williams, S. M., Dancik, D. B., and Allard, R. W. (1987). Chloroplast DNA polymorphisms in lodgepole and jack pines and their hybrids. *Proceedings of the National Academy of Science, USA*, **84**, 2097–100.

Westoby, M. and Rice, B. (1982). Evolution of the seed plants and inclusive fitness of plant tissues. *Evolution*, **36**, 713–24.

Whatley, J. M. (1982). Ultrastructure of plastid inheritance: green algae to angiosperms. *Biological Reviews*, **57**, 527–69.

White, T. L., Harris. H. G., and Kellison, R. C. (1977). Conelet abortion in longleaf pine. *Canadian Journal of Forest Research*, **7**, 378–82.

Willson, M. F. and Burley, N. (1983). *Mate choice in plants*. Princeton University Press.

5

Cannibalism in lower eukaryotes

David R. Waddell

Cannibalism is taxonomically widespread among lower eukaryotes (Sandon 1932), although most reports have been anecdotal. The first report of cannibalism in lower eukaryotes was in the ciliate *Oxytricha* (Haime 1853). It was subsequently reported in several other species of ciliates, including *Onychodromus grandis* (Maupas 1888), *Pleurotricha laceolata* (Joukowsky 1898), *Stentor coeruleus* (Gelei 1925), *Blepharisma undulans* (Dawson 1929), and *Stylonychia curvata* (Giese and Alden 1938). Although generally widespread among the ciliates, it is notable that cannibalism has not been reported for the most commonly studied smaller ciliates *Tetrahymena* and *Paramecium*, which generally feed upon bacteria. However, two species, *Tetrahymena vorax* (Buhse 1966; Buhse *et al.* 1978) and *T. patula* (Williams 1960) can undergo a transformation from a form which feeds upon bacteria to a carnivorous form which feeds on other cilates. The transition to the large form capable of feeding upon larger prey involves the resorption of the oral apparatus and the formation of a larger one by a process known as oral replacement. Cannibalism does occur in the form with the larger oral apparatus (N. E. Williams, personal communication).

Cannibalism occurs in several different contexts in the life cycle of cellular slime moulds. In several species, it occurs during the aggregation stage of fruiting body formation and during the formation of a sexual cyst structure known as the macrocyst. One species in particular, *Dictyostelium caveatum*, which is predatory upon other cellular slime moulds, exhibits a particularly voracious form of cannibalism which, in many respects, resembles that reported for ciliates. As noted above, it is the larger species or the larger forms of ciliates, in which predatory behaviour has been observed that also exhibit cannibalism. This suggests that predation and cannibalism are different manifestations of the same behaviour.

Cannibalism has been reported in the amoebae, *Amoeba vespertilio* (Lapage 1922), *A. verrucusa* (Ivanic 1927), and *A. gjorgjevici* (Ivanic 1927). However, one must be cautious in interpreting many reports of cannibalism in lower eukaryotes because it is often very difficult to establish separate species by morphological criteria alone. Therefore, cannibalism should probably be ascribed only to cloned stocks or to stocks in which a close relationship has been established by other criteria such as cell surface antigens or sequence homology. Lapage's (1922) report of cannibalism in *A. vespertilio* suffers from this problem, as this study was performed using relatively crude cultures started from a 10-year-old hay infusion containing a rich fauna. Lapage (1922) was cautious in proposing cannibalism as it was consistent with most of his observations. He rejected the idea that the ingested amoebae were parasites since they neither resembled any previously reported parasites nor replicated within the host amoeba. It was also not clear whether the cannibals obtained any nutrition in this way since ingested amoebae were usually later ejected from the cell.

Ivanic (1927) reported that the frequency of cannibalism in *A. verrucusa* was closely associated with the appearance of 'giant cells' in older cultures (about one-year-old). Fresh cultures showed no signs of cannibalism. Only the giant cells in older cultures fed upon sister cells and cysts, and the ingested amoebae were digested. This behavioural change suggests that the amoebae are capable of self or non-self recognition. Defects or changes in the recognition mechanism could thereby account for cannibalism. Ivanic (1927) proposed that cannibalism was due to disturbances in the division process that led to giant cells and subsequently to cannibalism. The absence of alternative food did not appear to be necessary to maintain the cannibalistic giant cells.

Cannibalism has also been reported to occur in both laboratory cultures and *in vivo* in the termite intestinal protozoan, *Trichonympha campanula* (Cleveland 1925). It is interesting that extensive cannibalism (50 per cent) was reported to occur between the intestinal protozoans of only a few of the termites examined (3 of 500), suggesting that the intestines of these animals may have been populated with cannibalistic variants. Cannibalism was directly observed in laboratory cultures of the protozoans but it was rare. Unfortunately no results were reported on laboratory cultures of the cannibalistic protozoans from the three termites showing extensive cannibalism in their gut fauna.

CANNIBALISM IN THE CILIATES

In order to understand certain characteristics of cannibalism in ciliates a few facts about their biology would be helpful. Ciliates are unique in having distinct nuclei specialized for transcription and sexual exchange. The micronuclei are relatively conventional nuclei with chromosomes and a replication cycle but exhibit little or no transcriptional activity. After sexual conjugation, micronuclei give rise to the macronuclei, which are the transcriptionally-active nuclei of ciliates. The formation of macronuclei from micronuclei involves the production of polytene chromosomes and, in many species, their subsequent processing into small molecules which are capable of autonomous replication by the removal of most non-coding DNA and addition of terminal palindromic repeats (Lawn *et al.* 1978; Gall 1986). The macronuclei divide during cell division without an elaborate segregation apparatus found in conventional eukaryotic nuclei. There are multiple copies of all transcribed genes so that segregation at cell division is assured statistically. The cell size and DNA content of practically all eukaryotic cells are highly correlated (Cavalier-Smith 1985) and ciliates are not an exception to this rule (Berger 1979). The specialization of nuclei in ciliates allows them to increase their cell size relatively easily and maintain a balance of DNA to cell volume.

The first detailed study of cannibalism in ciliates was published by Dawson (1919). He observed the presence of giant cells in cultures of an amicronucleate ciliate, *Oxytricha hymenostoma*. Dawson (1919) observed giant cells devouring clone mates. One particularly voracious cell contained the remnants of 16 clone mates. Dawson (1919) suggested that these giant cells continued to feed until it was physically impossible to ingest additional cells. Cannibalism was likely to occur shortly after bacteria are exhausted in these cultures (Dawson 1919). At this time, many pairs could be observed which were attached at their anterior ends in a way suggestive of sexual conjugation. However, these strains cannot complete mating because they lack micronuclei. After a cannibal had ingested its first cell, additional cannibalistic encounters rapidly followed leading to the formation of a 'giant'.

Since amicronucleate strains, like the one Dawson (1919) studied, cannot exchange genetic material during mating, this raised the possibility that cannibalism occurs when an attempt at exchanging micronuclei fails. However, Dawson (1929) later reported cannibalism in *Blepharisma undulans*, which has micronuclei. Giese (1938) extended the observations on *Blepharisma* and on another ciliate, *Stylonychia* (Giese and Alden 1938). Therefore, the inability to complete mating due to the lack of micronuclei cannot be the only cause of cannibalism.

Cannibalism may be a response to unsuccessful mating between individuals that were not suited to each other for other reasons (e.g. mating type) or may, in the absence of appropriate mates, be induced between clone mates or related strains. Since Dawson's (1919, 1929) observations, a considerable amount of information has accumulated over the mating process in ciliates especially in *Paramecium* (Watanabe 1990). Mating in ciliates begins during the early stationary phase as the cells begin to starve. Cells of compatible mating types adhere to each other via their cilia. The phenomenology of the genetics of the mating type system indicates that the adhesion is due to a polymorphic, 'lock-and-key-like' system. Each cell in a pair of compatible mating types provide only one type of molecule (i.e. either the lock or the key). In such a system, clone mates would normally be unable to mate since they would have either all 'lock' or all 'key' adhesion molecules. However, these adhesion molecules have not been biochemically characterized. In this system clone mates should not adhere. Since cannibalism occurred between clone mates in Dawson's (1919, 1929) studies, some type of homotypic interaction or mating type switching should occur if agglutination is the first step leading to cannibalism. High frequency 'selfers' which mate with clone mates are common in *Blepharisma* and the mechanism for this appears to be mating type switching (Miyake and Harumoto 1990). The switching occurs by asymmetrical cell divisions in which daughter cells are of opposite mating type. It is possible that this system is involved in producing cannibals.

The most striking difference between cannibals and normal cells is the much larger size of the cannibals. The presence of giants led to the discovery of cannibalistic strains and giant cells appear to accompany most reported cases of cannibalism in lower eukaryotes (Table 5.1). It is clear that the increase in cell size is more a function of prey size and prey type than cannibalism *per se* (Giese 1938; Giese and Alden 1938; Hewett 1980; Ricci *et al.*1989). Ciliates have rigid and elaborate cell membranes complete with specialized structures for ingestion known as oral apparati. It is also clear that cell size would strongly influence the outcome of predatory events in such cells. As cannibals ingest sister cells and grow, they increase more in breadth than in length; Dawson (1919) aptly described the effect as comparable to inflating a rubber hot water bottle. However, the cell size increase in giants was not due simply to the volume of the ingested cells, but was accompanied by increases in the size of macronuclei and of cortical structures (Tulchin and Hirshfield 1962). Giants, once produced, rapidly return to their normal size by accelerating the pace of the cell cycle when they are starved or fed bacteria (Giese 1938; Giese and Alden 1938). Therefore, cannibal giants are not distinct genetically-defined strains but alternative physiological forms of the same strain.

Giese (1938, 1973) hypothesized that it was the attainment of larger size which greatly increased the chances of successful cannibalism, and led to cannibal giants. He was able to alter the size of his *Blepharisma* by feeding them with other species of ciliates instead of bacteria. *Blepharisma* feeding on larger prey increased in size and often became cannabilistic. This seems to imply that cell recognition was not involved in the development of cannibals and that *Blepharisma* are omnivorous predators. However, this does not rule out the possibility that cannibal cells possess some type of recognition system. When bacterially-fed *Blepharisma* were given to *Tetrahymena*-fed *Blepharisma* (which are much larger), cannibalism would not take place unless the bacterially-fed animals had been starved for one to two weeks and the ciliate-fed cells had been starved for a few hours. Cannibalism thus depended upon a period of starvation even in a case where an obvious size difference between cells already existed. Therefore, increased cell size alone did not induce cannibalism, and the recognition systems that are induced by starvation in ciliates may play a role in this process.

One would expect that if the cells were omnivorous, then cannibalistic giants would always be present in cultures, albeit at frequencies dependent upon the availability of prey. Bacteria may be able to saturate phagocytic recognition receptors due to their greater numbers. However, *Tetrahymena*-fed cultures (which where smaller in size than the cannibal giants) should have exhibited cannibalism if the cells were indeed omnivorous. Since *Blepharisma* has a characteristic pink to red colour, it should have been possible to distinguish the cannibalistic events from predatory events in *Tetrahymena* by the colour of the phagosomes. Giess (1938, 1973) did not comment on this possibility, and although his conclusions seem to imply otherwise, the interesting question of whether self or non-self recognition was involved was not addressed. Instead, Giese (1938) simply mentioned that 'cannibal blepharismas are merely carnivorous individuals feeding upon

Table 5.1 Published accounts of cannibalism in lower eukaryotes

Taxa	Giant cells	Reference
Ciliates		
Oxytricha hymenostoma	yes	Haime (1853); Dawson (1919)
Onychodromus grandis	yes	Maupas (1888)
Stentor coeruleus	no	Gelei (1925)
Blepharisma (7 species)	yes	Dawson (1929); Giese and Alden (1938); Tulchin and Hirshfield (1962); Giese (1938); Christie and Hirshfield (1967); Hirshfield *et al.* (1963); Pierce et al. (1978)
Onychodromus quadricornutus	yes	Wicklow (1988, 1990)
Gastrostyla steinii	?	Weyer (1930)
Stylonychia curvata	yes	Giese and Alden (1938)
Frontonia leucas	no	Devi (1964)
Amoebae		
Amoeba vespertilio	yes	Lapage (1922); Ivanic (1927)
A. verrucosa	yes	Ivanic (1927)
A. gjorgjevici	yes	Ivanic (1927)
Dictostelium caveatum	yes	Waddell and Duffy (1986)
D. discoideum	yes	Olive et al. (1961); Huffman and Olive (1964); Huffman (1967); Erdos *et al.* (1972); Filosa and Dengler (1972); Fukui (1976); Saga and Yanigsawa (1982)
Chamydophrys schaudinni	no	Bélar (1921)
Other		
Trichonympha campanula	?	Cleveland (1925)

smaller members of their own species and can be produced at will by feeding bacterially-fed animals first with a diet of starved tetrahymenas and then with a diet of small blepharismas.' Padmavathi (1961) reached similar conclusions from a subsequent study on a different race of *Blepharisma*, but also did not directly address the question of self-recognition.

More recent work indicates that diffusible factors can induce gigantism and cannibalism. Vitamin E (a-tocopheryl succinate) induces changes in *Blepharisma americanum* which lead to giant formation (Lennarz and Bovee 1980). This vitamin is abundant in several of the grains used to prepare media for ciliates and also induces gigantism associated with cannibalism in the metazoan rotifer *Asplanchana sieboldi* (Gilbert 1973).

Diffusible factors also appear to be involved in an interesting developmental polymorphism that has been described in the hypotrich ciliate *Onychodromus quadricornutus* (Wicklow 1988). Well-fed clonal cultures of this species exhibit only one morph, called 'basic', which has a unimodal size distribution. Two distinct size classes exist in starving cultures: cannibalistic giants and a smaller morph called 'lanceolate' that differs from the 'basic' morph by having four large dorsal spines. The cannibalistic giants induce the dramatic increase in the size of the four dorsal spines of smaller cells. The induction is due to a diffusible factor which can pass through a membrane with a molecular weight cut-off of 10 000Da (Wicklow 1990). The larger spines on the smaller cells prevent intraspecific predation by the giants. Another predatory species, *Lembadion lucens,* also releases factors which induce cell transformations in prey ciliates (Kuhlmann and Heckmann 1985). The *Lembadion* factor(s) also induce the increase in dorsal spine size in *O. quadricornutus.* Whether the same factors are released by the conspecific giants of *O. quadricornutus* and *Lembadion* is not clear.

A different form of cannibalism which does not lead to giant formation has been reported in *Frontonia leucas* (Devi 1964) and *Stentor coeruleus* (Gelei 1925). In *Frontonia,* moribund cells are phagocytosed, and starvation was not necessary for cannibalism to occur. Therefore, in this case, the cannibals may be recognizing their prey through changes due to the moribund state of the cells. It is not known whether the ingestion of dead cells in strains that form giant cells leads to giant formation.

Whether or not starvation was necessary to induce cannibalism in *Stentor coeruleus* was not directly addressed (Gelei 1925). However, cannibalism occurred in only one of six identically-treated aquariums, and in this one only a small fraction of the cells were cannibalistic (16 of 963). The cannibals seemed to constitute a separate race distinct from the more 'tame' cells. Cannibalism was heritable: isolated cannibals divided to produce sister cells that then cannibalized each other. Cannibal cells appeared to be specialized in feeding upon other *S. coeruleus* as they did not contain *Paramecium* or *Rotaria* unlike the 'tame' race from the same aquarium. Why the aquarium containing the cannibal race did not eventually become dominated by the cannibals is not clear. Perhaps the more sessile nature of *S. coeruleus* prevented a high rate of encounters between *S. coeruleuss* and thereby allowed the 'tame' race to maintain its numbers.

Finally, it should be noted that in one of the species of ciliates in which giants occur–the *Stylonichia*–the animals also encyst in response to starvation. Cannibalism and encysting appear to be alternative responses to starvation stress. After starvation most of the cells formed cysts. Shortly thereafter giants appeared in the cultures which then proceeded to consume the cysts (Giese 1938). The availability of more than one response to starvation also exists in the cellular slime moulds. The implications of this will be discussed below.

CANNIBALISM IN THE CELLULAR SLIME MOULDS

Early reports of cannibalism in the cellular slime moulds centred around a controversy about the existence of a sexual cycle during fruiting body formation (Olive *et al.* 1961; Huffman and Olive 1964; Huffman 1967). Like ciliates, the common foodstuff of cellular slime moulds is bacteria. When feeding, they exist as autonomous amoebae. The multicellular stage of the life cycle is induced by starvation. When amoebae are starved for a few hours in the light, single cells and small groups of cells initiate chemotactic signals. These signals are propagated outwards by the relayed secretion of chemo-attractant by responding cells. Streams of amoebae move to the nascent aggregation centres. The aggregates thus formed undergo a series of morphological events that result in the formation of a simple fruiting body, composed of a stalk and a spore case.

Cytological evidence was presented for plasmogamy (cell fusion) which led to nuclear fusion and meiosis (Wilson and Ross 1957) during early stages of fruiting body formation. Subsequent studies (Olive *et al.* 1961; Huffman 1967) were not able to verify these early observations, and reported instead that cannibalism occurred. Since these studies sought a sexual cycle, reports of cannibalism were often that between different races (Olive *et al.* 1961). Huffman (1967) reported intraspecific cannibalism during the aggregation stage. Most strains showed low levels of intrastrain engulfment (<1 per cent). One strain in this study, *Dictyostelium discoideum* NC4-H, showed consistently higher levels of intrastrain engulfment (10–20 per cent). Such high levels of phagocytosis have not been noted in either electron micrographs of aggregating amoebae or between fluorescently labelled cells of a related strain (AX2 which was derived from NC4 after selection for growth in liquid media) (Waddell, unpublished observation).

When the presence of a sexual cycle was finally established, it became evident that it was accompanied by cannibalism (Erdos *et al.* 1972; Filosa and Dengler 1972). Like the formation of fruiting bodies, the sexual cycle which leads to the formation of macrocysts is initiated after starvation. Darkness and high humidity favour the formation of macrocysts over asexual fruiting bodies. The initial event in most cases is the fusion of amoebae of opposite mating types. However, homothallic strains which form macrocysts without a mating partner are known (Briscoe *et al.* 1987). Under certain conditions, additional fusions rapidly ensue which can lead to multinuclear giant cells containing as many as 50–60 nuclei (Saga and Yanigsawa 1982). Fusion of two of the nuclei leads, at least in some cases, to a zygote diploid nucleus. Each giant cell acts as the centre of an aggregation for other amoebae by secreting the chemo-attractant, cyclic AMP. Shortly after a thin sheath forms around these multicellular aggregates (12–14 hours after starvation), the other amoebae within the aggregate are phagocytosed by the giant cell.

At least in some cases the giant cell is a zygotic cell formed by the fusion of two of the nuclei. Cannibalism in these aggregates is asymmetric, with the giant cell feeding upon the smaller amoebae that were attracted to the giant cell. After phagocytosis of the other amoebae, the giant cell undergoes a highly-organized cell differentiation to form a large resistant structure with a cell wall, known as the macrocyst. This macrocyst is distinguishable from the smaller cysts formed by single amoebae in some species.

Giant cells that were experimentally freed by enzymatic digestion of the sheath exhibit a very interesting behaviour: they specifically migrate to other giant cells and attempt to ingest them. This indicates that only a single giant cell can exist in each macrocyst (Fukui 1976). The cannibalistic behaviour of these cells is restricted to the stage of macrocyst formation immediately after the formation of a sheath around the giant cells and its victims. The released giant cells strongly resemble the predatory form *D. caveatum*. However, the subsequent differentiation of the giant cell into a resistant cyst, distinguishs this type of cannibalism from that of *D. caveatum,* which will now be described.

CANNIBALISM IN *Dictyostelium caveatum*

The predatory slime mould *D. caveatum* was discovered in bat guano from a limestone cavern in Arkansas (Waddell 1982). Unlike other slime moulds which only occasionally attack cells of other species, *D. caveatum* is a specialist predator. Even when *D. caveatum* is mixed with prey amoebae of other species in ratios (1:10 000), that should result in aggregates which begin with a single *D. caveatum* amoeba, it is still able to phagocytose all of the amoebae of the other species, and construct its own fruiting bodies on the carcasses of the prey's slugs.

Although the amoebae of pure cultures of wild-type *D. caveatum* (BW1) often contain organelles suggesting cannibalism, these are probably autophagosomes formed by invaginations of the cell's own membrane, since double membranes have rarely been observed and do not themselves contains phagosomes (Waddell 1982). In any case, the wild-type strain does not exhibit dramatic changes in cell number during the first eight hours after starvation (Waddell and Vogel 1985). In this study, phagocytosis of *D. discoideum* prey amoebae by *D. caveatum* was observed by employing mutants, which were resistant to a lytic toxin from *Amanita,* to distinguish the prey and predator cells in mixing experiments. These experiments were usually performed with a control in which a lysis-resistant *D. caveatum* strain was mixed with a lysis-sensitive D. *caveatum* strain. Normally, the cell number of both cell types in this control remained constant. In later experiments this control failed and the numbers of cells of both types decreased significantly. The reason was that one of the *D. caveatum* strains had become cannibalistic (Waddell and Duffy 1986).

Cannibalistic strains also differ from wild-type strains in colony morphology. The principal difference is that they no longer produce fruiting bodies (Waddell and Duffy 1986). Multicellular aggregates form after starvation for bacteria at the same time as wild-type strains. However, these aggregates become smaller and eventually the amoebae in the aggregate disperse. The amoebae arising from such aggregates can be very large. Examination under phase microscopy confirmed that they were single amoebae.

The cannibalistic strains arise spontaneously in the population. A number of independent cannibalistic strains were obtained by growing the wild-type strain on large bacterial carpets and selecting areas of the growing front which did not develop beyond the aggregate stage (Waddell and Duffy 1986). Reversion to a strain exhibiting normal development was infrequent ($<10^{-4}$). A preliminary genetic study indicated that they were due to recessive mutations which probably fall into more than one complementation group (Duffy and Waddell 1988). Therefore, these cannibals, unlike those reported for most ciliates, with the possible exception of *S. coeruleus*, are the result of heritable changes.

Cannibalism in these strains is very marked. Suspension cultures exhibit half-lives of approximately four hours. Electron micrographs revealed cells filled with phagosomes which themselves often contain phagosomes. Under identical conditions, cultures of non-cannibalistic strains form tight aggregates and show unequivocal signs of cell differentiation (Waddell and Duffy 1986).

The cannibalistic strains also form very large cells. Since these cells lack the dual nuclei system of ciliates, they increase cell size by becoming multinucleate. As with ciliates, their large size does not appear to be due to cannibalism *per se* since a similar size increase occurs after predation on the amoebae of other species by non-cannibalistic strains. *D. caveatum* amoebae have also developed the capacity to feed upon prey cells of similar or larger size by nibbling pieces (Waddell and Vogel 1985). After the predatory amoebae have become large and multinucleate, they can then feed on the whole amoebae. As in the ciliates, when prey amoebae are exhausted (in the case of non-cannibalistic strains), or bacteria are provided, the cells divide rapidly and return to the normal mononucleate size.

Self/non-self recognition in *D. caveatum*

Wild-type amoebae are apparently able to distinguish self from non-self since under identical starvation conditions they exhibit strikingly different behaviour in the presence and absence of prey amoebae of other species. In the absence of prey, morphogenesis is completed in approximately 24 hours. When mixed with prey, the onset of the final stages of morphogenesis are delayed. Longer delays occur if the mixing ratio of predator to prey is decreased. The extra time is required to prey upon the amoebae of other species since the time between exhaustion of prey amoebae and fruiting body formation remained constant. Therefore, the wild-type strains recognize when prey amoebae are no longer present in an aggregate and then cooperate with each other to form a fruiting body.

Cannibalistic strains seem to have lost the ability to make the distinction between self and non-self, and they feed upon all available amoebae. However, most of the cannibalistic strains undergo the initial aggregation stages leading up to fruiting body formation. It seems that there is a normal response to bacterial starvation which activates the chemotactic machinery. Instead of cooperating to form a fruiting body, the aggregating amoebae feed upon each other. This implies that the wild-type cells have the capacity for self or non-self recognition, and that it is this recognition that is lost in the cannibal strains.

There are a number of mechanisms that would provide the appropriate recognition of self or non-self. A self-marking molecule could exist on the cell surface which prevents phagocytosis when it is recognized by receptors on wild-type cells. This would imply that the cannibal mutants either lack the self-marking molecule or that some aspect of the recognition pathway is altered so that cells with the self-marking molecules are now phagocytosed. Such a system would be similar in mechanism to the self-recognition system proposed by Burnet (1971) for colonial tunicates. However, since only a single isolate of *D. caveatum* exists it is not possible to establish if a polymorphic system exists that could distinguish clone mates from more distantly related *D. caveatum* strains.

A polymorphic non-self recognition system like that in higher vertebrates is difficult to envisage for an organism which exists as autonomous amoebae for much of its life cycle. Alternatively, one could hypothesize that cannibal cells acquire a phagocytosis receptor, not present on wild-type cells, that now allows them to phagocytose other *D. caveatum* amoebae. However, this would seem to imply that all prey were recognized by fairly specific receptors, which seems unlikely. Every species of cellular slime mould and soil amoebae that has been tested is phagocytosed by *D. caveatum*. Such a mechanism would also imply that cannibalism is adaptive under some conditions in order for such a self-phagocytosis receptor to have evolved. Proteins apparently exist in the cell membranes of wild-type *D. caveatum* which are not present in cannibalistic strains (Waddell, unpublished observations), which is consistent with the presence of a self-recognition system.

Mechanism of giant cell formation

The diet-related cell size increase in *D. caveatum*, similar to that found in ciliates, appears to depend upon a mechanism which balances DNA content and cell size. This mechanism is distinct from that found in yeast (Russell and Nurse 1987) which determines the size at which cells undergo division. This type of size control alters the ratio of DNA content to cell volume. In the cellular slime mould, the size increase occurs via an uncoupling of cytokinesis from the nuclear division cycle. Since both phagocytosis and cytokinesis utilize actin-based motility systems, competition for a limiting cell membrane component used by both systems may regulate cell size in this system (Waddell 1988). When cells attempt to ingest large cells, much of this component would be tied up by phagocytosis so that not enough would be available to build an efficient cytokinetic furrow for cell division. Since surface area increases at a slower rate than volume, a growing cell will eventually reach a size at which enough of the limiting component will become available for cytokinesis. This model was developed to explain the properties of pleiotropic mutants of the related slime mould, *Dictyostelium discoideum*, known as *phg* mutants (Waddell *et al.* 1987). These mutants exhibit defects in phagocytosis, adhesion, and cytokinesis. The proteins (currently unidentified) coded for by the *phg* genes may be parts of the limiting component in this model. Similar mutations have been isolated in *D. caveatum* and, as the model would predict, they exhibit an even larger cell size when feeding upon other amoebae (Waddell, unpublished observations).

The striking similarity between the prey-induced cell size increase in *D. caveatum* and the ciliates suggests that a common mechanism may underlie the phenomena in these very different cells. However, this is unlikely. Cytokinesis in ciliates, because of their complex cortex, is fundamentally different from that in cellular slime moulds and may not employ an actin-based system. Nevertheless, a logically similar model can be constructed for the ciliates. A cannibal gorged with its fellows must use much of its cell membrane to form the phagosomes. This may affect the availability of a critical membrane component which is necessary for cell division. Ingestion of latex particles by *Blepharisma* also induces an increase in cell size (Pierce *et al.* 1978). These authors suggest that membrane shortage itself may lead to gigantism. A global patterning system exists in ciliates which apparently measures distance along the cortex, as opposed to the interior of the cell (Frankel 1984). Perhaps a part of this system could be the limiting component in this case. Cell size in ciliates could also be regulated by the amount of DNA replication in the macronucleus that occurs during each cell cycle. This mechanism of size regulation would be analogous to increasing ploidy in organisms with normal mitosis. Although the mechanisms by which giants are formed may be very different between ciliates and cellular slime moulds, it is evident that over a very broad taxonomic distance, cannibalism, predatory behaviour, and gigantism are associated.

ALTERNATIVE RESPONSES TO STARVATION: ENCYSTMENT, MOTILITY AND CANNIBALISM

Cannibalism is associated with starvation in both ciliates and the cellular slime moulds. In both groups of organisms alternative responses to starvation other than cannibalism exist – in the ciliates cells can encyst or in some cases develop into fast-swimming (Nelson 1978) or other forms (Wicklow 1988).

By its very nature, rabid cannibalism seems to preclude some of the other types of responses. As mentioned above, cannibalistic ciliates also feed upon cysts formed by clone mates. It is not clear what factors influence the decision between encystment or cannibalism. Apparently both responses are possible in the same culture under identical conditions. Perhaps the size of cells at starvation has some influence on the decision.

In cells which cannot recognize self, the decision may be between encystment and becoming predatory in order to feed on any prey that is available. In the latter case, a predatory strategy might be disadvantageous if most of the neighbours were clone mates that had encysted because it would greatly reduce their numbers when new sources of bacteria became available. Other factors, such as differences in survivorship between the two strategies may be important, although encysting intuitively seems to be a much better strategy. In any case, most of the ciliates discussed so far prefer bacteria, if they are available, to eating other ciliates. If the cost of predation on clone mates was high then one would expect that a self or non-self recognition system would develop. Ciliates are certainly capable of the level of recognition required for self or non-self recognition since similar recognition occurs during mating (Hiwatashi 1981, Watanabe 1990). However, the development of a polymorphic recognition system solely to prevent predation upon close relatives may not be evolutionarily stable (Crozier 1986).

The dorsal spine induction in *Onychodromus quadricornutus* described above may be one way to prevent predation upon clone mates or close relatives while allowing a larger predator giant to feed upon other large cells. The formation of the large dorsal spines entails a cost since this morph grows more slowly than the basic morph (Wicklow 1990). If this induction is due to signals that can reflect genetic relationships, then this could be considered a type of self-recognition system. However, it is not yet clear how specific the signals are, especially since other predator species can also induce the change. Similarly, it is not clear whether this system has evolved because of selection due to intraspecific or interspecific predation (Wicklow 1988).

The cellular slime moulds also exhibit a variety of responses to starvation. Single cells of many species can encyst. All species can form the asexual multicellular fruiting body with a stalk and spore case. In most species, this cycle involves a migratory slug phase which exhibits sensitive photo- and thermotaxis. The sexual macrocyst cycle is also available to most species when cells of the appropriate mating types occur together. In these cases the lack of food results in the formation of a resistant or migrating structure. In the case of the fruiting body, the resistant structure may also serve as a means of dispersal.

In the cellular slime mould, developmentally-regulated cannibalism that occurs during macrocyst formation seems to combine both strategies. It is possible to interpret such cannibalism as a form of altruism since the cells participating in the formation of the macrocyst were previously autonomous and not necessarily the descendents of a single cell as in the embryos of higher organisms. Whether or not the cannibalized cells are willing participants in this process is not clear since the giant cells secrete the same chemo-attractant as that used for aggregation during fruiting body formation. In any case, the cannibalized cells probably contribute to the survivorship of the macrocyst. Since cannibalism in macrocysts is developmentally regulated and macrocysts usually have only one zygotic giant cell, this form of cannibalism allows the development of a resistant cyst structure and is clearly distinct from that of ciliate cannibals which devour cysts and *D. caveatum*.

The association between the timing of prey disappearance with the onset of the fruiting body construction suggests that wild-type *D. caveatum* strains undergo two forms of starvation: one for bacteria and one for amoebae. The cannibalistic strains only undergo starvation for bacteria and because they continue to feed upon each other, they are never able to complete multicellular morphogenesis. Consequently, cannibals have sacrificed the capacity to form a resistant structure when they opt for cannibalism. This would appear to entail a high cost since cellular slime mould amoebae cannot survive above 30°C. Perhaps it is not accidental that this species was isolated from a cavern in which the temperature was a constant 15°C.

Since *D. caveatum* amoebae do not form either microcysts or macrocysts, starving cells choose between cannibalism and the formation of an asexual fruiting body containing spores. *D. caveatum* appears to provide a model system in which to study the choice between an uncooperative cannibalistic response to starvation, and the cooperative response of fruiting body formation.

What could be the advantages of cannibalism? The large cannibalistic amoebae may be able to withstand starvation conditions for a longer period of time and also migrate further to find food than single non-cannibalistic amoebae. However, a migration stage also exists in the fruiting body mode. The spores formed in this mode can withstand much longer periods of starvation and are more resistant to high temperature and dessication than cannibalistic amoebae. Furthermore, the exponential reduction of the number of individuals caused by cannibalism would seem to be a very high cost to pay. In complex mixtures of slime moulds that may be undergoing food competition, predation would appear to be a good strategy. Temporary starvation for bacteria would allow the *D. caveatum* amoebae to consume and grow upon its competitors for bacteria, a double bonus. The additional costs necessary to maintain a self recognition system, necessary for fruiting body formation, that could not be invaded by cannibal mutants may be too high. No other species appears to have specialized as a predator and *D. caveatum* is not a common species, so the strategy must have its disadvantages. Perhaps one of them is the cost of preventing cannibalism.

It cannot be assumed that the amoebae in an aggregate are genetically identical, because the multicellular stage of the life cycle of cellular slime moulds begins with the aggregation of autonomous amoebae. The amoeba of related species, *Dictyostelium discoideum* and *D. mucoroides*, that use the same chemo-attractant will co-aggregate (see Raper 1940). However, an additional level of recognition must then come into play since amoebae of different species segregate and form separate fruiting bodies. This recognition system is not sufficient to prevent mutations in amoebae of the same species that allow the mutant amoebae to preferentially occupy the part of the structure which contributes to the next generation (Filosa 1962). Cannibalism, like these mutations, may be considered a selfish response. In a homogeneous environment without frequent clonal growth *D. caveatum* cultures would be dominated by cannibalistic mutants. However, it is curious that the multicellular stage of the life cycle is so stable. Periodic fluctuations in temperature or humidity would seem to favour the fruiting body strategy. Other factors such as the frequency at which clonal growth occurs may also be involved. However, *D. caveatum* was isolated from a cavern with a very homogeneous environment, which would seem to favour cannibalism. Perhaps some other features such as variation in the availability of bacteria selects against cannibalistic variants.

In both the ciliates and the cellular slime moulds, selection may have favoured those species which pursued a variety of responses to starvation. If the temperature increases, the cannibals would die but the encysted forms would survive. If, on the other hand, new food resources became available, then the more motile cannibals would be able to exploit them immediately without having to germinate. Laboratory cultures of *D. caveatum* that are not frequently cloned seem to be a mixture of cannibals and wild-type cells. Low levels of cannibalistic cells appear to allow fruiting body formation of wild-type *D. caveatum*. The tendency appears to be always toward cannibalism. Therefore, perhaps the periodic dying out of cannibals may be necessary to maintain the fruiting body mode in this species.

Since *D. caveatum* has only been isolated once and this was from a single clone, we do not know whether cannibalistic strains exist in natural populations. The frequency with which they arise in laboratory cultures strongly suggests that they do. It should also be noted that cannibalistic strains could not be isolated using the present procedures for isolating slime moulds, as these procedures depend upon observing the multicellular phase of the life cycle. Cannibalistic strains would therefore be difficult to distinguish from other soil amoebae. Some soil amoebae may in fact be former slime moulds!

SUMMARY

Cannibalism in lower eukaryotes is well established in ciliates and cellular slime moulds. The cannibalism exhibited in both groups of organisms has many similar properties: it occurs under starvation conditions, it is associated with predation, it leads to the formation of giant cannibalistic cells, and it is only one of many responses to starvation available to these organisms. Further studies are required to establish whether self/non-self recognition exists in ciliates and whether they distinguish between prey ciliates and cells of their own species. It is apparent that wild-type amoebae of the predatory slime mould do exhibit self/non-self recognition. However, the mechanism of this recognition is not yet established. A number of questions remain to be answered about why cannibalism is present in both groups of organisms: What conditions favoured the evolution of cannibalism? Is it adaptive or simply a cost associated with predation, and to what extent and under what conditions does it exist in natural populations?

ACKNOWLEDGEMENTS

I wish to thank Joseph Frankel, John Bonner, Barry Wicklow, David Haig and the editors for criticism of earlier drafts of this chapter and the Deutsche Forschungsgemeinschaft for a grant which supported some of the work reported here (Vo 348/1-2).

REFERENCES

Bélar, K. (1921). Untersuchungen über Thecamben der *Chlamydophrys*-Gruppe. *Archiv für Protistenkunde*, **43**, 287–354.

Berger, D. (1979). Regulation of macronuclear DNA content in *Paramecium tetraurelia*. *Journal of Protozoology*, **26**, 18–28.

Briscoe, D.A., Gooley, A.A., Bernstein, R.L., McKay, G.M., and Williams, K.L. (1987). Genetic diversity in cellular slime molds: allozyme electrophoresis and a monoclonal antibody reveal cryptic species among *Dictyostelium discoideum* strains. *Genetics*, **117**, 213–20.

Buhse, H.E. (1966). Oral morphogenesis during transformation from microstome to macrostome and macrostome to mecrostome in *Tetrahymena vorax* strain V2 type S1. *Transactions of the American Microscopical Society*, **85**, 305–13.

Buhse, H.E., Smith, H.E., Sonkin, D.J., and Fleming, J.J. (1978). Some factors which affect the maintenance and shape of the cytopharyngeal pouch in carnivorous forms of *Tetrahymena vorax*. *Transactions of the American Microscopical Society*, **97**, 350–62.

Burnet, F.M. (1971). 'Self-recognition' in colonial marine forms and flowering plants in relation to evolution of immunity. *Nature*, **232**, 230–5.

Cavalier-Smith, T. (1985). Cell volume and the evolution of eukaryotic genome size. In *The evolution of genome size*,(ed. T. Cavalier-Smith), pp. 105–84. John Wiley & Sons, New York.

Christie, S. L. and Hirshfield, H. I. (1967). Macronuclear variations during the life cycle of *Blepharisma dawsoni* n sp. and *B. wardsi*. *Journal of Protozoology*, **14**, 759–762.

Cleveland, L. R. (1925). The method by which *Trichonympha campanula*, a protozon in the intestine of termites, ingests solid particles of wood for food. *Biological Bulletin*, **48**, 282–8.

Crozier, R.H. (1986). Genetic clonal recognition abilities in marine invertebrates must be maintained by selection for something else. *Evolution*, **40**, 1100–1.

Dawson, J.A. (1919). An experimental study of an amicronucleate *Oxytricha* I. Study of the normal animal, with an account of cannibalism. *Journal of Experimental Zoology*, **29**, 473–513.

Dawson, J.A. (1929). Cannibalism in a ciliate, *Blepharisma*. *Proceedings of the Society for Experimental Biology & Medicine*, **26**, 335.

Devi, R.V. (1964). Cannibalism in *Frontonia leucas* Ehr. *Journal of Protozoology*, **11**, 304–7.

Duffy, K.T.I. and Waddell, D.R. (1988). Genetic characterization of cannibalism in *Dictyostelium caveatum*. *Current Microbiology*, **17**, 203–8.

Erdos, G.W., Neckerson, A.W., and Raper, K.B. (1972). Fine structure of macrocysts in *Polyspondylium violaceum*. *Cytobiology*, **6**, 351–66.

Filosa, M.F. (1962). Heterocytosis in cellular slime molds. *American Naturalist*, **96**, 79–91.

Filosa, M.F. and Dengler, R.E. (1972). Ultrastructure of macrocyst formation in the cellular slime mold, *Dictyostelium mucoroides*: extensive phagocytosis of amoebae by a specialized cell. *Developmental Biology*, **29**, 1–16.

Frankel, J. (1984). Pattern formation in ciliated protozoa. In *Pattern formation a primer in developmental biology* (ed. G.M. Malacinski and S.V. Bryant), pp. 163–96. Macmillan, New York.

Fukui, Y. (1976). Enzymatic dissociation of nascent macrocysts and partition of the liberated cytophagic giant cells in *Dictyostelium mucoroides. Development Growth and Differentiation*, **18**, 145–55.

Gall, J.G. (1986). *The molecular biology of ciliated protozoa*, (ed. J.G. Gall). Academic Press, New York.

Gelei, J. V. (1925). Über der Kannibalismus bei Stentoren. *Archiv für Protistenkunde*, **52**, 404–17.

Giese, A.C. (1938). Cannibalism and gigantism in *Blepharisma. Transactions of the American Microscopical Society*, **57**, 245–55.

Giese, A.C. (1973). Cannibalism. In *Blepharisma, the biology of a light-sensitive protozoan* (ed. A.C. Giese), pp. 123–34. Stanford University Press.

Giese, A.C. and Alden, R.H. (1938). Cannibalism and giant formation in *Stylonychia. Journal of Experimental Zoology*, **78**, 117–34.

Gilbert, J.J. (1973). Induction and ecological significance of gigantism in the rotifer *Asplanchna sieboldi. Science*, **181**, 63–6.

Haime, J. (1853). Observations sur les métamorphoses et l'organisation de la *Trichoda lynceus. Annales des Sciences Naturelles*, **3S, T.19**, 220.

Hewett, S.W. (1980). Prey-dependent cell size in a protozoan predator. *Journal of Protozoology*, **27**, 311–3.

Hirshfield, H.I., Chunosoff, L. and Bhandary, A.V. (1963). Macronuclear variability of *Blepharisma* associated with growth. In *Cell growth and cell division, Symp. Int. Soc. Cell Biology*, No. 11 (ed. R.J.C. Harris), pp. 27–56. New York.

Hiwatashi, K. (1981). Sexual interaction of the cell surface in *Paramecium.* In *Sexual interactions in eukaryotic microbes* (ed. D. O'Day and P.A. Horgen), pp. 351–78. Academic Press, New York.

Huffman, D.M. (1967). The role of engulfment in the *Dictyosteliaceae. Journal of Protozoology*, **14**, 762–4.

Huffman, D.M. and Olive, L.S. (1964). Engulfment and anastomosis in the cellular slime molds (*Acrasiales*). *American Journal of Botany*, **51**, 465–71.

Ivanic, M. (1927). Über den Kannibalismus bei *Amoeba verrucosa* (Ehrbg.), nebst Bemerkungen ber Kannibalismus bei Protozoen im allgemeinen. *Zoologischer Anzeiger*, **74**, 313–21.

Kuhlmann, H.W. and Heckmann, K. (1985). Interspecific morphogens regulating prey-predator relationships in protozoa. *Science*, **227**, 1347–9.

Lapage, G. (1922). Cannibalism in *Amoeba vespertilio. Quartarly Journal of Microscopical Science*, **66**, 669–710.

Lawn, R.M., Heumann, J.M., Herrick, G., and Prescott, D.M. (1978). The gene-size DNA molecules in *Oxytricha. Cold Spring Harbor Symposium of Quantitative Biology*, **42**, 483–92.

Lennartz, D.C. and Bovee, E.C. (1980). Induction of macrostome formation in *Blepharisma amiericanum* (Suzuki, 1954) by alpha-tocopheryl succinate. *Transactions of the American Microscopical Society*, **99**, 310–7.

Maupas, E. (1888). Recherches expérimentales sur la multiplication des Infusoires Ciliés. *Archives de Zoologie Exprimentale et Générale*, **2S.T6**, 165.

Miyake, A. and Harumoto, T. (1990). Asymmetrical cell division in *Blepharisma japonicum*: difference between daughter cells in mating-type expression. *Experimental Cell Research*, **190**, 65–8.

Nelson, E.M. (1978). Transformation in *Tetrahymena thermophila*. *Developmental Biology*, **66**, 17–31.

Olive, L.S., Dutta, K., and Stoianovitch, C. (1961). Variation in the cellular slime mold *Acrasis rosea*. *Journal of Protozoology*, **8**, 467–72.

Padmavathi, P.B. (1961). Giant-cannibals in *Blepharisma undulans*. *Archiv für Protistenkunde*, **105**, 341–4.

Pierce, E., Isquith, I.R., and Repak, A.J. (1978). Quantitative study of cannibal-gigantism in *Blepharisma*. *Acta Protozoologica*, **17**, 493–501.

Raper, K. B. (1940). The communal nature of the fruiting process in the Acrasiae. *American Journal of Botany*, **27**, 436–48.

Ricci, N., Banchett, R., and Pelamatt, R. (1989). Reversible carnivory in *Oxytricha bifaria*. A peculiar ecological adaptation. *Hydrobiologica*, **182**, 115–20.

Russell, P. and Nurse, P. (1987). Negative regulation of mitosis by wee1+ a gene encoding a protein kinase homolog. *Cell*, **49**, 559–67.

Saga, Y. and Yanigsawa, K. (1982). Macrocyst development in *Dicytostelium discoideum*. I. Induction of synchronous development by giant cells and biochemical analysis. *Journal of Cell Science*, **55**, 341–52.

Sandon, H. (1932). *The food of protozoa*. Misr-Sokkar Press, Cairo.

Tulchin, N. and Hirshfield, H.I. (1960). Studies of cannibal giant *Blepharisma undulans*. *Biological Bulletin*, **119**, 345.

Tulchin, N. and Hirshfield, H.I. (1962). Nucleic acid digestion and macronuclear form in cannibal *Blepharisma*, New York University strain. *Journal of Protozoology*, **9**, 200–3.

Waddell, D.R. (1982). A predatory slime mould. *Nature*, **298**, 464–6.

Waddell, D.R. (1988). Cell size in *Dictyostelium*. *Developmental Genetics*, **9**, 673–81.

Waddell, D.R. and Duffy, K.T.I. (1986). Breakdown of self/nonself recognition in cannibalistic strains of the predatory slime mold, *Dictyostelium caveatum*. *Journal of Cell Biology*, **102**, 298–305.

Waddell, D. R. and Vogel, G. (1985). Phagocytic behaviour of the predatory slime mold, *Dictyostelium caveatum*: cell nibbling. *Experimental Cell Research*, **159**, 323–34.

Waddell, D.R., Duffy, K.T.I., and Vogel, G. (1987). Cytokinesis is defective in *Dictyostelium* mutants with altered phagocytic recognition, adhesion and vegetative cell cohesion properties. *Journal of Cell Biology*, **105**, 2293–300.

Watanabe, T. (1990). The role of ciliary surfaces in mating in *Paramecium*. In *Ciliary and flagellar membranes* (ed. R.A. Bloodgood), pp. 149–71. Plenum Press, New York.

Weyer, G. (1930). Untersuchungen über die Morphologie und Physiologie des Formwechsels der *Gastrostyla steinii* Engelmann. *Archiv für Protistenkunde*, **71**, 140–228.

Wicklow, B.J. (1988). Developmental polymorphism induced by intraspecific predation in the ciliated protozoan *Onychodromus quadricornutus*. *Journal of Protozoology*, **35**, 137–41.

Wicklow, B.J. (1990). Intraspecific predator-released morphogens regulate defensive spine development and population growth rates in *Onychodromus quadricornutus*. *Journal of Protozoology*, **37**, 5a.

Williams, N.E. (1960). The polymorphic life history of *Tetrahymena patula*. *Journal of Protozoology*, **7**, 10–7.

Wilson, C.M. and Ross, I.K. (1957). Further cytological studies in the *Acrasiales*. *American Journal of Botany*, **44**, 345–50.

6

Cannibalism in gastropods

Bruno Baur

Gastropods (snails and slugs) constitute by far the largest and most diverse class of Mollusca, with an estimated 40 000 marine, 3000 freshwater and 24 000 terrestrial species (Solem 1974). They are characterized by a distinct head with eyes and tentacles, and a usually coiled shell. Slugs are snails in which the shell is lost or reduced to a remnant.

Gastropods occupy a wide variety of habitats including intertidal rocks, the deep sea bottom, mountain streams, tree stems, leaf litter and stone deserts (Fretter and Graham 1962; Hyman 1967; Solem 1974; Cain 1983; Calow 1983). The adaptive radiation of the gastropods included divergence in feeding structures associated with different feeding habits (Purchon 1977). Several species of gastropods are well-suited for detailed studies of feeding behaviour, but their potential as experimental organisms for examining major issues in the evolution of feeding behaviour has only recently begun to be exploited (Audesirk and Audesirk 1985). Hyman (1967), Hughes (1980), Taylor *et al.* (1980), Kohn (1983) and Audesirk and Audesirk (1985) reviewed gastropod feeding biology, including behavioural aspects, ingestion, neural integration of feeding behaviour, and ontogenetic development of feeding mechanisms.

This chapter examines cannibalism in gastropods. Cannibalism is defined as intraspecific predation, that is the killing and consumption of conspecifics at any stage of their life cycle. Detailed information on cannibalism in gastropods is sparse and often a by-product of studies with other aims. Consequently, this chapter is not a complete summary of all known cases of cannibalism in gastropods. Its purposes are rather

(1) to describe different types of cannibalism that occur among gastropods;

(2) to review egg cannibalism in hatchlings of a land snail species and evaluate the adaptive significance of this behaviour on the basis of a cost-benefit analysis;

(3) to predict what patterns of feeding behaviour may be expected in hatchlings of other species of herbivorous land snails for which information is still lacking;

(4) to evaluate mechanisms for avoiding cannibalism in mating gastropods; and

(5) to summarize the available evidence on the ecological and evolutionary consequences of cannibalism in gastropods.

TYPES OF CANNIBALISM IN GASTROPODS

Cannibalism in gastropods can be classified into three categories:

(1) adult and juvenile gastropods preying on conspecific adults and juveniles;
(2) adult gastropods preying on conspecific eggs; and
(3) larvae and newly hatched snails preying on conspecific eggs.

In this section I present examples for each type of cannibalism that occur in various gastropod groups. A systematic summary of gastropod species for which cannibalism has been reported is given in Table 6.1.

Adult and juvenile gastropods preying on conspecifics

The predatory prosobranch *Natica unifasciata* forages at low tide on sand beaches and mud flats (Hughes 1985). It detects its prey within about 5 cm, either by scent trails or by vibrations generated by the prey through the substratum. The predator quickly extends its propodium (anterior part of the foot partly covering the head) over a contacted prey and envelops it for one to four minutes, during which time the prey is subdued. The immobile prey, enclosed in a mucus envelope, is then towed 10–30 cm before the predator buries into the sand and commences to drill through the shell and ingest the contents. Smaller conspecifics are often consumed, but experiments demonstrate that *N. unifasciata* feeds opportunistically rather than selectively (Hughes 1985).

Oxychilus draparnaudi, *Aegopinella nitidula* and other zonitid terrestrial gastropods are predators on land snails (Boycott 1934; Frömming 1954; Mason 1970). In *A. nitidula*, predation almost invariably conforms to a characteristic pattern (Mordan 1977). The initial attack is made through the shell aperture of the prey and results in much of the head-foot being eaten. A second attack is then made in which the remaining visceral mass is reached by radulation of an irregularly shaped hole in the prey's shell, into which the predator inserts its head. The whole process normally takes between one and two hours in the laboratory. Analysis of *A. nitidula* faeces collected in the field indicated that about 10 per cent of the animals had eaten other gastropods. Food choice experiments revealed that the zonitid snail *Nesovitrea hammonis*, was the preferred prey, but conspecifics were also important. Predators were always larger than their prey, and only snails smaller than 4.5 mm in shell diameter were consumed. Mordan (1977) concluded that predation involving drilling through the shell of the prey is restricted to certain species of *Aegopinella*, since he did not observe it in any other zonitid genus. Frömming (1954) and Rondelaud (1980) recorded cannibalism in the zonitid snail *Oxychilus*

Table 6.1 Gastropod species for which cannibalism has been reported. The systematic order follows Vaught (1989), but nomenclature of species is that given by the authors cited.

Taxa	Type of cannibalism[1]	Habitat[2]	Type of study[3]	Source
Prosobranchia				
123 species (including all orders)	nu, he	m	fi, la	Fioroni (1988); Smith *et al.* (1989) (see also Table 6.2)
Mesogastropoda				
Ampullariidae				
Pila sp.	he	f/t	fi	Keawjam (1986)
Pomacea urceus	he	f/t	fi	Burky (1973)
Carinariidae				
Cardiapoda placenta	ac	m	fi	Hamner *et al.* (1975)
Carinaria cristata	ac	m	fi	Seapy (1980)
Naticidae				
Natica unifasciata	ac	m	fi, la	Hughes (1985)
Neogastropoda				
Muricidae				
Murex trunculus	ac	m	la	Spanier (1981)
Melongenidae				
Buscycon spiratum	ac	m	fi	Paine (1963*a*)
Fasciolariidae				
Fasciolaria tulipa	ac	m	fi	Paine (1963*a*)
Opisthobranchia				
Cephalospidea				
Aglajidae				
Navanax inermis	ac, jc	m	fi, la	Paine (1963*b*, 1965); Leonard and Lukowiak (1984*a*)
Sacoglossa				
Stiligeridae				
Calliopaea oophaga	ae	m	la	Jensen (1986)
Notaspidea				
Pleurobranchidae				
Pleurobranchaea californica	ac	m	la	Audesirk and Audesirk (1985)

Table 6.1 continued

Nudibranchia				
Chromodoridae				
Chromodoris zebra	ac	m	fi	Crozier (1918)
Tritoniidae				
Tritonia hombergi	ac	m	la	Thompson (1961)
Tergipedidae				
Precuthona peachii	ac	m	la	Christensen (1977)
Embletonia fuscata	ac	m	la	Chambers (1934)
Coryphellidae				
Coryphella stimpsoni	ac	m	la	Morse (1971)
Glaucidae				
Glaucus atlanticus	ac	m	la	Lalli and Gilmer (1989)
Facelinidae				
Hermissenda crassicornis	ac	m	la	Zack (1975); Rutowski (1982, 1983)
Aeolidiidae				
Aeolidia papillosa	ac	m	la	Rutowski (1983)
Pulmonata				
Basommatophora				
Ancylidae				
Ancylus fluviatilis	nu	f	la	Bondesen (1950)
Stylommatophora				
15 species from 6 families	he	t	fi, la	(see Table 6.3)
Spiraxidae				
Euglandina rosea	jc	t	la	Cook (1985*a*)
Succineidae				
Succinea putris	jc	t	la	Szabo and Szabo (1930)
Arionidae				
Arion subfuscus	ac, jc	t	fi, la	Rollo and Wellington (1979)
Zonitidae				
Zonitoides nitidus	ac, jc, ae, he	t	fi, la	Didier and Rondelaud (1983, 1989*a*,*b*)
Oxychilus draparnaudi	ac, jc, ae	t	fi, la	Frömming (1954); Rondelaud (1980)
O. alliarius	ac, jc, ae, he	t	fi, la	B. Baur (unpubl. data)

Table 6.1 continued

Aegopinella nitidula	ac, jc	t	fi, la	Frömming (1954); Mordan (1977)
Daudebardia langi	ac, ae	t	fi, la	Grossu (1968)
D. rufa	ac, ae	t	fi, la	Frömming (1954)
Limacidae				
Limax maximus	ac	t	fi, la	Rollo and Wellington (1979)
Deroceras caruanae	ac	t	fi, la	Rollo and Wellington (1979)
D. laeve	ac	t	fi, la	Rollo and Wellington (1979)
Testacellidae				
Testacella haliotidea	ac	t	fi, la	Geyer (1927)
Sphincterochilidae				
Sphincterochila boisseri	jc	t	la	Yom-Tov (1971)

[1] Types of cannibalism: ac = adults on conspecifics, jc = juveniles on conspecifics, ae = adults on eggs, nu = embryos on nurse eggs, he = embryos or newly-hatched snails on fertilized, conspecific eggs.
[2] Habitat: m = marine, f = freshwater, t = terrestrial.
[3] Type of study: fi = field, la = laboratory.

draparnaudi, and both report that the victims were always smaller than the cannibals, which invariably attacked the prey through the shell aperture.

Feeding in the carnivorous land snail *Euglandina rosea* usually commences with a period of trail-following, though randomly encountered immobile prey are also consumed (Cook 1983, 1985*a*, *b*, 1989). Upon contact, the prey is either swallowed whole or its soft parts are extracted from the shell. Which of these feeding techniques is employed depends on several factors, including the relative size of both predator and prey, and probably also the predator's hunger level. In general, this carnivore discriminates between prey species and rarely consumes conspecifics (Cook 1985*a*). Predators usually attack prey smaller than themselves.

Paine (1963*a)* studied a community of eight predatory marine snails and found that individuals of some species feed indiscriminately on gastropods including conspecifics as they encounter them. Other species that showed food preferences were not cannibalistic. Throughout, larger individuals killed and consumed smaller prey (see also Seapy 1980).

Paine (1963*b*) reported that the marine opisthobranch slug *Navanax inermis,* locates prey by following slime trails. However, trail directionality cannot be detected, therefore, trail following can lead *N. inermis* away from its prey (Blair and Seapy 1972). Recently, Leonard and Lukowiak (1984*a*) concluded that feeding on conspecifics in *N. inermis* is usually the result of chance encounters rather than active trail following, since most observed cannibalistic attacks occurred head-on or from the side. Upon contact with its prey, *N. inermis* rapidly withdraws its head, directs the labial palps toward the prey, arches the body just posterior to the head and strikes by protracting and rapidly expanding the pharynx. Striking is so rapid that an entire small prey can be captured and swallowed within 100 msec. The elastic body wall and the lack of a shell permit the ingestion of large prey. For example, a 60 g *N. inermis* has been observed to swallow a 70 g conspecific (Leonard and Lukowiak 1984*a*).

Like *N. inermis*, the carnivorous terrestrial semi-slug *Daudebardia langi* swallows conspecifics nearly as large as itself (Grossu 1968). The body of this largely subterranean species is very elastic and not retractable into the reduced shell, characters that allow the ingestion of large prey.

In several species of gastropods, cannibalism only occurs among juveniles; for example in hatchlings of the desert snail *Sphincterochila boisseri* (Yom-Tov 1971) and in *Succinea putris* (Szabo and Szabo 1930).

Living and decaying plant material, lichens, and fungi make up the staple diet of most land snails (exceptions are a few families of predatory land snails; see above and Table 6.1) (Frömming 1954; Cain 1983). In addition, most of these largely herbivorous land snails frequently scavenge on high-protein food, such as mammalian faeces and carrion, including conspecifics (e.g. *Arion* sp., *Opeas gracile, Macrochlamys indica, Arianta arbustorum*; Frömming 1954; Mitra and Biswas 1974; Raut and Ghose 1982; B. Baur, personal observations). The strong preference of the snails for high-nitrogen food in the field suggests that the availability of nitrogen in the natural diet may limit their growth and reproduction. However, since all these herbivorous gastropods invariably feed on dead conspecifics, they are not strictly cannibalistic. Feeding on carrion, including conspecifics, has also been observed in various species of predatory and non-predatory gastropods in marine and freshwater habitats (e.g. *Lymnaea stagnalis*; Merényi (1980)).

Adult gastropods preying on conspecific eggs

Oophagy by adults has been reported for a few species of marine opisthobranch slugs (Haefelfinger 1962; Crane 1971; Chia and Skeel 1973; Jensen 1986). Among them, *Calliopaea oophaga* eats conspecific eggs (Jensen 1986). Several species of carnivorous terrestrial gastropods are known as snail-egg predators, including *Oxychilus cellarius* (Wolda 1963; Cain and Currey 1968). Adult *Daudebardia* sp. and *O. alliarius* consume conspecific eggs (Grossu 1968; B. Baur, unpublished data).

Larvae and newly hatched snails preying on conspecific eggs

Two different types of embryonic food supply occur in marine prosobranchs: embryos may hatch as free-moving larvae (veligers) or they may metamorphose in the egg capsule. Whereas free-swimming veligers feed on a variety of planktonic

Table 6.2 Number of nurse eggs and hatchlings per egg capsule for several species of prosobranch gastropods

Taxa	Type of cannibalism	Egg size (μm)	Eggs per capsule	Nurse eggs per embryo	Hatching larvae or hatchlings	Hatching stage
Archaeogastropoda						
Neritidae						
Theodoxus fluviatilis	on	110–150	4–165	3–164	1	cw
Mesogastropoda						
Rissoidae						
Rissoa membranacea	fn	90—130	18–400	16	7–25	vl, cw
Calyptraeidae						
Crepidula cerithicola	on	160–180	105–300	21	5–14	cw
C. plana	fn	130–140	80	~3	25	vl
Cassididae						
Cassidaria echinophora	on	263–280	155–193	11	7–23	cw
Neogastropoda						
Muricidae						
Murex brandaris	on	130–330	109–196	24	6–8	cw
M. incarnatus	on	180–200	1036–2511	50	30–41	vl, cw
Nucella crassilabrum	on	204–293	134–1116	90	16–122	cw
N. lapillus	on, c	180–200	400–1000	10–30	2–40	cw
Buccinidae						
Volutopsis norvegicus	on	200–220	10^5	2.5×10^4–10^5	1–4	cw
Buccinum undatum	on, c	200–450	500–2000	110	10–30	cw
Fasciolariidae						
Fasciolaria tulipa	on, c	250	600–2300	200–500	2–16	cw
Pleuroploca australasia	on, c	250	300–900	100–900	1–3	cw

(Data from Fioroni (1988) and Smith *et al.* (1989)).
Type of cannibalism: fn = facultative feeding on nurse eggs, on = obligatory feeding on nurse eggs, c = cannibalism (preying on fertilized eggs and/or larvae). Hatching stage: vl = planktotrophic veliger larvae, cw = crawling juvenile.

Table 6.3 Species of herbivorous terrestrial gastropods (order Stylommatophora) for which egg cannibalism in hatchlings has been recorded and their breeding system

Taxon	Breeding system			Source
	out-crossing	self-fertilization	multiple mating	
Achatinidae				
Limicolaria martensiana	yes	no	yes	Owiny (1974)
Archachatina marginata	yes	no	yes	Plummer (1975)
Achatina achatina	yes	yes	yes	Hodasi (1979)
A. fulica	yes	yes	yes	Rees (1951); Ghose (1959); Pawson and Chase (1984)
Limacidae				
Deroceras sturanyi	yes	no	yes	Kosinska (1980)
D. agreste	yes	yes	yes	Carrick (1938)
Polygyridae				
Triodopsis albolabris	yes	yes[1]	?	Vail (1978); McCracken and Brussard (1980)
Sphincterochilidae				
Sphincterochila boissieri	yes	no	?	Yom-Tov (1971)
Hygromiidae				
Perforatella rubiginosa	yes	no	yes	Frömming (1954)
Helicidae				
Arianta arbustorum	yes	no	yes	Baur and Baur (1986)
Cepaea nemoralis	yes	no	yes	B. Baur (unpubl. data)
C. hortensis	yes	no	yes	B. Baur (unpubl. data)
Helix pomatia	yes	no	yes	Pollard (1975)
H. aspersa	yes	no	yes	Elmslie (1988)
Eobania vermiculata	yes	no	yes	P. Fioroni (pers. comm.)

[1] self-fertilizes only if a mate is not found after several months (McCracken and Brussard 1980).

cells, encapsulated embryos are provided with albumen and, additionally, with a large yolk supply either within each egg, or externally in the form of 'nurse' or 'food' eggs. Nurse eggs are sterile eggs, or eggs that stop developing at an early stage, which varies among species (Fioroni and Schmekel 1976; West 1983). The mechanisms controlling nutritive egg determination and formation are not clear (for discussions see Gallardo and Garrido 1987; Fioroni 1988).

Fioroni (1988) summarizes data for 113 prosobranch species producing and feeding on nurse eggs, and Smith *et al.* (1989) provide data for further 10 species: nurse eggs occur in two genera (one family) of the order Archaeogastropoda, in 18 genera (12 families) of Mesogastropoda and in 38 genera (nine families) of Neogastropoda. Franc (1943 cited in Fioroni 1988) assumes that 75 per cent of all neogastropods produce nurse eggs. Unfortunately, the available information is meagre; the reproductive biology for most of the species is still unknown.

The consumption of nurse eggs can be facultative (i.e. it may not occur in all egg capsules of a species) or obligate. In several species the larvae consume still unhatched embryos as well as 'abnormally' developing embryos. Furthermore, cannibalism between embryos within an egg capsule occurs both in species with and without nurse eggs.

Embryos only have access to nurse eggs that occur in the same egg capsule. Nurse eggs, initially indistinguishable in size from normally-developing eggs, can occur in large numbers. The eggs of *Colus islandicus* are 200–210 μm in diameter, but of 7,350 in a capsule only one to five hatch (Thorson 1935); *Neptunea antiqua* has larger eggs (300 μm) and produces about 5,000 per capsule, of which one or two hatch (Pearce and Thorson 1967) (for further examples see Table 6.2).

When a species provides nurse eggs to its offspring, some embryos typically reach much larger sizes than others. For example, hatchling size ranges from 0.54 to 1.83 mm in *Murex virgineus* (Natarajan 1957) and from 0.66 to 1.84 mm in *M. brandaris* (Fioroni 1966). Interspecific comparisons indicate that hatchling size in prosobranch species with nurse eggs is more variable than in those without nurse eggs, and that shell length of hatching snails is positively correlated with the average nurse-egg supply for each embryo (Fioroni 1966, 1988).

An individual embryo can attain a relatively large hatchling size only if it is able to acquire more than the average number of nurse eggs during its embryonic feeding period or if it cannibalizes its capsule-mates. An unequal distribution of nurse eggs can be obtained by three means. First, nurse-egg supplies may vary between egg capsules; when one capsule is more well-stocked with nurse eggs than others are, each of its embryos will acquire more than the average nurse-egg supply. Second, some capsules may contain many more embryos than others even though all capsules are about equally well-stocked with nurse eggs. Third, some embryos within each capsule may eat more than their 'share' of the nurse eggs supplied, thus attaining large size at the expense of (e.g. out-competing) their capsule-mates.

The ratio of viable eggs to nurse eggs varies considerably both between and within species (Table 6.2; Fioroni (1988), Smith *et al.* (1989)). For example, in *Thais emarginata* some capsules contain only one embryo which becomes large, while others contain as many as 33 embryos, each of which remains relatively small (Spight 1976). Compared with the number of embryos, nurse eggs are distributed more regularly in egg capsules of *T. emarginata*. Embryos tend to reach the same size when nurse eggs are present in excess (Spight 1976).

Larger (older) females of *Buccinum undatum* produce larger egg capsules that contain more nurse eggs but fewer embryos than smaller (younger) females (Giese 1978). Furthermore, the very last capsules of *Buccinum* sp. produced at the end of the reproductive period contain very few nurse and fertilized eggs (Thorson 1935; Giese 1978). In several species, egg size, the presence or absence of nurse eggs, the ratio of nurse eggs to embryos, the total number of eggs per capsule, and hatching stage (veliger larvae or crawling juveniles) depend on various habitat characteristics including salinity and geographic location (Thorson 1950; Fioroni 1988). In general, the number of nurse eggs per capsule increases towards the Arctic; this trend is found both between species and, in some cases, within species (Thorson 1950). With low abundances of food, planktonic larvae are at a disadvantage in the polar regions, and therefore greater maternal nutrition is required.

In summary, embryonic food supply in marine prosobranch gastropods varies from a lack of nurse eggs, to various ratios of viable eggs to nurse eggs with facultative or obligate consumption of nurse eggs, eating of unhatched runts, and cannibalism among capsule-mates.

Prosobranch apple snails of the genera *Pila* and *Pomacea*, that occur in tropical freshwaters and are amphibious, produce calcareous-shelled eggs which they deposit in batches in the soil out of water (Keawjam 1986). In *Pomacea urceus*, the earliest snails to hatch feed upon the remaining eggs in the brood chamber and thus kill unhatched siblings within the batch (Burky 1973).

Hatchlings of various species of herbivorous terrestrial gastropods cannibalize eggs (Table 6.3). Newly-hatched snails consume unhatched sibling eggs, including those with fully-developed embryos (Baur and Baur 1986). In natural populations, hatchlings of the land snail *Arianta arbustorum* also eat eggs from neighbouring batches (Baur 1987*a*). Thus, in stylommatophoran land snails, egg cannibalism occurs within and between clutches, in contrast to marine prosobranch gastropods, in which nurse-egg consumption and cannibalism are restricted to embryos within a capsule.

EGG CANNIBALISM IN HATCHLINGS OF A LAND SNAIL

Reproduction and egg cannibalism

The land snail *A. arbustorum* is a simultaneously hermaphroditic, self-incompatible helicid land snail that is common in moist habitats of northwestern and central Europe. Snails become sexually mature at an age of two to four years and live on average three to four years as adults, but lifespans of up to 14 years have been recorded (Baur and Raboud 1988). Adult snails, measuring 16–20 mm in shell diameter, lay one to three batches of 20–80 eggs per year (Baur and Raboud 1988). The nearly-spherical eggs are approximately 3 mm in diameter (Baur 1984). Egg batches are deposited into small holes in the soil, in moss, and under decaying grass or leaves (Baur 1988*a*). Batches of *A. arbustorum* eggs consist mainly of half-siblings as a result of repeated mating and sperm storage (B. Baur, unpublished data). Hatching takes place two to four weeks after oviposition, depending on ambient temperature. Developing embryos become visible through the egg shell a few days before eclosion. At this stage fertilized and unfertilized eggs can easily be distinguished. Hatchlings emerge by eating a hole in the egg shell. Then they devour the remains of their egg shell and consume unhatched sibling eggs, including those with fully-developed embryos (Baur and Baur 1986). Hatchlings do not discriminate between unfertilized eggs and those containing still unhatched embryos (B. Baur, unpublished data).

Age specificity of egg cannibalism

Egg cannibalism in *A. arbustorum* is restricted to the hatchling stage. The hatchlings' propensity for egg cannibalism decreases with increasing age (Baur 1987*b*). In food choice tests, newly-hatched snails with no prior feeding experience fed exclusively on eggs, four-day-old snails ate eggs and lettuce in equal proportions, while 16-day-old animals preferred lettuce. Individuals older than four weeks fed exclusively on lettuce; even in the absence of vegetable food, individuals will starve rather than eat conspecific eggs. Interestingly, the food preference of young snails is not influenced by either cannibalistic or vegetarian early feeding experience (Baur 1987*b*). Egg cannibalism in *Helix pomatia* and *Cepaea nemoralis* also occurs exclusively during the hatchling stage, suggesting that this behaviour is commonly age-specific in herbivorous terrestrial gastropods (Baur 1988*b*, B. Baur, unpublished data).

The absence of egg cannibalism by older snails may be associated with an age-specific occurrence of digestive enzymes. Hatching snails may require specific enzymes to digest the egg shell, which consists primarily of protein, mucopolysaccharides, and calcium (see below), while the digestion of vegetable food requires cellulase. In *H. pomatia*, the digestive enzymes cellulase and chitinase are partly produced by symbiotic bacteria (Owen 1966; Grassé 1968). It is possible that some time is required for newly-hatched snails to accumulate those enzymes necessary for digesting vegetable food. Such an age-specific occurrence of digestive enzymes has been found in the land snail *Achatina fulica* (Grassé 1968). This might be a proximate explanation for both the age-specific decline in egg preference and the increase in preference for vegetables by *Arianta arbustorum* and *H. pomatia*.

Species specificity of egg cannibalism

Egg-choice tests demonstrated that cannibalistic hatchlings of *A. arbustorum* and

H. pomatia showed a significant preference for conspecific eggs (Baur 1988*c*), and hatchlings of *C. nemoralis* and *C. hortensis* did not consume eggs of the helicid land snails *A. arbustorum* and *Perforatella incarnata* (B. Baur, unpublished data). Thus, egg consumption by newly-hatched snails probably results from active food choice. Indeed, gastropods are known to use chemical cues in locating food and can discriminate between food types on the basis of taste and smell (Croll 1983).

Nutritional and energetic benefits of egg cannibalism

Significant benefits accrue to cannibalistic hatchlings of *A. arbustorum*. Laboratory experiments demonstrated that newly-hatched snails fed a cannibalistic diet during their first 10 days of life increased in wet weight 2.6 times as much as siblings fed on lettuce (Baur 1990*a*). In this experiment, egg consumption within 10 days ranged from 0.7 to 4.0 eggs per individual and weight increase of cannibalistic hatchlings was positively correlated with the number of eggs consumed. Diet did not affect hatchling survival during the first 10 days, but did influence future survival: 67 per cent of the individuals initially fed on eggs attained adulthood compared with 38 per cent of those fed on lettuce. Cannibalistic hatchlings completed shell growth more rapidly and thus became sexually mature earlier than non-cannibalistic ones, but the two groups did not differ in adult shell size. Thus, a cannibalistic diet during the hatchling stage will give accelerated growth and higher survival.

Snail eggs are rich in energy and nutrients (proteins, mucopolysaccharides, calcium) (Duncan 1975; Tompa 1984). For example, the protein concentration in freshly laid eggs of *H. pomatia* averages 14 per cent (Alyakrinskaya 1981). Eggs of *A. arbustorum* contain crystals of calcium carbonate in the jelly layer of the eggshell (cf. Tompa 1976). The function of the calcified egg shell, in addition to mechanical support of egg contents, is to supply the developing embryo with enough calcium to form the embryonic shell by the time of hatching (Tompa 1975, 1980). The availability of calcium and protein may limit growth in newly-hatched *A. arbustorum* (cf. Crowell 1973). The accelerated growth and enhanced survival of cannibalistic hatchlings is probably due to the increased consumption of calcium and protein provided by conspecific eggs. A vegetarian diet, in contrast, is difficult for newly-hatched snails to digest, probably due to their initial lack of cellulase (see section on age specificity).

Geographic and individual variation in propensity for cannibalism

The extent of within-batch egg cannibalism depends primarily on the hatching asynchrony of the eggs and on the hatchlings' propensity for cannibalism (Baur and Baur 1986; Baur 1987*a*). In particular, egg cannibalism is more likely when hatching is more asynchronous. Under laboratory conditions, intrinsic hatching success of *A. arbustorum* from three different populations averaged 89, 90 and 87 per cent (Baur and Baur 1986), and the corresponding hatching asynchrony was 2.4, 2.2 and 2.3 days, respectively. When the hatching asynchrony was experimentally lengthened by a further four days (by keeping half of the eggs of each batch at a temperature of 9°C for seven days), the extent of egg cannibalism increased, resulting in hatching successes being reduced to 58, 62 and 76 per cent, respectively (Baur and Baur 1986). The differential hatching success was a result of differences in cannibalistic propensity; snails that emerged from relatively smaller eggs of lowland populations exhibited a higher propensity for cannibalism than hatchlings from relatively larger eggs of a mountain population. Furthermore, in two populations the propensity for cannibalism varied between batches (progeny from different mothers) (Baur and Baur 1986). Similarly, in egg-choice tests (see below), hatchlings from different mothers and, above all, from different populations varied enormously in their propensity for cannibalism; in some batches all the hatchlings were cannibalistic, while in others none or only a few hatchlings were cannibalistic (Baur 1987*a*). In three different populations from Central Sweden, 40.5, 68.9 and 74.0 per cent of the hatchlings consumed at least one conspecific egg within four days. Within populations, the extent of cannibalism was not influenced by clutch size. The large variation between populations in propensity for cannibalism suggests different costs and benefits of egg cannibalism in different habitats.

Under natural conditions, hatching asynchrony and the consequent extent of egg cannibalism will also depend on the type of oviposition (batches or scattered), the spatial heterogeneity of egg-laying sites, and climatic conditions. For example, in egg batches of the slug *Deroceras agreste*, hatching asynchrony increases as the temperature decreases (Carrick 1938, 1942).

Selective forces acting on the mother

Cannibalism among offspring can be influenced by parental manipulation (Alexander 1974; Polis 1981). A mother snail could manipulate the extent of within-batch cannibalism by altering the degree of hatching asynchrony in the batch, for example by her choice of the oviposition site. An oviposition site that provides equal humidity and temperature conditions for all eggs in the batch synchronizes hatching (cf. Bayne 1968*a*). The presence of another egg batch in an oviposition site may render it less suitable if between-batch cannibalism occurs (see below). Thus ovipositing snails may increase the survival prospects of their offspring if they are able to detect the presence of other egg batches (which will probably hatch earlier) and lay their eggs elsewhere.

The extent of between-batch cannibalism depends on several factors, including the spatial distribution and density of egg batches, the hatching asynchrony between batches and the distances travelled by newly-hatched snails (Baur 1987*a*, 1988*a*). In a field study including 11 natural populations of *A. arbustorum* in central Sweden, the density of egg batches found in August ranged from 0.2 to 21.0 per m^2 (Baur 1988*d*). In populations with very high batch densities (7.8 and more batches per m^2), 21–39 per cent of all batches were 5 cm or less from their nearest neighbour batch. These eggs run a high risk of being eaten by earlier hatched conspecifics from neighbour batches, since the distance covered by newly-hatched *A. arbustorum* within one day averages 5.8 cm (Baur 1988*d*).

Surprisingly, *A. arbustorum* does not react to the presence of conspecific eggs when starting to oviposit, thus exposing its eggs to the risk of between-batch cannibalism (Baur 1988*a*). The snails prefer to oviposit in soil of higher moisture content and thus reduce the risk of later egg desiccation. Both the type and spatial distribution of oviposition sites in natural populations suggest that abiotic factors, such as permanent moisture, exert a stronger influence on the choice of the oviposition site than does between-batch egg cannibalism.

Selection may favour egg cannibalism if the potential reduction of the mother snail's total fecundity due to sibling cannibalism is compensated for by the increased survivorship of her cannibalistic progeny. In *A. arbustorum*, not all eggs within a batch hatch; indeed, some eggs are not fertilized and in others the embryos die during development. Under natural conditions, eggs that fail to hatch can serve as 'trophic' eggs. When intra- and interspecific egg predation were prevented, hatching success averaged 51.4 per cent in 1,047 egg batches of *A. arbustorum* kept in cages in the field; in only 221 batches (21.1 per cent) was hatching success 90 per cent or greater (see Baur and Raboud 1988).

The proportion of fertilized eggs within batches in *A. arbustorum* decreases towards the end of the reproductive season (Baur 1990*c*), which means enhanced opportunities of egg cannibalism for newly hatched snails. Due to the short time until first hibernation, snails hatching late in autumn may have the lowest prospects of survival and, consequently, cannibalism may be most profitable for these hatchlings. It is not known whether an ovipositing snail is able to manipulate the proportion of fertilized eggs within a batch or whether the lower hatching success observed at the end of the breeding season is simply a result of various other environmental factors.

Selective forces acting on the offspring

Cannibalistic hatchlings of *A. arbustorum* and *H. pomatia* do not distinguish between sib and non-sib eggs (Baur 1987*a*, 1990*b*). In four series of choice tests, newly-hatched *A. arbustorum* showed no egg preference when offered sibling eggs, eggs from other relatives (= kin eggs) and non-kin eggs (conspecific eggs from another population) (Baur 1987*a*).

Multiple mating is common in helicid land snails, and sperm can be stored for more than a year (Murray 1964). The eggs of a single batch of *A. arbustorum* are often fertilized by sperm from two or more individuals and consequently, batches consist mainly of half-siblings (B. Baur, unpublished data).

Under which circumstances can we expect that egg cannibalism in *A. arbustorum* might evolve? Using Hamilton's rule (1964), the necessary benefits of egg cannibalism in relation to the degree of relatedness between cannibal and victim can be estimated. According to Hamilton, the inclusive fitness of an individual will be increased, and the selfish trait selected for, if $-k < 1/r$. In this equation k is the change in fitness of the egg eaten divided by the gain in fitness of the cannibal, and r is the coefficient of relatedness between the two individuals ($r = 0.5$ for siblings and $r = 0.25$ for half-sibs). The function k can be expressed as $-p/p'$, where p is the probability of survival from egg to adulthood and p' is the increase in the cannibal's survival probability (Eickwort 1973). The cannibalistic trait will be selected for if $p/p' < 2$ for sibs or < 4 for half-sibs.

Andreassen (1981) found a survivorship from egg to maturity of 0.69 per cent in a natural population of *A. arbustorum* near Bergen (Norway), whereas juvenile survival was 30.5 per cent for the first two months of life. For the cannibalistic trait to be selected, a cannibal would have to increase its probability of survival from 0.69 per cent to just over 1.04 per cent according to Hamilton's rule (see above), assuming relatedness to be 0.5 (or to 0.86 per cent when $r = 0.25$). The shell diameter of a newly hatched snail increases from 2.7 to 3.4 mm within three days following the consumption of one conspecific egg (Baur 1986, 1990*a*), whereas a non-cannibalistic snail requires at least 21 days for a comparable increase in size. As a result of this actual gain in time, the cannibal's probability of survival to adulthood rises from 0.69 to 0.98 per cent, which indicates that within-batch cannibalism is adaptive if the victims are half-siblings.

Due to multiple mating and sperm storage, batches of *A. arbustorum* consist mainly of half-siblings (B. Baur, unpublished data). Moreover, the survival rate of 0.69 per cent used in the example above is rather high. Under certain environmental conditions, such as hot and dry weather, the probability of egg survival may be much lower, and the necessary gain of the cannibal's probability of survival is correspondingly reduced. For example, Wolda (1963) found a survivorship from egg to maturity of 0.30 per cent in a natural population of *C. nemoralis*. If such low survival rates also prevail in *A. arbustorum*, egg cannibalism may be adaptive even among full-siblings.

Experimental evidence indicates that cannibalistic *A. arbustorum* have a higher survivorship than non-cannibalistic individuals (see above). Furthermore, cannibalistic hatchlings tend to complete shell growth more rapidly than non-cannibalistic ones, which may lead to a greater seasonal and lifetime reproductive output in cannibalistic snails. On the other hand, snails fed a cannibalistic diet during the hatchling stage did not differ in adult shell size from non-cannibalistic ones, indicating that egg cannibalism does not affect fecundity through snail size; egg size, clutch size, and number of clutches produced per year are significantly correlated with adult shell size in *A. arbustorum* (Baur 1984; Baur and Raboud 1988).

EVOLUTION OF EGG CANNIBALISM IN HATCHLINGS

Why are the hatchlings of herbivorous land snails cannibalistic in some species, but not in others? Comparisons of certain reproductive and morphological traits (structure of the egg shell, clutch size, egg size, and breeding system) between cannibalistic and non-cannibalistic species might provide some clues.

Eggshell structure

After emergence, hatchlings of some terrestrial gastropods consume the remains of their eggshells, while other species do not (Tompa 1980). Available evidence indicates that the latter group consists of species with non-cannibalistic hatchlings (e.g. *Punctum pygmaeum, Vallonia costata*), whereas hatchlings of all species known to be cannibalistic eat their own eggshells (Table 6.3). This suggests that the consumption of the eggshell may be a prerequisite for the occurrence of egg cannibalism.

Following Tompa (1976, 1984), eggshells of terrestrial gastropods can be classified into three basic types: (1) uncalcified, with only an organic, jelly-like shell layer (e.g. *Succinea putris*), (2) partly calcified, with discrete crystals of calcium carbonate in a jelly matrix (e.g. *H. pomatia*), and (3) heavily calcified, with an outer shell consisting of a layer of fused calcium carbonate crystals, making the eggshell hard and brittle like a bird's egg (e.g. *Cepaea nemoralis*; see also Bayne 1968*b*). Eggshells containing calcium carbonate occur in at least 36 of the 65 families of stylommatophoran land snails (Tompa 1976). All species with cannibalistic hatchlings listed in Table 6.3 have partly or heavily calcified eggshells. In addition to providing proteins and mucopolysaccharides, a calcified eggshell may supply cannibalistic hatchlings with calcium for shell growth (see above).

Clutch size and egg size

Reproductive modes of terrestrial gastropods comprise numerous morphological, physiological and behavioural adaptations for nonaquatic egg deposition and subsequent juvenile development (Tompa 1984). A number of species deposit their eggs in batches (e.g. *A. arbustorum, H. pomatia*), while others deposit their eggs singly; this latter type of oviposition occurs mainly in very small species (e.g. *Punctum pygmaeum*; Baur 1989). Still other species retain their eggs in the oviduct, giving birth to miniature snails (e.g. *Balea perversa*; Tompa 1979). Presumably this prevents both desiccation and predation of eggs.

Deposition of eggs in batches (i.e. brood aggregation) may be a prerequisite for egg cannibalism. Brood aggregation occurs in all species with cannibalistic hatchlings listed in Table 6.3, while egg cannibalism has not been reported for species that deposit their eggs singly (e.g. *P. pygmaeum, Vallonia costata*; B. Baur, unpublished data).

Snail species that deposit their eggs singly produce larger eggs in relation to their shell size than species that lay their eggs in batches (Baur 1989). This difference in relative egg size is highly significant in a comparative study on the life history of land snail species from northwestern and central Europe (Baur and Bengtsson, in preparation). Thus, egg cannibalism tends to occur in species that provide relatively little maternal nutrition in a propagule.

Breeding system

Stylommatophoran land snails are generally described as self-incompatible, simultaneous hermaphrodites (Duncan 1975). Recent studies, however, indicate that self-fertilization may be more widespread than usually assumed (Foltz *et al.* 1984; Baur 1987*c*). The three species known to be non-cannibalistic have self-fertilization (*Vallonia costata*, *Punctum pygmaeum* and *Vertigo pusilla*) (B. Baur, unpublished data). In contrast, species with cannibalistic hatchlings tend to be self-incompatible and mate repeatedly (Table 6.3). Their eggs may be fertilized by sperm from several individuals, and consequently egg cannibalism may more often affect half-siblings than full-siblings (see above). This may have facilitated the evolution of egg cannibalism.

In summary, present evidence suggests that egg cannibalism tends to occur in species of terrestrial gastropods that have obligate out-crossing and multiple mating, produce small eggs in relation to their shell size, deposit their eggs in batches, and have hatchlings that eat their own calcified eggshells.

CANNIBALISM DURING COURTSHIP AND MATING

The fact that cannibalism in soft-bodied, carnivorous gastropods is not restricted to smaller conspecifics means that species having obligate reciprocal mating are faced with the problem that encounters may often be agonistic as well as sexual. Hence, cannibalism can be an important factor moulding courtship and mating behaviour in carnivorous gastropods (see also Elgar, this volume). In general, copulatory duration in gastropods lasts from many minutes to several hours (e.g. Fretter and Graham 1962; Hyman 1967; Kupfermann and Carew 1974).

The carnivorous nudibranch *Hermissenda crassicornis* engages in reciprocal copulation, which is very rapid; intromission lasting only about six seconds (Rutowski 1983). As the partners move apart, they may bite one another. Because individuals of *H. crassicornis* are cannibalistic (Zack 1975; Rutowski 1982), Longley and Longley (1982) suggested that the brief duration of sexual contact is an adaptation to minimize time spent with a sexual partner that is also a potential predator. The speed with which copulation occurs results in a high failure rate: only 38 per cent of the pairwise intromission attempts were successful (Rutowski 1983). The nudibranch *Aeolidia papillosa*, which frequently attacks conspecifics, also has very brief copulations in comparison to non-cannibalistic species (Kupfermann and Carew 1974; Longley and Longley 1982; Rutowski 1983).

However, rapid intromission times are not characteristic for all cannibalistic slugs. For example, *Navanax inermis* exhibits prolonged sexual encounters (copulation duration 30–120 min), during which cannibalistic behaviour is in some way suppressed (Leonard and Lukowiak 1984*a*,*b*). *Pleurobranchaea californica* is another cannibalistic opisthobranch in which mating encounters are prolonged; the animals will interrupt mating to feed, but not on each other (Audesirk and Audesirk 1985). On the other hand, Lalli and Gillmer (1989) observed that a smaller individual (15 mm in body length) of *Glaucus atlanticus* was cannibalized after copulation by its larger mate (18 mm).

CANNIBALISM AND LOCAL POPULATION SIZE

Indiscriminate predation that includes conspecifics can act as a density-dependent factor affecting local population size. A variety of factors such as abundance of alternative prey, quality of food, availability of specific nutrients, and stress are thought to influence the extent of cannibalism within a population (Fox 1975; Polis 1981). The role of cannibalism in the dynamics of natural snail populations has been largely ignored. Paine (1965) found that cannibalism makes a significant contribution to the natural diet of the predatory marine slug *Navanax inermis*, and demonstrated experimentally that cannibalism can act as a density-dependent factor affecting population size. As local population density increases, slugs encounter more conspecifics, leading to an increased frequency of cannibalism. Cannibalism may similarly affect local population size in the predatory land snail *Oxychilus draparnaudi* (Rondelaud 1980).

Between-batch cannibalism in hatchlings of herbivorous land snails may also limit local population size in natural populations of *Arianta arbustorum* in central Sweden; the density of egg batches was positively correlated with local population density and between-batch cannibalism was observed in batches deposited close together (Baur 1987*a*, 1988*d*) (see section on selective forces acting on the mother). The extent of between-batch egg cannibalism in *A. arbustorum* under laboratory conditions increased with decreasing distance between batches (Baur 1988*d*), suggesting that between-batch cannibalism is density-dependent in natural populations (Baur 1987*a*; Baur 1988*d*).

DISCUSSION

Cannibalism between adult gastropods is usually incidental and could be considered as a by-product of the species' feeding behaviour. The few reports on intraspecific predation in gastropods suggest that cannibalism may depend on the degree of feeding specialization. Cannibalism has not been observed in highly specialized predators (e.g. *Conus* sp. Kohn 1959, 1983; Kohn and Nybakken 1975). Among generalist predators, cannibalism seems to occur opportunistically (e.g. Paine 1963*a*; Seapy 1980; Hughes 1985), whereas scavenging on carrion, including conspecifics, can frequently be observed in herbivorous species. Thus, cannibalism among adult gastropods appears to be a consequence of the lack of specificity of predation, linked perhaps with food deprivation or other environmental constraints, such as a limited period for foraging activity. For example, Hughes (1985) suggested that the opportunistic feeding behaviour of *Natica unifasciata* may be associated with both the low prey densities in the habitat and foraging periods constrained by environmental hazards during low tide. Under such conditions a cannibal is just another occasional predator.

The scattered reports of cannibalism indicate that potential prey size is limited by the presence and shape of the predator's shell. Cannibalistic individuals of both marine and terrestrial snail species with a retractable body are able to swallow only small conspecifics or consume only the soft body of the prey in parts, while slug species lacking a shell or having a greatly reduced shell can ingest large conspecifics.

Obligate or facultative egg cannibalism by larvae and newly hatched snails seems to be widespread among marine prosobranchs and stylommatophoran land snails. In both groups of gastropods there is a graduation in the consumption of conspecific eggs by embryos or hatchlings that ranges from no eggs, to infertile (nurse) eggs, unhatched runts, and finally any unhatched sibs.

In an evolutionary context, two different solutions to the problem of increasing maternal nutrition can be recognized: increasing egg size and producing nurse eggs (Thorson 1950). The probability of survival of newly hatched gastropods is likely to increase with hatchling size (Spight 1976), which in turn is positively correlated with egg size, except in the case of nurse-egg feeders (Fioroni and Schmekel 1976; Fioroni 1988). Hence, the evolution of large eggs may be a selective response to high post-hatching mortality. However, large eggs of marine snails take longer to develop (Perron 1981) and, as Spight (1975) points out, the selective cost of prolonged embryonic development is high pre-hatching mortality. Snail eggs are subject to intense predation; Spight (1975) reported that the gastropod *Thais lamellosa* may incur up to 42 per cent mortality from predation and physical stress before hatching. Furthermore, egg size may be limited by morphological and physiological constraints on the mother (e.g. by the width of the shell aperture; B. Baur, unpublished data).

Another solution is to provide hatchlings with food. Alexander (1974) suggested that if parents are unable to increase investment in young through increasing egg size, an alternative strategy is to increase clutch size and allow some siblings to consume others. The optimum clutch size can be found by calculating the clutch size leading to maximum brood productivity, taking into account the effects of sibling cannibalism and possible trade-offs. The mother may, however, be able to regulate the amount of sibling cannibalism by laying eggs sequentially, rather than simultaneously. Sequential egg production, which leads to a size hierarchy among the young, is very common among birds, and several explanations have been proposed (Lack 1968). Lack suggested that it is advantageous to have a size hierarchy in view of uncertainty of food supplies; in a poor year, competition between siblings of different sizes leads to the death of the runt. However, it is unknown whether gastropods can regulate the amount of egg cannibalism by the way in which they deposit their eggs or by producing different proportions of unfertilized eggs within a batch.

Maternal provision of nurse eggs of different types to hatchlings may be more widespread than currently realized. Numerous species of marine prosobranch snails normally produce nurse eggs, which serve as the first food for their progeny (Fioroni 1966, 1988; Fioroni and Schmekel 1976; Smith *et al.* 1989). In some other species the larvae additionally consume siblings (Hoagland 1979, 1986; Fioroni 1988). In terrestrial gastropods, the extent of egg cannibalism mainly depends on the hatching asynchrony of the young within a batch, which in turn is influenced by proximate factors (e.g. drought and temperature) and probably genetic factors. Under favourable conditions, all young of a clutch may hatch. However, under less favourable conditions, hatching asynchrony may increase and the snails that emerge first will eat the remaining unhatched sibs. In this way a few juveniles may have a chance for survival even under severe conditions.

ACKNOWLEDGEMENTS

I thank A. Baur, A. J. Cain, A. Cook, B. J. Crespi, M. L. Crump, M. A. Elgar, P. Schmid-Hempel, S. C. Stearns, and S. Ulfstrand for discussion and comments on the manuscript and D. Geiger and W. Neuckel for help in literature search. The Swiss National Science Foundation and the Freiwillige Akademische Gesellschaft, University of Basel, supported my research and preparation of this review.

REFERENCES

Alexander, R. D. (1974). The evolution of social behaviour. *Annual Review of Ecology and Systematics*, **4**, 325–83.

Alyakrinskaya, I. O. (1981). Egg nutrient content in gastropods. *Doklady Akademii Nauk SSSR*, **260**, 245–8.

Andreassen, E. M. (1981). Population dynamics of *Arianta arbustorum* and *Cepaea hortensis* in western Norway. *Fauna norvegica Serie A*, **2**, 1–13.

Audesirk, T. and Audesirk, G. (1985). Behavior of gastropod molluscs. In *The Mollusca*, Vol. 8, *Neurobiology and behavior*, Part 1, (ed. A. O. D. Willows), pp. 1–94. Academic Press, London.

Baur, B. (1984). Shell size and growth rate differences for alpine populations of *Arianta arbustorum* (L.) (Pulmonata: Helicidae). *Revue Suisse de Zoologie*, **91**, 37–46.

Baur, B. (1986). The influence of cannibalistic egg eating on the growth of young *Arianta arbustorum* (L.) (Helicidae). *Snail Farming Research*, **1**, 9–17.

Baur, B. (1987*a*). Can cannibalistic hatchlings of the land snail *Arianta arbustorum* distinguish between sib and non-sib eggs? *Behaviour*, **103**, 259–65.

Baur, B. (1987*b*). Effects of early feeding experience and age on the cannibalistic propensity of the land snail *Arianta arbustorum*. *Canadian Journal of Zoology*, **65**, 3068–70.

Baur, B. (1987*c*). The minute land snail *Punctum pygmaeum* (Draparnaud) can reproduce in the absence of a mate. *Journal of Molluscan Studies*, **53**, 112–13.

Baur, B. (1988*a*). Do the risks of egg cannibalism and desiccation influence the choice of oviposition sites in the land snail *Arianta arbustorum*? *Journal of Zoology (Lond.)*, **216**, 495–502.

Baur, B. (1988*b*). Age-specific food preferences in hatchlings of *Helix pomatia* (L.). *Snail Farming Research*, **2**, 14–19.

Baur, B. (1988*c*). Egg-species recognition in cannibalistic hatchlings of the land snails *Arianta arbustorum* and *Helix pomatia*. *Experientia*, **44**, 276–7.

Baur, B. (1988*d*). Population regulation in the land snail *Arianta arbustorum*: density effects on adult size, clutch size and incidence of egg cannibalism. *Oecologia (Berl.)*, **77**, 390–4.

Baur, B. (1989). Growth and reproduction of the minute land snail *Punctum pygmaeum* (Draparnaud). *Journal of Molluscan Studies*, **55**, 383–7.

Baur, B. (1990*a*). Possible benefits of egg cannibalism in the land snail *Arianta arbustorum*. *Functional Ecology*, **4**, 679–84.

Baur, B. (1990*b*). Egg cannibalism in hatchlings of the land snail *Helix pomatia:* nutritional advantage may outweigh lack of kin recognition. *Malacological Review*, **23**, 103–5.

Baur, B. (1990*c*). Seasonal changes in clutch size, egg size and mode of oviposition in *Arianta arbustorum* L. (Gastropoda) from alpine populations. *Zoologischer Anzeiger*, **225**, 253–64.

Baur, B. and Baur, A. (1986). Proximate factors influencing egg cannibalism in the land snail *Arianta arbustorum* (Pulmonata, Helicidae). *Oecologia (Berl.)*, **70**, 283–7.

Baur, B. and Raboud, C. (1988). Life history of the land snail *Arianta arbustorum* along an altitudinal gradient. *Journal of Animal Ecology*, **57**, 71–87.

Bayne, C. (1968*a*). A study of the desiccation of egg capsules of eight gastropod species. *Journal of Zoology (Lond.)*, **155**, 401–11.

Bayne, C. (1968*b*). Histochemical studies on the egg capsules of eight gastropod molluscs. *Proceedings of the Malacological Society of London*, **38**, 199–212.

Blair, G. M. and Seapy, R. R. (1972). Selective predation and prey location in the sea slug *Navanax inermis*. *Veliger*, **15**, 119–24.

Bondesen, P. (1950). A comparative morphological-biological analysis of the egg capsules of freshwater pulmonate gastropods. *Natura Jutlandica*, **3**, 1–208.

Boycott, A. E. (1934). The habitats of land mollusca in Britain. *Journal of Ecology*, **22**, 1–38.

Burky, A. J. (1973). Organic content of eggs and juveniles of an amphibious snail, *Pomacea urceus* (Müller), from the Venezuelan savannah and their ecological significance. *Malacological Review*, **6**, 59.

Cain, A. J. (1983). Ecology and ecogenetics of terrestrial molluscan populations. In *The Mollusca, Ecology*, Vol. 6, (ed. W. D. Russell-Hunter), pp. 597–647. Academic Press, London.

Cain, A. J. and Currey, J. D. (1968). Studies on *Cepaea*. III. Ecogenetics of a population of *Cepaea nemoralis* (L.) subject to strong area effects. *Philosophical Transactions of the Royal Society of London, Series B*, **253**, 447–82.

Calow, P. (1983). Life-cycle patterns and evolution. In *The Mollusca, Ecology*, Vol. 6, (ed. W. D. Russell-Hunter), pp. 649–78. Academic Press, London.

Carrick, R. (1938). The life-history and development of *Agriolimax agrestis*, the grey field slug. *Transactions of the Royal Society of Edinburgh*, **59**, 563–97.

Carrick, R. (1942). The grey field slug, *Agriolimax agrestis* L., and its environment. *Annals of Applied Biology*, **29**, 43–55.

Chambers, L. A. (1934). Studies on the organs of reproduction in the nudibranchiate mollusks, with special reference to *Embletonia fuscata* Gould. *Bulletin of the American Museum of Natural History*, **66**, 599–641.

Chia, F.-S. and Skeel, M. (1973). The effect of food consumption on growth, fecundity and mortality in a sacoglossan opisthobranch, *Olea hansineensis*. *Veliger*, **16**, 153–8.

Christensen, H. (1977). Feeding and reproduction in *Precuthon peachii* (Mollusca, Nudibranchia). *Ophelia*, **16**, 131–42.

Cook, A. (1983). Feeding by the carnivorous snail *Euglandina rosea* Férussac. *Journal of Molluscan Studies*, Supplement 12A, 32–35.

Cook, A. (1985*a*). Functional aspects of trail following by the carnivorous snail *Euglandina rosea*. *Malacologia*, **26**, 173–81.

Cook, A. (1985*b*). The organisation of feeding in the carnivorous snail *Euglandina rosea*. *Malacologia*, **26**, 183–9.

Cook, A. (1989). The basis of food choice by the carnivorous snail, *Euglandina rosea*. In *Slugs and snails in world agriculture*, Monograph No. 41, (ed. I. F. Henderson), pp. 367–72. British Crop Protection Council, Thornton Heath.

Crane, S. (1971). The feeding and reproductive behavior of the sacoglossan gastropod *Olea hansineensis* Agersborg, 1923. *Veliger*, **14**, 57–9.

Croll, R. P. (1983). Gastropod chemoreception. *Biological Review*, **58**, 293–319.

Crowell, H. H. (1973). Laboratory study of calcium requirements of the brown garden snail, *Helix aspersa* Müller. *Proceedings of the Malacological Society of London*, **40**, 491–503.

Crozier, W. J. (1918). Assortative mating in a nudibranch, *Chromodoris zebra* Heilprin. *Journal of Experimental Zoology*, **27**, 247–92.

Didier, B. and Rondelaud, D. (1983). Les effets d'un aliment composé sur la fécondité et la croissance de *Zonitoides nitidus* Müller. *Haliotis*, **13**, 25–8.

Didier, B. and Rondelaud, D. (1989*a*). Premiéres données sur le régime alimentaire de *Zonitoides nitidus* Müller. *Bulletin de la Société d'Histoire Naturelle de Toulouse*, **125**, 55–60.

Didier, B. and Rondelaud, D. (1989*b*). Les caractéristiques des proies consommées par le Mollusque *Zonitoides nitidus* Müller et leur dynamique en juin, juillet et août. *Bulletin de la Société d'Histoire Naturelle de Toulouse*, **125**, 111–17.

Duncan, C. J. (1975). Reproduction. In *Pulmonates*, Vol. 1, (ed. V. Fretter and J. Peake), pp. 309–65. Academic Press, London.

Eickwort, K. R. (1973). Cannibalism and kin selection in *Labidomera clivicollis* (Coleoptera: Chrysomelidae). *American Naturalist*, **107**, 452–3.

Elmslie, L. J. (1988). Studies on the feeding of newly hatched *Helix aspersa*. *Snail Farming Research*, **2**, 45–8.

Fioroni, P. (1966). Zur Morphologie und Embryogenese des Darmtraktes und der transitorischen Organe bei Prosobranchiern (Mollusca, Gastropoda). *Revue Suisse de Zoologie*, **73**, 621–876.

Fioroni, P. (1988). Die Prosobranchier-Entwicklung mit Nähreiern. *Zoologischer Anzeiger*, **221**, 201–47.

Fioroni, P. and Schmekel, L. (1976). Die nährstoffreiche Gastropoden-Ontogenese. *Zoologisches Jahrbuch Abteilung Anatomie und Ontogenie der Tiere*, **96**, 74–171.

Foltz, D. W., Ochman, H., and Selander, R. K. (1984). Genetic diversity and breeding systems in terrestrial slugs of the families Limacidae and Arionidae. *Malacologia*, **25**, 593–605.

Fox, L. R. (1975). Cannibalism in natural populations. *Annual Review of Ecology and Systematics*, **6**, 87–106.

Fretter, V. and Graham, A. (1962). *British prosobranch molluscs: their functional anatomy and ecology*. Ray Society, London.
Frömming, E. (1954). *Biologie der mitteleuropäischen Landgastropoden*. Duncker and Humblot, Berlin.
Gallardo, C. S. and Garrido, O. A. (1987). Nutritive egg formation in the marine snails *Crepidula dilatata* and *Nucella crassilabrum*. *International Journal of Invertebrate Reproduction and Development*, **11**, 239–54.
Geyer, D. (1927). *Unsere Land- und Süsswasser-Mollusken*. Lutz Verlag, Stuttgart.
Ghose, K. C. (1959). Observations on the mating and oviposition of two land pulmonates, *Achatina fulica* Bowdich and *Macrochlamys indica* Godwin-Austen. *Journal of the Bombay Natural History Society*, **56**, 183–7.
Giese, K. (1978). Zur Embryonalentwicklung von *Buccinum undatum* L. (Gastropoda, Prosobranchia, Stenoglossa). *Zoologisches Jahrbuch Abteilung Anatomie und Ontogenie der Tiere*, **100**, 65–117.
Grassé, P.-P. (1968). Traité de zoologie. Tome 5/3. *Mollusques, Gastéropodes et Scaphopodes*, pp. 377–8. Masson et Cie, Paris.
Grossu, A. V. (1968). Cannibalisme chez *Daudebardia*. *Archiv für Molluskenkunde*, **98**, 139–41.
Haefelfinger, H.-R. (1962). Quelques faits concernant la nutrition chez *Favorinus branchialis* (Rathke 1802) et *Stiliger vesiculosus* (Deshayes 1864), deux Mollusques Opisthobranches. *Revue Suisse de Zoologie*, **69**, 311–16.
Hamilton, W. D. (1964). The genetical evolution of social behaviour I, II. *Journal of Theoretical Biology*, **7**, 1–51.
Hamner, W. M., Madin, L. P., Alldredge, A. L., Gilmer, R. W., and Hamner, P. P. (1975). Underwater observations of gelatinous zooplankton: sampling problems, feeding biology and behaviour. *Limnology and Oceanography*, **20**, 907–17.
Hoagland, K. E. (1979). The behavior of three sympatric species of *Crepidula* (Gastropoda: Prosobranchia) from the Atlantic, with implications for evolutionary ecology. *Nautilus*, **94**, 143–9.
Hoagland, K. E. (1986). Patterns of encapsulation and brooding in the Calyptraeidae (Prosobranchia: Mesogastropoda). *American Malacological Bulletin*, **4**, 173–83.
Hodasi, J. K. M. (1979). Life-history studies of *Achatina* (*Achatina*) *achatina* (Linné). *Journal of Molluscan Studies*, **45**, 328–39.
Hughes, R. N. (1980). Optimal foraging theory in the marine context. *Oceanography and Marine Biology Annual Review*, **18**, 423–81.
Hughes, R. N. (1985). Predatory behaviour of *Natica unifasciata* feeding intertidally on gastropods. *Journal of Molluscan Studies*, **51**, 331–5.
Hyman, L. H. (1967). *The invertebrates*, Vol. 6, *Mollusca* I. McGraw-Hill, New York.
Jensen, K. R. (1986). Obervations on feeding, copulation and spawning of the ascoglossan opisthobranch *Calliopaea oophaga* Lemche. *Ophellia*, **25**, 97–106.
Keawjam, R. S. (1986). The apple snails of Thailand: distribution, habitats and shell morphology. *Malacological Review*, **19**, 61–81.
Kohn, A. J. (1959). The ecology of Conus in Hawaii. *Ecological Monographs*, **29**, 47–90.
Kohn, A. J. (1983). Feeding biology of gastropods. In *The Mollusca, Physiology*, Part 2, Vol. 5, (ed. A. S. M. Saleuddin and K. M. Wilbur), pp. 1–63. Academic Press, London.
Kohn, A. J. and Nybakken, J. W. (1975). Ecology of *Conus* on Eastern Indian Ocean fringing reefs: diversity of species and resource utilization. *Marine Biology*, **29**, 211–34.
Kosinska, M. (1980). The life-cycle of *Deroceras sturanyi*. *Zoologica Poloniae*, **28**, 3–155.

Kupfermann, I. and Carew, T. J. (1974). Behavior patterns of *Aplysia californica* in its natural environment. *Behavioral Biology*, **12**, 317–37.

Lack, D. (1968). *Ecological adaptations for breeding in birds.* Methuen, London.

Lalli, C. M. and Gilmer, R. W. (1989). *Pelagic snails. The biology of holoplanktonic gastropod mollusks.* Stanford University Press.

Leonard, J. L. and Lukowiak, K. (1984*a*). An ethogram of the sea slug, *Navanax inermis* (Gastropoda, Opisthobranchia). *Zeitschrift für Tierpsychologie*, **65**, 327–45.

Leonard, J. L. and Lukowiak, K. (1984*b*). Male–female conflict in a simultaneous hermaphrodite resolved by sperm trading. *American Naturalist*, **124**, 282–6.

Longley, R. D. and Longley, A. J. (1982). *Hermissenda:* agonistic behavior or mating behavior? *Veliger*, **24**, 230–1.

Mason, C. F. (1970). Food, feeding rates and assimilation in woodland snails. *Oecologia (Berl.)*, **4**, 358–73.

McCracken, G. F. and Brussard, P. F. (1980). Self-fertilization in the white-lipped land snail *Triodopsis albolabris*. *Biological Journal of the Linnean Society*, **14**, 429–34.

Merényi, L. (1980). An observed case of cannibalism among the aquarium reared specimens of *Lymnaea stagnalis* (L.). *Soosiana*, **8**, 3–4. (in Hungarian, with English summary.)

Mitra, T. R. and Biswas, S. K. (1974). Necrophagous habit in *Opeas gracile* (Stylommatophora: Subulinidae). *Malacological Review*, **7**, 136.

Mordan, P. B. (1977). Factors affecting the distribution and abundance of *Aegopinella* and *Nesovitrea* (Pulmonata: Zonitidae) at Monks Wood National Nature Reserve, Huntingdonshire. *Biological Journal of the Linnean Society*, **9**, 59–72.

Morse, M. P. (1971). Biology and life history of the nudibranch mollusc, *Coryphella stimpsoni* (Verill 1879). *Biological Bulletin*, **140**, 84–94.

Murray, J. J. (1964). Multiple mating and effective population size in *Cepaea nemoralis*. *Evolution*, **18**, 283–91.

Natarajan, A. V. (1957). Studies on the egg masses and larval development of some prosobranchs from the Gulf of Mannar and the Palk Bay. *Proceedings of the Indian Academy of Sciences*, **46**, 170–228.

Owen, G. (1966). Digestion. In *Physiology of Mollusca*, Vol. II, (ed. K. M. Wilbur and C. M. Yonge), pp. 53–96. Academic Press, London.

Owiny, A. M. (1974). Some aspects of the breeding biology of the equatorial land snail *Limicolaria martensiana* (Achatinidae: Pulmonata). *Journal of Zoology (Lond.)*, **172**, 191–206.

Paine, R. T. (1963*a*). Trophic relationships of eight sympatric predatory gastropods. *Ecology*, **44**, 63–73.

Paine, R. T. (1963*b*). Food recognititon and predation on opisthobranchs by *Navanax inermis* (Gastropoda: Opisthobranchia). *Veliger*, **6**, 1–9.

Paine, R. T. (1965). Natural history, limiting factors and energetics of the opisthobranch *Navanax inermis*. *Ecology*, **46**, 603–19.

Pawson, P. A. and Chase, R. (1984). The life-cycle and reproductive activity of *Achatina fulica* (Bowdich) in laboratory culture. *Journal of Molluscan Studies*, **50**, 85–91.

Pearce, J. B. and Thorson, G. (1967). The feeding and reproductive biology of the red whelk, *Neptunea antiqua* (L.) (Gastropoda, Prosobranchia). *Ophelia*, **4**, 277–314.

Perron, F. E. (1981). The partitioning of reproductive energy between ova and protective capsules in marine gastropods of the genus *Conus*. *American Naturalist*, **118**, 110–18.

Plummer, J. M. (1975). Observations on the reproduction, growth and longevity of a laboratory colony of *Archachatina (Calachatina) marginata* (Swainson) subspecies ovum. *Proceedings of the Malacological Society of London*, **41**, 395–413.

Polis, G. A. (1981). The evolution and dynamics of intraspecific predation. *Annual Review of Ecology and Systematics*, **12**, 225–51.

Pollard, E. (1975). Aspects of the ecology of *Helix pomatia* L. *Journal of Animal Ecology*, **44**, 305–29.

Purchon, R. D. (1977). *The biology of the Mollusca*. Pergamon, New York.

Raut, S. K. and Ghose, K.C. (1982). Cannibalism in the garden snail *Macrochlamys indica* Godwin-Austen (Stylommatophora: Mollusca). *Journal of the Bombay Natural History Society*, **79**, 706–7.

Rees, W. J. (1951). The giant African snail. *Proceedings of the Zoological Society of London*, **120**, 577–98.

Rollo, C. D. and Wellington, W. G. (1979). Intra- and inter-specific agonistic behavior among terrestrial slugs (Pulmonata: Stylommatophora). *Canadian Journal of Zoology*, **57**, 846–55.

Rondelaud, D. (1980). Données complémentaires sur l'éco-éthologie d'*Oxychilus draparnaudi* Beck (Mollusque, Pulmoné): Proposition d'une nouvelle technique d'élevage. *Haliotis*, **10**, 45–8.

Rutowski, R. L. (1982). The temporal structure of behavioral interactions in *Hermissenda crassicornis* (Opisthobranchia). *Veliger*, **24**, 227–9.

Rutowski, R. L. (1983). Mating and egg mass production in the aeolid nudibranch *Hermissenda crassicornis* (Gastropoda: Opisthobranchia). *Biological Bulletin*, **165**, 276–85.

Seapy, R. R. (1980). Predation by the epipelagic heteropod mollusk *Carinaria cristata* forma *japonica*. *Marine Biology*, **60**, 137–46.

Smith, B. J., Black, J. H., and Shepherd, S. A. (1989). Molluscan egg masses and capsules. In *Marine invertebrates of Southern Australia*, Part II, (ed. S. A. Shepherd and I. M. Thomas), pp. 841–91. South Australian Government Printing Division, Adelaide.

Solem, A. (1974). *The shell makers introducing mollusks*. Wiley, New York.

Spanier, E. (1981). Aspects of the biology and behaviour of the purple snail *Murex trunculus*. *Israel Journal of Zoology*, **30**, 106–7.

Spight, T. M. (1975). On a snail's chances of becoming a year old. *Oikos*, **26**, 9–14.

Spight, T. M. (1976). Hatching size and the distribution of nurse eggs among prosobranch embryos. *Biological Bulletin*, **150**, 491–9.

Szabo, J. and Szabo, M. (1930). Todesursachen und pathologische Erscheinungen bei Pulmonaten. *Archiv für Molluskenkunde*, **62**, 123–30.

Taylor, J. D., Morris, N. J., and Taylor, C. N. (1980). Food specialization and the evolution of predatory prosobranch gastropods. *Palaeontology*, **23**, 375–409.

Thompson, T. E. (1961). The structure and mode of functioning of the reproductive organs of *Tritonia hombergi* (Gastropoda, Opisthobranchia). *Quarterly Journal of Microscopical Science*, **102**, 1–14.

Thorson, G. (1935). Studies on the egg-capsules and development of Arctic marine prosobranchs. *Meddelser om Grønland*, **100**, 1–71.

Thorson, G. (1950). Reproductive and larval ecology of marine bottom invertebrates. *Biological Review*, **25**, 1–45.

Tompa, A. S. (1975). Embryonic use of egg shell calcium in a gastropod. *Nature*, **255**, 232–3.

Tompa, A. S. (1976). A comparative study of the ultrastructure and mineralogy of calcified land snail eggs (Pulmonata: Stylommatophora). *Journal of Morphology*, **150**, 861–88.

Tompa, A. S. (1979). Oviparity, egg retention and ovoviviparity in pulmonates. *Journal of Molluscan Studies,* **45**, 155–60.

Tompa, A. S. (1980). Studies on the reproductive biology of gastropods: part 3. Calcium provision and the evolution of terrestrial eggs among gastropods. *Journal of Conchology,* **30**, 145–54.

Tompa, A. S. (1984). Land snails (Stylommatophora). In *The Mollusca*, Vol. 7, (ed. A. S. Tompa, N. H. Verdonk, and J. A. M. van den Biggelaar), pp. 47–140. Academic Press, London.

Vail, V. A. (1978). Laboratory observations on the eggs and young of *Triodopsis albolabris major* (Pulmonata: Polygyridae). *Malacological Review,* **11**, 39–46.

Vaught, K. C. (1989). *A classification of the living Mollusca.* American Malacologists Inc., Melbourne, Florida.

West, D. L. (1983). Reproductive biology of *Colus stimpsoni* (Prosobranchia: Buccinidae). 5. Nutritive egg formation. *Veliger*, **25**, 299–305.

Wolda, H. (1963). Natural populations of the polymorphic land snail *Cepaea nemoralis* (L.). *Archives Néerlandaises de Zoologie,* **15**, 381–471.

Yom-Tov, Y. (1971). The biology of two desert snails *Trochoidea* (*Xerocrassa*) *seetzeni* and *Sphincterochila boissieri. Israel Journal of Zoology,* **20**, 231–48.

Zack, S. (1975). A description and analysis of agonistic behavior patterns in an opisthobranch mollusc, *Hermissenda crassicornis. Behaviour*, **53**, 238–67.

7

Sexual cannibalism in spiders and other invertebrates

Mark A. Elgar

A variety of animals, but most notably certain species of praying mantids, scorpions, and spiders, have gained some notoriety for attempting to consume their mates, with descriptions of this behaviour making good copy for popular articles on courtship and mating in animals. However, the evolutionary basis of this form of cannibalism has received little empirical or theoretical attention in the biological literature, and descriptions are usually anecdotal or associated with general descriptions of courtship and mating. For example, many textbooks state that cannibalism of sexual partners has been responsible for the evolution of elaborate courtship behaviour in many spiders (e.g. Manning 1972; Gould 1982; Cloudsley-Thompson 1988), despite the lack of either within-species or comparative evidence. Cannibalism of sexual partners may have been ignored in the past because it did not accord with the traditional view of courtship and mating as largely cooperative and harmonious ventures, a view that is being replaced by the idea that there are conflicts of interest between the sexes (e.g. Krebs and Davies 1987). Because cannibalism of sexual partners represents an extreme form of the conflict of interests between males and females, it offers an interesting model system for understanding animal mating and courtship behaviour.

Cannibalism of sexual partners, or more specifically sexual cannibalism, refers to the killing and consumption of males by females at some stage during either courtship, copulation, or shortly thereafter. There are infrequent reports of cannibalism of adult females by males, and it usually occurs under laboratory conditions. In this chapter, sexual cannibalism relates specifically to the consumption of males by females. It is frequently thought that sexual cannibalism occurs only after the female has copulated with the male. However, in many species of spiders, sexual cannibalism more frequently occurs *before* the male has managed to transfer sperm to the female. The timing of sexual cannibalism in relation to sperm transfer is important because the implications for males and females are very different. For example, a female that consumes the male after sperm transfer may increase her expected reproductive success, whereas it will be decreased if he is eaten before sperm transfer takes place. Similarly, the function of male courtship behaviour is unlikely to be to reduce the risk of cannibalism that occurs during or after copulation.

This chapter has two aims: first, to examine the evolution of sexual cannibalism, particulary in the context of the conflicts of interest between males and females, and second to describe the various behavioural and morphological characteristics of males that may have evolved as a result of sexual cannibalism. The first section will briefly describe the taxonomic distribution of cannibalism, based on reports from the literature. The following section considers the evolution of sexual cannibalism. Particular attention will be directed towards two models that specifically account for the evolution of sexual cannibalism; one considers sexual cannibalism after copulation (Buskirk *et al.* 1984), and the other considers sexual cannibalism before copulation (Newman and Elgar 1992). The various explanations for the evolution of sexual cannibalism will be assessed against the available evidence. The final section discusses the influence of sexual cannibalism on courtship behaviour and sexual dimorphism, particularly within the orb-weaving spiders.

TAXONOMIC DISTRIBUTION OF SEXUAL CANNIBALISM

The taxonomic distribution of sexual cannibalism was collated from published records of courtship and mating behaviour, and is given in Table 7.1. Several observations can be made from this survey. First, sexual cannibalism is widely distributed among invertebrates, including gastropods, copepods, insects, and many species of arachnids, indicating that the behaviour has evolved at least several times. Second, it occurs frequently in the arachnids, contrary to some opinions (e.g. Kaston and Kaston 1953; Main 1976; Savory 1977), and probably represents an important component of their courtship and mating behaviour. Third, despite its widespread distribution, most species records are concentrated within the scorpions and the spiders. Contrary to popular perception, sexual cannibalism is absent in many species of mantids (see Liske & Davis 1987), and recent field and laboratory experiments show that decapitation of the male mantid by the female is not a necessary prerequisite for successful copulation (Roeder 1935; Liske and Davis 1984, 1987; Birkhead *et al.* 1988; Lawrence 1992). Records of sexual cannibalism in spiders are concentrated principally among the orbweaving families; in other families it is reported in only one and sometimes two species per family. In contrast, sexual cannibalism is more broadly distributed among scorpions (see Table 7.1 and below). Finally, many of the incidences of sexual cannibalism refer to the female capturing and consuming the male before copulation has occurred.

The list in Table 7.1 is not an exhaustive nor definitive account of the frequency of sexual cannibalism. Most records come from anecdotal accounts that may either under- or overestimate the frequency of cannibalism for a particular species. Some observations of the courtship behaviour may represent very few observations of a full courtship and mating sequence, and cannibalism is unlikely to be reported if it happened infrequently.

Table 7.1 The frequency and timing of sexual cannibalism (before, during or after mating) of the male by the female during courtship and mating

Taxa	Timing[1]	Frequency[2]	Place[3]	Source
Insecta				
Mantidae				
Sphodromantis lineola	both	often	captive	Edmunds (1975)
Polyspilota spp.	before	–	wild	Lawrence (1992)
Tenodera aridifolia	after	rare	captive	Liske and Davis (1987)
Mantis religiosa	after	often	field	Lawrence (1992)
Stagmonantis carolina	before	–	captive	Lawrence (1992)
Hestiasula brunneriana	before	–	captive	Lawrence (1992)
Tarachodes afzelii	both	often	captive	Edmunds (1975)
Oxypilus hamatus	both	often	captive	Edmunds (1975)
Miomantis paykullii	before	–	captive	Lawrence (1992)
Catasigerpes occidentalis	before	–	captive	Lawrence (1992)
Humbertiella similis	after	–	captive	Lawrence (1992)
Ameles decolor	before	–		Lawrence (1992)
Coptopteryx viridis	before	–		Lawrence (1992)
Miomantis caffra	after	–		Lawrence (1992)
Orthodera ministralis	after	–		Lawrence (1992)
Hierodula membranacea	after	rare	captive	Birkhead et al. (1988)
Coleoptera				
Carabidae				
Carabus auratus				Thornhill (1976)
Diptera				
Ceratopogonidae				
Serromyia femorata	during	often	field	Edwards (1920)
Mallochohelea nitida	during	often	field	Downes (1978)
Bezzia modocensis	during	often	field	Downes (1978)
Probezzia concinna	during	often	field	Downes (1978)
Palpomyia spp.	during	often	field	Downes (1978)
Orthoptera				
Stenopelmatidae				
Stenopelmatus nigrocapitatus	before	anecdote		Field and Sandlant (1983)
S. fuscus	before	anecdote		Field and Sandlant (1983)
Copepoda				
Cyclopoida				
Acanthocyclops viridis	after	anecdote		Smyly (1970)

Table 7.1 (*continued*)

Taxa	Timing[1]	Frequency[2]	Place[3]	Source
Gastropoda				
Opisthobranchia				
Aeolidae				
Hermissenda crassicornis	after	anecdote	captive	Rutowski (1983)
Navanax inermis	after	anecdote	captive	Paine (1965)
Glaucidae				
Glaucus atlanticus	after	anecdote		Lalli and Gilmer (1989)
Arachnida				
Scorpiones				
Bothriuridae				
Bothriurus bonariensis	after	anecdote	–	Polis (1990)
Urophonius brachycentrus	after	anecdote	–	Polis (1990)
Buthidae				
Buthus occitanus	after	anecdote	–	Polis (1990)
Isometrus maculatus	after	anecdote	–	Polis (1990)
Leiurus quinquestriatus	after	anecdote	–	Polis (1990)
Tityus fasciolatus	after	anecdote	–	Polis (1990)
Scorpionidae				
Heterometrus scaber	after	anecdote	–	Polis (1990)
Pandinus imperator	after	anecdote	–	Polis (1990)
Vaejovidae				
Paruroctonus luteolus	after	rare	captive	Polis and Farley (1979)
P. mesaensis	after	anecdote	–	Polis (1990)
Vejovoidus longiunguis	after	anecdote	–	Polis (1990)
Araneae				
Araneidae				
Araneus quadratus	before	anecdote	field	Elgar (pers. obs.)
A. pallidus	during	anecdote	captive	Grasshoff (1964)
A. diadematus	before	often	captive	Elgar and Nash (1988)
Argiope florida	before	anecdote	field	Robinson and Robinson (1980)
A. cuspidata	before	anecdote	field	Robinson and Robinson (1980)
A. bruennichi	before	anecdote	field	Bristowe (1958)
A. ocyaloides	before	anecdote	field	Robinson and Robinson (1980)

Taxa	Timing[1]	Frequency[2]	Place[3]	Source
A. radon	before	anecdote	field	Robinson and Robinson (1980)
A. reinwardti	before	anecdote	field	Robinson and Robinson (1980)
A. aemula	during	always	field	Sasaki and Iwashi (ms)
A. argentata	before	anecdote	field	Robinson and Robinson (1980)
A. aetheria	before	anecdote	field	Robinson and Robinson (1980)
A. picta	before	anecdote	field	Robinson and Robinson (1980)
A. aurantia	before	anecdote	field	Robinson and Robinson (1980)
Cyrtophora moluccensis	after	anecdote	field	Berry (1987)
C. citricola	before	anecdote	field	Lubin (1974)
Eriophora transmarinus	before	anecdote	field	Elgar (pers. obs.)
E. fuliginea	before	anecdote	field	Robinson and Robinson (1980)
Gea eff	before	anecdote	field	Robinson and Robinson (1980)
Nephila clavipes	before	anecdote	field	Robinson and Robinson (1980)
N. edulis	before	anecdote	field	Austin and Anderson (1978)
Scoloderus cordatus	before	anecdote	field	Stowe (1978)
Zilla sp Wau #2	before	anecdote	field	Robinson and Robinson (1980)
Z. sp Wau #1	before	anecdote	field	Robinson and Robinson (1980)
Clubionidae				
Supunna picta	before	anecdote	captive	Jackson and Poulsen (1990)
Dictynidae				
Dictyna calcarata	before	rare	captive	Jackson (1979)
Dysderidae				
Dysdera crocata	before	rare	captive	Jackson and Pollard (1982)
Gnaphosidae				
Herpyllus blackwalli	before	anecdote	–	Bristowe (1958)
Heteropodidae				
Heteropoda venatoria	after	anecdote	field	Ross *et al.* (1982)

Table 7.1 (*continued*)

Taxa	Timing[1]	Frequency[2]	Place[3]	Source
Linyphiidae				
Frontinella pyramitela	before	rare	captive	Austad (1982; pers. comm.)
Lycosidae				
Lycosa scutulata	after	anecdote	–	Montgomery (1910)
L. rabida	after	anecdote	captive	Rovner (1972)
Metidae				
Meta segmentata	before	anecdote	field	Rubenstein (1987)
Pisauridae				
Pisaura mirabilis	before	rare	captive	Austad and Thornhill (1986)
Dolomedes fimbriatus	before	anecdote	–	Gerhardt (1926)
D. triton	before	anecdote	captive	Roland and Rovner (1983)
Salticidae				
Marpissa marina	both	rare	captive	Jackson *et al.* (1990)
Phidippus audax	before	anecdote	captive	Austad (pers. comm.)
P. clarus	before	anecdote	captive	Austad (pers. comm.)
P. johnsoni	before	rare	captive	Jackson (1980)
P. morsitans	before	anecdote	–	Peckham and Peckham (1889)
P. rufus	before	anecdote	–	Peckham and Peckham (1889)
Portia schultzi	before	rare	captive	Jackson and Hallas (1986)
P. labiata	before	rare	captive	Jackson and Hallas (1986)
Simaetha paetula	before	rare	captive	Jackson (1985)
Tetragnathidae				
Leucage sp	before	anecdote	captive	Elgar (pers. obs.)
Theridiidae				
Latrodectus hesperus	after	rare	captive	Ross and Smith (1979)
L. hesselti	after	often	captive	Forster (1992)

Taxa	Timing[1]	Frequency[2]	Place[3]	Source
Thomisidae				
Xysticus sp	before	anecdote	field	Bristowe (1958)

[1] Timing of sexual cannibalism; before, during or after copulation.
[2] Rare: < 10 per cent of observations; often: > 10 per cent; anecdote: not determined.
[3] captive: observed under laboratory conditions; field: observed under field conditions; otherwise not specified.

There are few records of the frequency of sexual cannibalism in natural populations: sexual cannibalism occurred only once in 1150 observations of pairings of the dome spider *Linyphia litigiosa,* and in this case the male failed to court the female in the usual fashion (Watson, personal communication); all of over 100 pairings in the orb-weaver *Argiope aemula* (T. Sasaki and O. Iwahashi, personal communication); and 30 per cent of 13 pairings in the praying mantis *Mantis religiosa* (Lawrence 1992). Other studies provide details of the frequency of sexual cannibalism under laboratory conditions. In the orb-weaving spider *Araneus diadematus,* sexual cannibalism occurred in 40 per cent of 52 mating experiments (Elgar and Nash 1988); over half of 19 pairings in *Latrodectus hasselti* (Forster 1992); one per cent of 1834 interactions between male and female *Phiddipus johnsoni* (Jackson 1980); two per cent of 43 interactions in *Jacksonoides queenslandica* (Jackson 1988), although the latter two studies do not indicate how many different individuals are represented. In mantids, cannibalism occurred in 55, 83 and 30 per cent of the courtship sequencies of *Oxypilus hamatus, Tarachodes afzelii* and *Sphodromantis lineola* respectively (Edmunds 1975).

EVOLUTION OF SEXUAL CANNIBALISM

Sexual cannibalism is surprisingly uncommon relative to the other types of cannibalism (e.g. see Fox 1975; Polis 1981; Nentwig 1986; this volume). An obvious prerequisite for sexual cannibalism is that the female must be predatory and capable of capturing the male, but these conditions are met by many organisms that do not practice sexual cannibalism. Why are females of some species sexually cannibalistic while others are not? Explanations of sexual cannibalism fall into two categories according to whether it is considered from the male or female perspective. This depends in part upon the timing of sexual cannibalism. Not surprisingly, sexual cannibalism before mating is considered from the female's perspective and may be explained in terms of mistaken identity, female choice of partner, or more simply an economic foraging decision. Sexual cannibalism after mating is considered more from the male perspective, and therefore may represent a form of paternal investment or ensuring paternity of the female's eggs. These various explanations are considered in the next section.

Sexual cannibalism and mistaken identity

Discussions of sexual cannibalism, particularly with respect to courtship and copulation, focus on the predatory nature of the female, and it is often implied that males fall victim to her as a consequence of mistaken identity. If sexual cannibalism occurs through mistaken identity, then it has little adaptive significance and the behaviour would not have evolved in response to some selective advantage to either males or females (e.g. Gould 1984; Jamieson 1986).

Females of orb-weaving spiders may mistake the identity of a courting male in two ways. First, these spiders are typically intolerant of other conspecific individuals, and will usually engage in physical combat with another spider that ventures onto the web. In *Araneus diadematus*, the initial reaction of the female to a courting male is identical to her reaction to an intruding female (author's personal observations). The female shudders the entire web, by bouncing up and down, and sometimes will run directly toward the male. Thus, a male must rapidly establish his credentials as a potential mate, rather than as an intruding female. Nevertheless, cannibalism can take place at any time from when the male enters the female's web until he actually attempts to copulate with her, suggesting that sexual cannibalism is not a simple consequence of mistaken identity. Second, in many orb-weaving spiders, the male ventures across the orb-web to the centre of the web. Since the orb-web is designed for capturing prey, and females respond to the vibrations made by prey arrested by the web, the male may run a high risk of being mistaken for a prey item. It is not surprising that the approach of males of these species of orb-weaving spiders is usually described as cautious. Nevertheless, such cautious behaviour would also be expected if the female was cannibalistic and could distinguish between males and prey items.

There are several reasons to suppose that sexual cannibalism is not primarily the consequence of mistaken identity. First, despite its widespread occurrence among orb-web building spiders, sexual cannibalism is almost completely absent in other web building spiders. It is possible that the taxonomic distribution of sexual cannibalism among web-building spiders is related to the distribution of their prey types. For example, web-builders without sexual cannibalism might usually capture prey that do not fall into the same size distribution of males, while orb-weavers more frequently capture prey items that includes those similar in size to males. If this is the case, mistaken identity may be more common for orb-weaving spiders. However, such an explanation implies that vibratory signals along the web provide the owner with little information about the nature of foreign items on the web, which seems unlikely.

Second, sexual cannibalism occurs in many highly visual species, such as the praying mantis or the salticid spiders. The courtship behaviour of the latter relies almost exclusively on visual cues that may be remarkably simple. An approaching male making 'distracting movements' is safe, whereas one that runs away is more likely to be treated as prey, and mistaken identity is therefore unlikely (Forster 1982; 1985). Furthermore, the predatory behaviour of female salticids toward prey items contrasts markedly with that toward courting males (e.g. Jackson and Hallas 1986), indicating that the capture and consumption of males is unlikely to be the result of mistaken identity.

Finally, under certain circumstances (see below) there may be some costs to females that fail to distinguish potential mates from potential prey items, and it seems reasonable that selection will favour improved discrimination abilities if there is any possibility of a female remaining unmated and failing to reproduce.

Post-mating sexual cannibalism: paternal investment

Earlier theoretical considerations of sexual cannibalism *after* the male has transferred sperm to the female (post-mating sexual cannibalism) emphazise the contribution of the male's body to the production of offspring (e.g. Thornhill 1976; Parker 1979). Thus, explanations for the evolution of the provision of nuptial gifts (reviewed in Simmons and Parker 1989) may have some relevance for the evolution of this form of sexual cannibalism. An obvious difference between sexual cannibalism and the provision of nuptial gifts is that the costs to the cannibalized male are considerably greater than that of a male providing a gift. These costs may not be so great for those species, such as the linyphiid spider *Baryphyma pratense* (Blest 1987) and the haglid orthopteran *Cyphoderris* (Dodson *et al.* 1983), in which parts of the male's body are consumed by the female perhaps diminishing his future survival or mating success (Morris *et al.* 1989).

Thornhill (1976) suggested that if post-mating sexual cannibalism is a form of paternal investment, it will have implications for mating systems. He argues that in species that commonly practice post-mating sexual cannibalism, selection will favour a role reversal in the mating system because of the potentially larger parental investment of males. If female reproductive success is limited by her ability to induce males to sacrifice their lives for her consumption, then selection would favour greater male coyness during courtship (compared with other arthropods), and females would attempt to advertize their fitness through more elaborate courtship. The cautious courtship behaviour of males might be interpreted as a form of male mate choice.

Parker (1979) noted that post-mating sexual cannibalism could represent an extreme form of the conflict of interests between the sexes. However, this conflict would be reduced if the increased fitness of the offspring, as a result of the male donating his soma through female cannibalism, exceeded the value of the male's searching alternatives. Such a scenario would be particularly likely if females only mate once, or males suffer high mortality when moving between females.

Buskirk *et al.* (1984) expressed Parker's (1979) verbal ideas more formally in a model that attempted to account for the evolution of post-mating sexual cannibalism. The model explicitly predicts that selection will favour post-mating sexual cannibalism, if the proportion of additional offspring produced as a result of sexual cannibalism is greater than the number of offspring the male would expect to produce if he were not cannibalized, but instead attempted to find mating opportunities with other females. In other words, sexual cannibalism will be favoured if the expected number of matings for a male (E_{NM}) is low, and the proportional increase in offspring resulting from cannibalism (P_{MC}) is high.

The model assumes that there are no age specific mortality effects (either through cannibalism or other causes of death) or any relationship between male age and either male fertility or somatic contribution by cannibalized males. Unfortunately, data are unavailable to assess this assumption, but the inclusion of these age-related effects may well alter the decision of a male to sacrifice future mating opportunities according to his age.

What is the evidence for Buskirk *et al.*'s (1984) paternal investment model of post-mating sexual cannibalism? Buskirk *et al.* (1984) suggest estimates of E_{NM}, and P_{MC} for several groups of invertebrates and compare these with the occurrence of sexual cannibalism (see Table 7.2). While the evidence appears supportive, the estimates of E_{NM} and P_{MC} are entirely qualitative, and their conclusions should be interpreted cautiously. There are few data that provide quantitative estimates of E_{NM}, and P_{MC}; the addition of a single male to a female's diet has a significant effect on the body weight of the orb-weaving spider *Araneus diadematus* (Elgar and Nash 1988), but females of this spider invariably attack the male before copulation and therefore these data have little relevance to this model (but see below). The addition of a male in the diet of the praying mantis *Hierodula membranacea*, also significantly increases female fecundity (Birkhead *et al.* 1988). In *Mantis religiosa*, the sex ratio becomes increasingly female biased during the breeding season (Lawrence 1992), suggesting that females might expect to encounter fewer males toward the end of the season. Buskirk *et al.* (1984) estimate a low E_{NM} value for the black widow spider *Latrodectus*, because the tip of the embolis on the male's palp breaks off, apparently preventing any further successful copulations. However, Breene and Sweet (1985) show that the male black widow spider *L. mactans* is capable of successfully mating with at least three females, and utilizes an escape mechanism involving the possible spatial disorientation of the female. Although E_{NM} for this species will also depend upon the density of females, Breene and Sweet's (1985) observations suggest that it may be high, in which case the model would not predict sexual cannibalism.

Table 7.2 Comparison of the paternal investment model's predictions for sexual cannibalism with the evidence

Taxon	Encounter rate (E_{NM})	Male contribution (P_{MC})	Sexual cannibalism predicted	Sexual cannibalism observed
Paruroctonus mesacensis	low	high	yes	yes
Phiddipus johnsoni	high	low	no	rare
Lycosid spiders	high	high	no	rare
Praying mantids	low	high	yes	some
Latrodectus sp.	low	low	yes	yes

(From Buskirk *et al.* 1984) (see text for details)

The Australian orb-weaving spider *Phonognatha graeffii* appears to provide the conditions under which Buskirk *et al.*'s (1984) model would predict post-mating sexual cannibalism. This spider constructs a leaf retreat located near the centre of the orb-web. During the breeding season, males can be found with sub-adult and adult females in this leaf retreat. Although it is not known whether males move between retreats in natural populations, they do not do so under captive conditions and there is little size dimorphism in this species (Elgar, unpublished data). Accordingly, *P. graeffii* appears to have a low E_{NM} and a high P_{MC}, and post-mating sexual cannibalism is predicted. My observations show that while females are capable of cannibalizing males, sexual cannibalism rarely, if ever, occurs under natural conditions.

Gould (1984) draws a rather different interpretation of the published accounts of sexual cannibalism used by Buskirk *et al.* (1984) in support of their model. In particular, Gould (1984) raises the important issue that sexual cannibalism as paternal investment requires some degree of male complicity, yet he finds little behavioural evidence that males actively elicit sexual cannibalism and concludes that it has evolved rarely, if at all. Both Gould (1984) and Jamieson (1986) argue that if sexual cannibalism represents the simple failure to escape the female and is relatively rare, then it is a by-product of other phenomena and therefore not a selected trait in itself. In other words, sexual cannibalism simply reflects the inevitable odds of a dangerous game of attempting to mate with a rapacious female (Gould 1984). Of course, such an argument begs the question why these females are so rapacious, and why sexual cannibalism is confined to certain taxonomic groups.

What is the evidence of male complicity in post-mating sexual cannibalism? Most observations of the mating behaviour of sexually cannibalistic species note that the males usually move rapidly away from the female after mating is completed. For example, many sexually cannibalistic scorpions possess a specific post-mating escape behaviour (Polis 1990), apparently contradicting Buskirk *et al.*'s (1984) model. However, males should only sacrifice their lives if E_{NM} is low, and this parameter may vary during the course of a male's lifetime. Consequently, we expect a conflict of interest between males and females, resulting in both sexual cannibalism and escape behaviour. Perhaps younger males, with relatively high expectations of future matings, attempt to escape from the female, while older males are more willing partners of sexual cannibalism.

Complicity by one partner is not, however, always expected in sexually cannibalistic species; sexual cannibalism has been reported in some species of nudibranchs (e.g. Rutowski 1983; Lalli and Gilmer 1989, see also Baur, this volume). Mating in these species is very rapid and is followed by repeated attacks by both partners (Rutowski 1983). Complicity by one partner is not expected in these species because they are simultaneously hermaphroditic. Consequently, both partners are expected to attempt simultaneously both to eat the other partner and to avoid being eaten, which appears to be the case. Unfortunately, accounts of sexual cannibalism in *Hermissenda* are mostly anecdotal, and courtship, mating, and sexual cannibalism in these species deserves further consideration.

Courtship for orb-weaving spiders in the genus *Araneus* takes place on a specially constructed mating thread, and the pair mate by hanging from this thread. For these spiders, the male is sufficently large that he is able to clasp the female and then spring away after mating. However, the male of *A. pallidus* is so small that at the moment of mating, he somersaults through 180° and the female is able to clasp him with her chelicerae. While the male inserts his pedipalps, the female begins to feed on him (Grasshorf 1964), but it is not known whether males are always cannibalized during mating. In another orb-weaving spider *Cyrtophora cicatrosa*, the female grasps the copulating male with her chelicerae (Blanke 1983), although it is not clear whether she subsequently cannibalizes him.

A similar situation for the Australian red-back spider *Latrodectus hesselti* is recorded by Forster (1992). Copulation in the red-back spider takes place on the web; the female is suspended upside down, and the male is in the venter-to-venter position. However, as soon as sperm is being transferred, the male raises himself up into a head-stand position, and turns a complete somersault bringing his posterior abdomen into the proximity of the female's mouthparts from which fluids are exuded almost immediately. As insemination proceeds, the female masticates the male's abdomen and may penetrate it with her fangs. After copulation is complete, the male withdraws a few centimetres from the female. Shortly afterwards, the male briefly re-enacts several courtship components, returns to the quiescent female and engages his second palp, despite the rear of his abdomen being mutilated. Again, he somersaults onto his back and the female continues digesting his abdomen. As the male makes his second withdrawal, the female gathers her legs around him and wraps him with her silk. However, not all males are consumed; about one-third of the males that made at least one insertion managed to escape.

The most suggestive evidence of male complicity in sexual cannibalism is provided from experiments by T. Sasaki and O. Iwahashi (personal communication). Like *Latrodectus*, the male of *Argiope aemura* makes one insertion, moves away from the female and then returns to make a second insertion. However, Sasaki and Iwahashi found that the male was always killed sometime during the second insertion. Of particular interest is that while the male actively escapes from the female after the first insertion, he makes no attempt to do so after the second, and males always die without a struggle. It is impossible to assess the number of matings a male might expect if he were not cannibalized, because in Sasaki and Iwahashi's study of over 100 copulations, the male was always consumed. Males of *A. aemura* are very small compared with females, as is the case for other species of *Argiope* (see Elgar 1991). Thus, the male contribution to female fecundity is likely to be negligible, unless his body represents a source of important nutrients for the female. Why the male sacrifices his life is a mystery.

It is tempting to argue that sexual cannibalism in *Araneus pallidus*, *C. cicatrosa*, *L. hesselti*, and *A. aemura* represents instances of male complicity, although it is possible that males are only able to copulate successfully if they are cannibalized. While this may be true of *A. pallidus* and *C. cicatrosa*, it is not the case for *Argiope aemura* because males can avoid cannibalism while making the first insertion. Of course, this may be because the female makes no attempt to cannibalize him at this stage. It also seems unlikely for *L. hesselti* because not all males were captured by the female. Interestingly, when *L. hesselti* males are cross-mated with the related *L. katipo*, the somersault still occurs but *L. katipo* females do not attempt to consume the male (Forster 1992).

Post-mating sexual cannibalism: mating effort

In insects, the provision of nuptial gifts by males to females has been explained either in terms of paternal investment (see above) or mating effort, where the nuptial gift effectively increases the proportion of eggs a male fertilizes from a given female. In their review of the various merits of these explanations, Simmons and Parker (1989) argue that the provision of somatic gifts (parts of the male's body and post-mating sexual cannibalism) has probably evolved primarily as a means of increasing paternal investment. In contrast, gifts that have been synthezised by the male have probably evolved as a means of increasing the male's mating effort. While their arguments are likely to be generally true, sexual cannibalism in the ceratopogonid midges appears to be a possible exception.

In many predaceous species of ceratopogonid biting midges, the female eats the male while mating is in progress (Downes 1978). Once the pair are in copula, the female pierces the male cuticle, and dissolves and sucks out the body contents as she would with any other small insect prey. It takes about half an hour for the female to empty the male, at which point she disengages from him. When the female disengages, the membrane cuticle in the abdominal segments of the male breaks, leaving the torn off terminalia attached to the female in the mating position. It is not known how long the terminalia remains attached to the female, but presumably it must be discarded before oviposition or subsequent matings. Furthermore, it is not clear whether the male terminalia always breaks off after mating, or whether this only occurs if the female cannibalizes the male.

Mating in these species usually occurs in swarms, suggesting intense competition between males for fertilization priority. Consequently selection may favour mechanisms that ensure that a male that mates with a female also fertilizes all her eggs. Sexual cannibalism for these midges may provide such a mechanism of ensuring paternity, rather than representing a form of paternal investment. Indeed, Downes (1978) doubted that the male's body makes much contribution to female fecundity.

Pre-mating sexual cannibalism: mating and foraging choices

If, as Gould (1984) concludes, sexual cannibalism is simply the result of a dangerous mating game, why should the female attempt to cannibalize her mate, particularly before copulation has taken place? Elgar and Nash (1988) concluded that pre-mating sexual cannibalism may allow females to exert a form of female choice. Their experiments showed that smaller males were more likely to be cannibalized than larger males, and that female size did not determine variation in male mating success. Furthermore, the mating success of males was not influenced by male age, contradicting the view that cannibalism is the result of male senility (e.g. Jones 1983). Elgar and Nash (1988) also showed that the addition of a single male to the female's diet significantly increased her body weight, thus increasing her fecundity. By attempting to cannibalize males, females expect to either increase their fecundity or mate with larger males. Large male body size might reflect greater ability at capturing prey or locating good foraging sites. However, such a strategy provides a potentially significant cost to females in areas of low male density because she risks remaining unmated. Elgar and Nash (1988) suggest that females may alter their cannibalistic behaviour according to their encounter rate with males.

An alternative, but not mutually exclusive, explanation is suggested by a stochastic dynamic programming model developed by Newman and Elgar (1992). Their model is primarily concerned with explaining pre-mating sexual cannibalism by virgin females. Although this behaviour may provide a virgin female with a nutritious meal that might subsequently increase her fecundity, she risks remaining unmated and therefore not reproducing.

The model considers a virgin female's expected reproductive output, which is assumed to be an increasing function of the number of pedipalp insertions she has received and her body weight, both of which will depend upon her decision to mate or cannibalize a courting male. Below a certain weight, the female will be unable to produce eggs, and virgin females will be unable to produce any viable eggs. The model futher assumes that the female has three options, each of which is assigned a certain probability:

(1) to try to cannibalize the male without mating first;
(2) to allow the male to make one insertion and then attempt to cannibalize him; or
(3) to allow the male to make two insertions and then attempt to cannibalize him. Incorporated in the model are varying encounter rates with males of different sizes (and thus different energetic gains) and a varying food intake rate from other prey items.

Newman and Elgar's (1992) model showed that pre-mating sexual cannibalism by virgin females can occur as a result of purely nutritional considerations. The most important factors influencing a female's decision to cannibalize the male were her encounter rate with males, and her expected food intake from other prey items. The model makes several explicit predictions that can be experimentally tested. First, females are less likely to cannibalize a male before mating if they expect to encounter fewer males, as originally suggested by Elgar and Nash (1988). Male encounter rate will depend upon a variety of ecological factors including female dispersion and density, male mate searching behaviour and the sex ratio. The sex ratio will be affected by male mortality, as a result of predation, starvation, and sexual cannibalism. Thus, virgin females may be less likely to be cannibalistic toward the end of the mating season when males may be less abundant. Second, females are less likely to cannibalize the male if they expect a high mean, or a low variance in their food intake rate from other sources. Unfortunately there are no experimental or comparative data with which to test these ideas. Nevertheless, it is interesting to note that the spiders used in the study by Elgar and Nash (1988), which showed high levels of sexual cannibalism, were collected from an area with a high density of spiders.

AVOIDING SEXUAL CANNIBALISM

Courtship behaviour

Several features of the courtship and post-copulatory behaviour of males appear to have evolved in response to reducing the risk of sexual cannibalism. For example, male orb-weaving spiders commonly attempt to mate with the female when she has just moulted, and Robinson and Robinson (1980) argue that this reduces the risk of cannibalism because the female is incapable of attack. In other species of orb-weaving spiders, males construct mating threads that may function to reduce the risk of cannibalism (Robinson and Robinson 1980; Elgar *et al.* 1990). A variety of morphological traits may also reduce the risk of cannibalism. For example, the jaws of female *Tetragnatha* orb-weaving spiders are held open by specialized curved spurs on the jaws of the males, apparently preventing the female from biting and killing the male (Bristowe 1958).

In many spiders, the male usually waits for the female to capture a prey item before starting to court her (Robinson and Robinson 1980; Foelix 1982), and he then attempts to mate with her while she is otherwise occupied with feeding. This delay in courting and mating is thought to reduce the risk of cannibalism, although there is no supporting evidence. The male nursery web spider *Pisaura mirabilis* (Pisauridae) first catches a prey item, wraps it in silk and then presents this nuptial gift to the female. If she is receptive, she will seize the fly and start feeding on it, at which stage the male begins copulating (Austad and Thornhill 1986). Bristowe (1958) suggested that the provision of a nuptial gift reduces the risk of cannibalism, but Austad and Thornhill (1986) found little supporting evidence, arguing that the provision of nuptial gifts may have evolved as a form of paternal investment (see Simmons and Parker 1989 for review).

Male spiders are often particularly vulnerable when they separate from the female. In many orb-weaving spiders, the male may gather in the mating thread and use the tension to spring away very rapidly from the female, thus reducing the chance of her capturing him. In some species of crab spiders, such as *Xysticus cristatus*, the male throws several strands of silk over the female before mating, which apparently makes it difficult for the female to disentangle her legs, preventing her from seizing him when he dismounts (Bristowe 1958). Similar behaviour has been observed in several other families of spiders (e.g. Ross and Smith 1979; Robinson and Robinson 1980; Bruce and Carico 1988; Sierwald 1988). The male lycosid *Lycosa rabida* pinches the female's carapace with his chelicerae as he leaps off her, which apparently allows him to escape unharmed (Rovner 1972). These various male traits indicate that sexual cannibalism is a substantial risk for the courting male, and may represent a potential cost of sexual reproduction that is not usually considered (see Lewis 1987, for review).

Courtship behaviour in spiders, scorpions and certain praying mantids is most aptly described as cautious (see Bristowe 1958; Robinson 1982; Liske and Davis 1987; Polis 1990), and in the majority of these species, females frequently attempt to capture the male, even in species where cannibalism is not recorded (e.g. Bristowe 1958; Robinson and Robinson 1980). It is not surprising, therefore, that an important function of courtship in these species was thought to be the suppression of the predatory behaviour of the female, thus allowing the male to copulate rather than be cannibalized (e.g. Darwin 1871; Montgomery 1910; Bristowe 1958; Platnick 1971; Manning 1972; Robinson and Robinson 1980; Polis 1981; Gould 1982; Foelix 1982; Cloudsley-Thompson 1988).

Certain aspects of male courtship behaviour may well be important in reducing the risk of sexual cannibalism. Courtship in spiders varies in both duration and apparent complexity; in some species, males mate with the female almost as soon as she is located with little, if any, courtship behaviour, while in others the male may wave his legs and perform complex dances (reviewed in Platnick 1971). It is often thought that such complex behaviour functions to reduce the predatory tendancies of the female and stimulate her sexual response (e.g. Bristowe 1958; Manning 1972; Cloudsley-Thompson 1988). Although many differences in specific courtship behaviours are common to higher taxonomic levels, there is still some variation among closely related species. For example, comparative studies of the salticid genus *Portia*, reveal some qualitative differences in courtship behaviour that may be attributable to the frequency of sexual cannibalism (Jackson and Hallas 1986). Similar patterns may apply to praying mantids; Liske and Davis (1987) argue that the unusual male 'abdominal flexion display' observed in the sexually cannibalistic mantis *Tenodera aridifolia* functions to reduce the risk of sexual cannibalism, and it would be interesting to know if similar displays are only found in other sexually cannibalistic species.

Many species of scorpions exhibit a characteristic post-mating escape behaviour, which may reduce the risk of sexual cannibalism (Polis 1990). If this were the case, then sexually cannibalistic species should be more likely to exhibit post-mating escape behaviour than non sexually cannibalistic species. However, the idea is not strongly supported by a simple analysis of comparative data (Table 7.3) assembled by Polis (1990). Fisher's exact test suggests that the two variables are independent ($p > 0.35$), if sexual cannibalism and post-mating escape behaviour are scored as absent if there are no records of it occurring. Such an analysis is clearly preliminary, and more data are required. In particular, it would be interesting to know whether there is intraspecific variation in post-mating escape behaviour that is correlated

Table 7.3 Sexual cannibalism, consumption of spermatophore by the female and postmating escape behaviour by the male in scorpions

Taxa	Sexual cannibalism	Post-mating escape
Bothriuridae		
Bothriurus asper	?	–
B. bonariensis	+	?
B. flavidus	?	+
Urophonius brachycentrus	+	+
Buthidae		
Buthus occitanus	+	+
Isometrus maculatus	+	?
Leiurus quinquestriatus	+	+
Parabuthus planicauda	?	–
Tityus (two species)	–	+
T. fasciolatus	+	?
T. trinitatis	?	–
Chactidae		
Euscorpius (three species)	?	
Diplocentridae		
Nebo hierichonticus	?	
Scorpionidae		
Heterometrus scaber	+	?
Opistophthalmus latimanus	–	+
Pandinus imperator	+	?
Vaejovidae		
Paruroctonus (two species)	+	+
Vejovoidus longiunguis	+	+

+ = behaviour observed; – = behaviour reported to be absent; ? = no records
(From Polis 1990).

with the male's expected future matings, as predicted by the models of post-mating sexual cannibalism (see above).

It is unlikely that the primary function of the often elaborate courtship behaviour of sexually cannibalistic species is to reduce the risk of female attack, since elaborate courtship behaviour is exhibited by many different taxonomic groups, including invertebrates such as butterflies, grasshoppers, and dipterans that do not practice sexual cannibalism (Savory 1977; Jackson 1982). Indeed, the salticids exhibit the most obviously elaborate courtship behaviour among spiders, yet sexual cannibalism is relatively uncommon in this family. Most species in this family are sexually dimorphic in colour and the males are probably the most colourful of all spiders. The combination of these factors strongly suggest that courtship behaviour in this group may have evolved through sexual selection (Peckham and Peckham 1889; see also Jackson 1982), and this is probably true of other taxonomic groups.

Sexual size dimorphism

Several comparative analyses have indicated that sexual cannibalism has been an important selective agent in moulding a variety of morphological and behavioural features of orb-weaving spiders (Elgar *et al.* 1990; Elgar 1991). These studies, and a comparative analysis across a broader taxonomic distribution of spiders, are described in the next section.

Sexual size dimorphism across all spiders

Pre-mating sexual cannibalism should always be actively resisted by the male, but the ability of a female to capture a male may depend on his size relative to the female. Cannibalism that is not associated with courtship and mating is common among spiders (see Nentwig 1986), and it seems likely that the victims of cannibalism would be smaller than the cannibals. Perhaps sexual cannibalism is common among spiders because males are generally smaller than females, and that the variation in the occurrence of sexual cannibalism within the Araneae is associated with sexual dimorphism; species in which the female is much larger than the male may be more likely to be sexually cannibalistic than species that show less dimorphism. This idea can be examined by comparing size dimorphism between sexually and non-sexually cannibalistic species.

Comparative studies frequently encounter the problem of distinguishing between those patterns that arise in response to similar selective forces, and those that are attributable to common ancestry (see Ridley 1983; Harvey and Pagel 1991, for reviews). The problem is less acute if the correct phylogeny is known, but this is frequently not the case. The results of any comparative study of the Araneae must be viewed with some degree of caution since not only is the phylogeny largely unknown, but the taxonomy of many groups is still a matter of contention (see Coddington 1986; Shear 1986). The choice of an appropriate taxonomic level of analysis is also important for comparative studies. For example, species can rarely be assumed to represent independent data points (see Ridley 1983, 1989; Harvey and Pagel 1991, for critiques of such an approach), and therefore may be inappropriate as the unit of analysis, particularly for comparative studies that use data from a broad taxonomic distribution.

Data on male and female body size, and the presence of sexual cannibalism (derived from conditions of courtship and mating behaviour) for 139 species were assembled from the literature. The complete data set, including references, can be obtained from the author on request. All measures of body size were logarithmically transformed before data manipulation and analysis, in order to normalize their distributions. Family means were calculated from constituent generic means, and generic means were calculated from constituent species values. Each family was assigned to three categories of sexual cannibalism according to the frequency with which it had been noted: frequent (two or more species per family), infrequent (one species per family), and absent. These data are shown in Table 7.4.

Table 7.4 Sexual cannibalism and size dimorphism in 26 families of spiders[1]

Family	Cannibalism		Female length (mm)	Male length (mm)	Size dimorphism (mm)
	Timing	Frequency			
Agelinidae	•	absent	10.62	9.11	0.177
Anyphaenidae	•	absent	6.00	5.00	0.061
Araneidae	before	frequent	11.03	4.85	–0.536
Atypidae	•	absent	12.50	8.00	0.003
Clubionidae	•	infrequent	7.69	6.44	0.085
Dictynidae	before	infrequent	4.78	3.74	–0.017
Dipluridae	•	absent	4.00	3.46	0.071
Dysderidae	before	infrequent	12.27	9.65	0.054
Eresidae	•	absent	12.50	9.50	0.021
Gnaphosidae	before	infrequent	11.89	10.04	0.123
Heteropodidae	after	infrequent	23.00	20.00	0.199
Linyphidae	•	absent	3.79	3.22	0.049
Lycosidae	after	infrequent	7.36	6.15	0.078
Nesticidae	•	absent	5.00	4.00	0.008
Pholicidae	•	absent	4.33	4.33	0.221
Pisauridae	before	frequent	17.24	11.48	–0.087
Oonopidae	•	absent	2.00	1.50	–0.122
Oxyopidae	•	absent	6.00	6.00	0.244
Salticidae	before	infrequent	5.50	4.76	0.093
Scytodidae	•	absent	5.00	4.00	0.008
Sparassidae	•	absent	13.50	9.00	–0.104
Tetragnathidae	before	infrequent	7.46	5.42	–0.059
Theriidae	after	infrequent	5.75	3.85	–0.159
Thomisidae	before	infrequent	7.77	4.33	–0.323
Uloboridae	•	absent	5.47	3.74	–0.143
Philodromidae	•	absent	10.00	8.00	0.057

[1] These data are derived from 139 species. Copies of the complete data set, and their sources, can be obtained on request from the author. Size dimorphism is measured as the residual from the regression of male size on female size.

There is a positive correlation between male body length and female body length across all families ($r^2 = 0.90$, slope = 0.93, $P < 0.001$, n = 26 families). Residuals from this regression curve were used as measures of sexual size dimorphism. Large negative values indicate highly sexually dimorphic species (i.e. the male is much smaller than the female), while positive values indicate little, if any, sexual dimorphism.

The comparative data offers little support to the idea that sexual cannibalism is associated with sexual dimorphism, and that the reason for its widespread occurrence among spiders is due to the difference in size between males and females. Sexual dimorphism in families for which pre-mating sexual cannibalism has been frequently recorded is significantly greater (mean = – 0.31, SE = 0.22, n = 2) than in those families for which cannibalism has been infrequently reported (mean = 0.04, SE = 0.03, n = 5) or is apparently absent (mean = 0.02, SE = 0.03; n = 19; F = 4.3, $P < 0.05$). However, this pattern may be primarily due to the extreme size dimorphism among species within the Araneidae, since there is no significant covariation between sexual size dimorphism and sexual cannibalism across genera within either the Salticidae or the Araneidae. There is a positive correlation between male body length and female body length across genera within the Salticidae ($r^2 = 0.86$, $P < 0.001$, n = 29 genera), but no significant difference between those genera that are sexually cannibalistic (mean = – 0.06, SE = 0.11, n = 3) and those that are not (mean = 0.01, SE = 0.03, n = 26; t = 0.85, $P < 0.3$). Similarly, there is a positive correlation between male body length and female body length across genera within the orb-weaving spiders ($r^2 = 0.34$, $P < 0.01$, n = 18 genera), but no significant difference between those genera that are sexually cannibalistic (mean = 0.08, SE = 0.10, n = 8) and those that are not (mean = - 0.02, SE = 0.17, n = 10; t = 0.77, $P < 0.4$). These analyses indicate that the patterns of sexual dimorphism across spiders are not associated with sexual cannibalism, and it seems likely that other factors are responsible for the diversity of sexual size dimorphism in this taxon.

Sexual cannibalism, size dimorphism, and courtship location in orb-weaving spiders

It has been frequently argued that the extreme size dimorphism of some species within the orb-weaving spiders (Araneidae) reduces the risk of cannibalism (Darwin 1871; Cambridge 1871; Robinson and Robinson 1980; but see Vollrath 1980). However, not all sexually cannibalistic species show such large degrees of sexual size dimorphism. Comparative data reveal an association between sexual size dimorphism and courtship behaviour, which may have arisen as a result of selection through sexual cannibalism (Elgar 1991).

Mating in orb-weaving spiders can take place at the central hub of the orb-web or on a mating thread specially constructed by the courting male. In the former case, the male initially traverses the orb-web to the centre, where he climbs onto the female and inserts each pedipalp (Robinson and Robinson 1980). In these species, the male approaches the female extremely slowly; should the female move during his approach, he will either remain still or even drop off the web. The advantage of being relatively small is that the female may either fail to detect the male, as is often the case for small prey items, or she may simply ignore the male because his small size provides little nutritional value (Darwin 1871; Cambridge 1871; Robinson and Robinson 1980; Newman and Elgar 1992).

In other species, the courting male constructs a mating thread, either within or outside the female's orb-web. In the former, the male first cuts a hole in the female's web and constructs the mating thread across it. Following vibratory courtship, the female moves onto the mating thread where mating takes place (Robinson and Robinson 1980). In the latter, the male rarely moves onto the orb-web, but instead constructs a mating thread from the edge of the orb-web to either the structural threads or the surrounding vegetation. Again, the female is attracted by vibratory courtship, and mating takes place on the mating thread (Bristowe 1958; Robinson and Robinson 1980; Elgar and Nash 1988). Selection by sexual cannibalism may favour relatively large males, since larger males in these species are less likely to be cannibalized, apparently as a result of their physical strength (Elgar and Nash 1988).

The comparative data of species from three subfamilies of orb-weaving spiders support this idea (Elgar 1991); species that mate at the hub are more dimorphic than those that mate on a mating thread within the orb-web, which are in turn more dimorphic than those that mate on a mating thread constructed outside of the orb-web (Elgar 1991). However, the pattern was confounded by a taxonomic association; all species within the subfamily Nephilinae mate at the hub, while all species within the subfamily Araneinae mate on a mating thread constructed outside the orb-web, and most of the Argiopinae mate on a mating thread constructed within the orb-web. In fact, the differences in size dimorphism, although significantly associated with differences in the location of courtship and mating, were significantly different between these three subfamilies and therefore might be related to other differences in behaviour or ecology.

However, the problem could be resolved by examining the association between size dimorphism and courtship location within the single genus *Argiope*, in which some species court and mate on a mating thread that is constructed within a hole cut out of the orb-web, while others mate at the hub. The pattern was the same; species that mate at the hub are more dimorphic than species that mate on a mating thread.

How did this pattern arise? There is some evidence that there has been an evolutionary trend toward smaller body size in female orb-weaving spiders (Craig 1987). The decrease in female size will lead to a decrease in male size, until a certain threshold below which males might be unable to survive. Any further reduction in female size would decrease the advantage of relatively small body size for males, in terms of their detectability by, or nutritional value to, the female. Selection may then favour alternative mechanisms for the male to avoid sexual cannibalism, such as different locations of courtship and mating (Elgar 1991). It seems unlikely that selection initially favoured different locations of courtship, thereby promoting changes in size dimorphism.

Sexual cannibalism may have further implications for patterns of sexual dimorphism in orb-weaving spiders. Males of species that court on a mating thread use their legs to both manipulate the female into a responsive position, and to spring away from the female after copulation. As previously noted, large males are less likely to be cannibalized than small males (Elgar and Nash 1988), perhaps because their longer legs increase their ability to physically manipulate the female. If the lower risk of sexual cannibalism associated with large body size is general across species that use mating threads, then selection may favour relatively longer leg length in males of species that are relatively smaller than females. Smaller males with longer legs can reduce the risk of cannibalism by manipulating the female into a position that is more responsive to copulation, while remaining at a safe distance from her.

This idea was examined by Elgar *et al.* (1990) in a comparative analysis of morphological data from three subfamilies of orb-weaving spiders. Each subfamily differed in either the location of courtship or the occurrence of sexual cannibalism. Mating in the Araneinae takes place on a mating thread and the male uses his leg to manipulate the female. Although males in the Gasteracanthinae construct a mating thread, they do not use their legs to manipulate the female, and pre-mating sexual cannibalism has not been observed in this subfamily. Males in the Tetragnathinae do not use mating threads, and many species have spurs on their chelicerae that apparently prevent the female from biting the male (Bristowe 1958). The data were analysed within each subfamily, and showed that in the Araneinae, there was a significant correlation between sexual dimorphism and leg length; males of species that were relatively small compared with the female, had relatively longer legs than were expected for their body size.

Several lines of evidence were provided to support the view that pre-mating sexual cannibalism was responsible for this pattern. First, there was no relationship between relative female size and leg length within each of the three subfamilies. This may be a consequence of the different lifestyles of males and females (Montgomery 1910), since males do not build webs. However, there was no relationship between relative male size and relative male leg length within either the Tetragnathinae or the Gasteracanthinae. Elgar *et al.* (1990) argue that this is expected because mating does not take place on a mating thread in the Tetragnathinae, males of the Gasteracanthinae do not use their legs to manipulate the female, and apparently the females are not sexually cannibalistic. Thus, there would be no selection through sexual cannibalism for longer leg length in these two subfamilies.

CONCLUDING REMARKS

Sexual cannibalism has been reported across a wide range of species. However, there are good reasons to believe that sexual cannibalism is not the occasional result of the male's failure to escape the female's rapacious behaviour, perhaps through mistaken identity. For many species, sexual cannibalism is a frequent hazard associated with courtship and mating, and a variety of morphological and behavioural traits of males apparently function to reduce the risk of sexual cannibalism. In addition, the comparative data provide evidence that the patterns of variation in sexual dimorphism and courtship behaviour, particularly in orb-weaving spiders, has arisen in response to the selective influence of sexual cannibalism. It seems reasonable to conclude, therefore, that sexual cannibalism is of some evolutionary significance.

The conditions under which sexual cannibalism might be favoured depend critically upon whether it occurs before or after copulation. The models developed by Buskirk *et al.* (1984) and Newman and Elgar (1992) address these different forms of sexual cannibalism, and make contrasting predictions about the relationship between sexual cannibalism and male encounter rate, clearly illustrating the important difference between pre- and post-mating sexual cannibalism. Newman and Elgar (1992) predict *pre-mating* sexual cannibalism if females expect a high encounter rate of males, while Buskirk *et al.* (1984) predict *post-mating* sexual cannibalism if males expect a low encounter rate with females. The former considers sexual cannibalism from the female's perspective, with sexual cannibalism having the potential to reduce her reproductive success, while the latter considers the male's perspective, with sexual cannibalism potentially increasing his reproductive success. The models indicate that the expected encounter rate of the opposite sex, and the nutritional value of the male in relation to other prey items, are the most important variables for the evolution of this behaviour. These models require further development and experimental analysis.

There is still much to be learned about sexual cannibalism in spiders and other invertebrates. The question of whether post-mating sexual cannibalism represents some form of adaptive behaviour by the male or whether it only benefits the female still remains unanswered. Experimental studies are required to examine whether females alter their sexually cannibalistic behaviour according to local conditions, and the evolutionary implications of this behaviour will remain poorly understood until there are data available on the frequency of sexual cannibalism in natural populations. While intriguing in its own right, sexual cannibalism has broader implications for understanding animal mating systems, because it represents possibly the most extreme example of the conflict of interests between males and females.

ACKNOWLEDGEMENTS

I am grateful for helpful comments and advice from Steve Austad, Bruno Baur, Bernie Crespi, Lyn Forster, Robert Jackson, Penny Kukuk, Sue Lawrence, Christian Peeters, Risa Rosenberg, Ken Ross, Anita Timso-Louie, Paul Watson, and Nina Wedell. Lyn Forster, Osamu Iwahashi, and Sue Lawrence kindly gave me copies of their manuscripts. Financial support has been provided by a Queen Elizabeth II Research Fellowship, the University of New South Wales, the Australian Research Council, the Association for the Study of Animal Behaviour, and the Christensen Research Institute.

REFERENCES

Austad, S. N. (1982). A game theoretical interpretation of male combat in the bowl and doily spider (*Frontinella pyramitela*). *Animal Behaviour*, **31**, 59–73.

Austad, S. N. and Thornhill, R. (1986). Female reproductive variation in a nuptial-feeding spider, *Pisaura mirabilis*. *Bulletin of the British Arachnological Society*, **7**, 48–52.

Austin, A. D. and Anderson, D. T. (1978). Reproduction and development of the spider *Nephila edulis* (Koch) (Araneidae: Araneae). *Australian Journal of Zoology*, **26**, 501–18.

Berry, J. W. (1987). Notes on the life history and behavior of the communal spider *Cyrtophora moluccensis* (Doleschall) (Araneae, Araneidae) in Yap, Caroline Islands. *Journal of Arachnology*, **15**, 309–19.

Birkhead, T. R., Lee, K. E., and Young, P. (1988). Sexual cannibalism in the praying mantis *Hierodula membranacea*. *Behaviour*, **106**, 112–18.

Blest, A. D. (1987). The copulation of a linyphiid spider, *Baryphyma pratense*: does a female receive a blood-meal from her mate? *Journal of Zoology, London*, **213**, 189–91.

Blanke, R. (1983). Behaviour as an indicator of taxonomic relationships in Arachnology (Arachnida: Araneae). *Verhandlugen des Naturwissenschaftlichen Vereins in Hamburg*, **26**, 217–26.

Breene, R. G. and Sweet, M. H. (1985). Evidence of insemination of multiple females by the male black widow spider, *Latrodectus mactans* (Araneae, Theridiidae). *Journal of Arachnology*, **13**, 331–5.

Bristowe, W. S. (1958). *The world of spiders*. Collins, London.

Bruce, J. A. and Carico, J. E. (1988). Silk use during mating in *Pisaurina mira* (Walckenaer) (Araneae, Pisauridae). *Journal of Arachnology*, **16**, 1–4.

Buskirk, R. E., Frohlich, C., and Ross, K. G. (1984). The natural selection of sexual cannibalism. *American Naturalist*, **123**, 612–25.

Cambridge, O. -P. (1871). Notes on some arachnida collected by Cuthbert Collingwood esq., M.D., during rambles in the China Sea, etc. *Proceedings of the Zoological Society of London*, (1871), 617-22.

Cloudsley-Thompson, J. L. (1988). *Evolution and adaptation of terrestrial arthropods*. Springer-Verlag, Berlin.

Coddington, J. (1986). The monophyletic origin of the orb web. In *Spiders: webs, behavior and evolution*, (ed. W. A. Shear), pp. 319-63. Stanford University Press.

Craig, C. L. (1987). The significance of spider size to the diversification of spider-web architectures and spider reproductive modes. *American Naturalist*, **129**, 47–68.

Darwin, C. (1871). *Sexual selection and the descent of man*. Murray, London.

Dodson, G. N., Morris, G. K., and Gwynne, D. T. (1983). Mating behavior of the primitive Orthopteran genus *Cyphoderris* (Haglidae). In *Orthopteran mating systems: sexual competition in a diverse group of insects*, (ed. D. T. Gwynne and G. K. Morris), pp. 305–17. Westview Press, Boulder.

Downes, J. A. (1978). Feeding and mating in the insectivorous ceratopogoninae (Diptera). *Memoirs of the Entomological Society of Canada*, **104**, 1–62.

Edmunds, M. (1975) Courtship, mating and possible sex pheromones in three species of mantodea. *Entomologist's Monthly Magazine*, **111**, 53–57.

Edwards, F. W., Jr. (1920). Some records of predaceous Ceratopogoninae (Diptera). *Entomologists' Monthly Magazine*, **56**, 203–5.

Elgar, M. A. (1991). Sexual cannibalism, size dimorphism and courtship behavior in orb-weaving spiders (Araneae). *Evolution*, **45**, 444–8.

Elgar, M. A. and Nash, D. R. (1988). Sexual cannibalism in the garden spider *Araneus diadematus*. *Animal Behaviour*, **36**, 1511–17.

Elgar, M. A., Ghaffar, N., and Read, A. (1990). Sexual dimorphism in leg length in orb-weaving spiders: a possible role for sexual cannibalism. *Journal of Zoology, London*, **222**, 455–70

Field, L. H. and Sandlant, G. R. (1983). Aggression and mating behavior in the Stenopelmatidae (Orthoptera, Ensifera). In *Orthopteran mating systems: sexual competition in a diverse group of insects*, (ed. D. T. Gwynne and G. K. Morris), pp. 120–46. Westview Press, Boulder.

Foelix, R. F. (1982). *Biology of spiders*. Harvard University Press, Cambridge, Mass.

Forster, L. (1982). Visual communication in jumping spiders (Salticidae). In *Spider communication: mechanisms and ecological significance*, (ed. P. N. Witt and J. S. Rovner), pp. 161–212. Princeton University Press.

Forster, L. (1985). Target discrimination in jumping spiders (Araneae: Salticidae). In *Neurobiology of arachnids*, (ed. F. G. Barth), pp. 249–74. Springer-Verlag, Berlin.

Forster, L. (1992). The stereotyped behaviour of sexual cannibalism in *Latrodectus hasselti*, Thorell (Araneae: Theridiidae), the Australian redback spider. *Australian Journal of Zoology* (in press).

Fox, L. R. (1975). Cannibalism in natural populations. *Annual Review of Ecology and Systematics*, **6**, 87–106.

Gerhardt, U. (1926). Weitere Untersuchungen zur biologie der spinnen. *Zeitschrift für Morphologie und Ökologie Tiere*, **6**, 1–77.

Gould, J. L. (1982). *Ethology: mechanisms and evolution of behaviour*. W. W. Norton, New York.

Gould, S. J. (1984). Only his wings remained. *Natural History*, **93** (9), 10–18.

Grasshoff, M. (1964). Die Kreuzspinne *Araneus pallidus*–ihr Netzbau und ihre Paarungsbiologie. *Natur und Museum*, **94**, 305–14.

Harvey, P. H. and Pagel, M. D. (1991). *The comparative method in evolutionary biology*. Oxford University Press.

Jackson, R. R. (1979). Comparative studies of *Dictyna* and *Mallos* (Araneae, Dictynidae) II. The relationship between courtship, mating aggression and cannibalism in species with differing types of social organisation. *Revue Arachnologique*, **2**, 103–32.

Jackson, R. R. (1980). Cannibalism as a factor in the mating strategy of the jumping spider *Phidippus johnsoni* (Araneae: Salticidae). *Bulletin of the British Arachnologial Society*, **5**, 129–33.

Jackson, R. R. (1982). The behavior of communicating in jumping spiders (Salticidae). In *Spider communication: mechanisms and ecological significance*, (eds. P. N. Witt and J. S. Rovner), pp. 213–47. Stanford University Press.

Jackson, R. R. (1985). The biology of *Simaetha paetula* and *S. thoracica*, web-building jumping spiders (Araneae, Salticidae) from Queensland: co-habitation with social spiders, utilization of silk, predatory behaviour and intraspecific interactions. *Journal of Zoology, London* (B), **1**, 175–210.

Jackson, R. R. (1988). The biology of *Jacksonoides queenslandica*, a jumping spider (Araneae, Salticidae) from Queensland: intraspecific interactions, web-invasion, predators, and prey. *New Zealand Journal of Zoology*, **15**, 1–37.

Jackson, R. R. and Hallas, S. E. A. (1986). Comparative biology of *Portia africana*,

P. albimana, P. fimbriata, P. labiata, and P. shultzi, araneophagic, web-building jumping spiders (Araneae: Salticidae): utilisation of webs, predatory versatility, and intraspecific interactions. *New Zealand Journal of Zoology*, **13**, 423–89.

Jackson, R. R. and Pollard, S. D. (1982). The biology of *Dysdera crocata* (Araneae, Dysderidae): intraspecific interactions. *Journal of Zoology, London*, **198**, 197–214.

Jackson, R. R., Pollard, S. D., MacNab, A. M., and Cooper, K. J. (1990). The complex communicatory behaviour of *Marpissa marina*, a New Zealand jumping spider (Araneae: Salticidae). *New Zealand Journal of Zoology*, **17**, 25–38.

Jackson, R. R. and Poulsen, B. A. (1990). Predatory versatility and intraspecific interactions of *Supunna picta* (Araneae: Clubionidae). *New Zealand Journal of Zoology*, **17**, 169–84.

Jamieson, I. G. (1986). The functional approach to behavior: is it useful? *American Naturalist*, **127**, 195–208.

Jones, D. (1983). *Spiders of Britain and northern Europe*. Hamlyn, London.

Kaston, B. J. and Kaston, E. (1953). *The spiders*. Brown, Dubuque.

Krebs, J. R. and Davies, N. B. (1987). *An introduction to behavioural ecology*, 2nd edition. Blackwell Scientific Publications, Oxford.

Lalli, C. M. and Gilmer, R. W. (1989). *Pelagic snails. The biology of Holoplanktonic gastropod mollusks*. Stanford University Press.

Lawrence, S. E. (1991). Sexual cannibalism in the praying mantid, *Mantis religiosa:* a field study. *Animal Behaviour*. (In press.)

Lewis, W. M. Jr. (1987). The cost of sex. In *The evolution of sex and its consequences*, (ed. S. C. Stearns), pp. 33–57. Birkhuser Verlag, Basel.

Liske, E. and Davis, W. J. (1984). Sexual behaviour of the Chinese praying mantis. *Animal Behaviour*, 32, 916–18.

Liske, E. and Davis, W. J. (1987). Courtship and mating behviour of the Chinese praying mantis *Tenodera aridfolia sinensis*. *Animal Behaviour*, **35**, 1524–38.

Lubin, Y. D. (1974) Adaptive advantages and the evolution of colony formation in *Cyrtophora* (Araneae: Araneidae). *Zoological Journal of the Linnean Society*, **54**, 321–39.

Main, B. Y. (1976). *Spiders*. Collins, Sydney.

Manning, A. (1972). *An introduction to animal behaviour*, 2nd edition. Edward Arnold, London.

Montgomery, T. H., Jr. (1910). The significance of the courtship and secondary sexual characters of Araneads. *American Naturalist*, **44**, 151–77.

Morris, G. K., Gwynne, D. T., Klimas, D. E., and Sakaluk, S. K. (1989). Virgin male mating advantage in a primitive acoustic insect (Orthoptera: Haglidae). *Journal of Insect Behaviour*, 2, 173–85.

Nentwig, W. (1986). The prey of spiders. In *Ecophysiology of spiders*, (ed. W. Nentwig), pp. 249–73. Springer-Verlag, Berlin.

Newman, J. A. and Elgar, M. A. (1992). Sexual cannibalism in orb-weaving spiders: an economic model. *American Naturalist*. (In press.)

Paine, R. T. (1965). Natural history, limiting factors and energetics of the opisthobranch *Navanax inermis*. *Ecology*, **46**, 603–19.

Parker, G. A. (1979). Sexual selection and sexual conflict. In *Sexual selection and reproductive competition in insects*, (eds. M. S. Blum and N. A. Blum), pp. 123–66. Academic Press, London.

Peckham, G. W. and Peckham, E. G. (1889). Observations on sexual selection in spiders of the family Attidae. *Occasional Papers of the Natural History Society of Wisconsin*, **1**, 1–60.

Platnick, N. (1971). The evolution of courtship behaviour in spiders. *Bulletin of the British Arachnological Society*, **2**, 40–47.

Polis, G. A. (1981). The evolution and dynamics of intraspecific predation. *Annual Review of Ecology and Systematics*, **12**, 225–51.

Polis, G. A. (ed.) (1990). *The biology of scorpions*, Stanford University Press.

Polis, G. A. and Farley, R. D. (1979). Behavior and ecology of mating in the cannibalistic scorpion, *Paruroctonus mesaensis* Stahnke. *Journal of Arachnology*, **7**, 33–46.

Ridley, M. (1983). *The explanation of organic diversity. The comparative method and adaptations for breeding*. Clarendon Press, Oxford.

Ridley, M. (1989). Why not to use species in comparative tests. *Journal of theoretical Biology*, **136**, 361–4.

Robinson, M. H. (1982). Courtship and mating behavior in spiders. *Annual Review of Entomology*, **27**, 1–20.

Robinson, M. H. and Robinson, B. (1980). Comparative studies of the courtship and mating behavior of tropical araneid spiders. *Pacific Insects Mongraphs*, **36**, 1–218.

Roland, C. and Rovner, J. S. (1983). Chemical and vibratory communication in the aquatic pisaurid spider *Dolomedes triton* (Araneae: Pisauridae). *Journal of Arachnology*, **11**, 77–85.

Roeder, K. D. (1935). An experimental analysis of the sexual behaviour of the praying mantis (*Mantis religiosa*). *Biological Bulletin*, **69**, 203–20.

Ross, J., Richman, D. B., Mansour, F., Trambarulo, A., and Whitcomb, W. H. (1982). The life cycle of *Heteropoda venatoria* (Linnaeus) (Araneae: Heteropididae). *Psyche*, **89**, 297–306.

Ross, K. and Smith, R. L. (1979). Aspects of the courtship behavior of the black widow spider, *Latrodectus hesperus* (Araneae, Theridiidae) with evidence for the existence of a contact sex pheromone. *Journal of Arachnology*, **7**, 69–77.

Rovner, J. S. (1972). Copulation in the lycosid spider (*Lycosa rabida* Walckenaer): a quantitative study. *Animal Behaviour*, **20**, 133–8.

Rubenstein, D. I. (1987). Alternative reproductive tactics in the spider *Meta segmentata*. *Behavioral Ecology and Sociobiology*, **20**, 229–37.

Rutowski, R. L. (1983). Mating and egg mass production in the aeolid nudibranch *Hermissenda crassicornis* (Gastropoda: Opisthobranchia). *Biological Bulletin*, **165**, 276–85.

Savory, T. (1977). *Arachnida*, 2nd edition. Academic Press, London.

Shear, W. A. (1986). The evolution of web-building behavior in spiders: a third generation of hypotheses. In *Spiders: webs, behavior and evolution*, (ed. W. A. Shear), pp. 364–400. Stanford University Press.

Sierwald, P. (1988). Notes on the behavior of *Thalassius spinosissimus* (Arachnidae: Araneae: Pisauridae). *Psyche*, **95**, 243–52.

Simmons, L. W. and Parker, G. A. (1989). Nuptial feeding in insects: mating effort versus paternal investment. *Ethology*, **81**, 332–49.

Smyly, W. J. P. (1970). Observations of rate of development, longevity and fecundity of *Acanthocyclops viridis* (Jurine)(Copepoda, Cyclopoida) in relation to type of prey. *Crustaceana*, **18**, 21–36.

Stowe, M. K. (1978). Observations of two nocturnal orbweavers that build specialized webs: *Scoloderus cordatus* and *Wizia ectypa* (Araneae: Araneidae). *Journal of Arachnology*, **6**, 141–6

Thornhill, R. (1976). Sexual selection and paternal investment in insects. *American Naturalist*, **110**, 153–63.

Vollrath, F. (1980). Male body size and fitness in the web-building spider *Nephila clavipes*. *Zeitschrift für Tierpsychologie*, **53**, 61–78.

8

Cannibalism in beetles

Lori Stevens

Beetles are holometabolous insects with egg, larval, pupal, and adult life stages. Cannibalism often involves mobile stages (adults and larvae) eating quiescent stages (eggs and pupae). Adult beetles are sclerotized and invulnerable to cannibalism, with the consequence that cannibalism among mobile stages usually involves larvae eating larvae (Tables 8.1 and 8.2); however, adults occasionally eat larvae and there are two reported cases of adults eating callow adults. Cannibalism in beetles can be classified into two categories (Hamilton 1964*a*,*b*; Fox 1975; Polis 1981, 1984): (1) cannibalism of kin, including sibling cannibalism where early hatching larvae cannibalize eggs or later hatching larvae from the same clutch and filial cannibalism where adults eat larval offspring, and (2) non-kin cannibalism in which cannibalism is not specifically directed toward or away from closely related individuals. Non-kin cannibalism in beetles apparently does not involve individuals specifically seeking out and eating one's own species (Tables 8.1 and 8.2), in fact, such cannibals do not necessarily discriminate between conspecifics and other prey. Indiscriminate or incidental cannibalism are terms that may be used to describe non-kin cannibalism, however, ecological factors and age specificity of cannibalism may bias indiscriminate cannibalism towards unrelated individuals (e.g. Mertz *et al.* 1984). I use the term non-kin cannibalism to refer to cannibalism when relatedness between victim and perpetrator is variable and/or unknown.

Cannibalism of kin has been observed only in predacious ladybeetles (Coccinellidae), leaf-eating beetles (Chrysomelidae) and burying beetles (Silphidae). Non-kin cannibalism also occurs in leaf-eating beetles (Chrysomelidae), and in several species of granivorous beetles (Bostrychidae), saw-toothed grain beetles (Cucujidae), bark beetles (Scolytidae), flour and darkling beetles (Tenebrionidae), ground beetles (Carabidae), whirligig beetles (Grynidae), rove beetles (Staphylinidae), carrion beetles (Silphidae), and ladybeetles (Coccinellidae) (Table 8.1). One species of ladybeetle displays both types of cannibalism (Kaddou 1960; Osawa 1989).

Studies of cannibalism in beetles vary from anecdotal notation, to measures of nutritional benefits and cannibalism rates, and multi-generation (>50) studies of the evolution of cannibalism. Cannibalism in beetles has been discussed previously by Park *et al.* (1965), King and Dawson (1972), Mertz (1972), Fox (1975), and Polis (1981, 1984). However, two recent important studies have not yet been

Table 8.1 Records of cannibalism of kin among coleopterans.

Taxa	Source
Coccinellidae (carnivorous beetles)	
Adalia bipunctata	Banks (1956); Dimetry (1974); Iperti (1966)
Adalia decempunctata	Dixon (1959)
Adonia (= Semiadalia) undecimnotata	Iperti (1966)
Adonia variegata	Iperti (1966)
Cheilomenes lunata	Brown (1972)
Coccinella axiridis	Tan (1934) (in Pienkowski 1965)
Coccinella septempunctata	Banks (1956); Iperti (1966)
Coleomegilla maculata lengi	Pienkowski (1965)
Harmonia axyridis	Kawai (1978); Osawa (1989)
Harmonia (= Propylea) quatturodecimpunctata	Banks (1956); Iperti (1966)
Hippodamia quinquesignata	Kaddou (1960)
Lioadalia flavmaculata	Brown (1972)
Chrysomelidae (herbivorous leaf beetles)	
Plagiodera versicolora	Wade and Breden (1986)
Labidomera clivicollis	Eickwort (1973)
Siliphidae (burying beetles)	
Nicrophorus vespilloides	Bartlett (1987)

reviewed. One investigates the influence of group structure in the evolution of cannibalistic behaviour, particularly sibling cannibalism in the imported willow leaf beetle *Plagiodera versicolora* (Chrysomelidae) (Breden and Wade 1985,1987, 1989; Wade and Breden 1986). The second is a long-term study (> 50 generations) of the evolution of cannibalism in four genetic strains of *Tribolium confusum* (Stevens 1989; Stevens and Mertz 1985).

Natural selection will favour cannibalistic behaviour if $-k < 1/r$, where k is the ratio of the change in fitness of the victim to the change in fitness of the perpetrator, and r is the degree of relatedness (Hamilton 1964*a*,*b*). The selective advantage to cannibalism will increase as either the benefit to the cannibal increases or the cost to the victim or relatedness decreases. Thus, cannibalism will be favoured by natural selection when it increases an individual's fitness through direct, immediate gains, usually nutritional, and as a result may be associated with poor environments and/or physiological stress.

Table 8.2 Coleopterans in which indiscriminate or incidental cannibalism behavior has been reported

Taxa	Description and source
Herbivorous beetles	
Bostrychidae (granivorous beetles)	
Rhizopertha dominica	Partly eaten pupae found in grains from which other insects had previously emerged as adults (Crombie 1944)
Chrysomelidae (leaf eating beetles)	
Leptinotarsa decemlineata	Adults eat eggs (Tamaki *et al.* 1982)
Leptinotarsa juncta	Adults of both sexes and larvae eat eggs (D. McCauley, personal communication)
Cucujidae (saw-toothed grain beetle)	
Oryzaephilus surinamensis	Cannibalism of eggs and pupae by adults and larvae (Crombie 1946)
Scolytidae (bark beetles)	
Ips avulsus *Ips calligraphus* *Scolytus scolytus* *Scolytus multistriatus* *Tomicus piniperda*	Larger larvae probably encounter and eat smaller larvae as they elongate their gallery (Wagner *et al.* 1987, 1988; Beaver 1974)
Tenebrionidae (flour beetles)	
Tribolium castaneum *Tribolium confusum* *Tribolium madens* *Tribolium brevicornis*	Cannibalism of pupae and eggs by larvae and adults (Chapman 1928; Park *et al.* 1965; Jillson and Costantino 1980; Teleky 1980); cannibalism of callow adults by sclerotized adults (Young 1970)
Tenebrionidae (darkling beetles)	
Zophobas atratus (= *rugipes*)	Cannibalism of pupae and eggs by larvae and of eggs by adults (Tschinkel 1978)
Carnivorous beetles	
Carabidae (ground beetles)	
Pterostichus oblongopunctatus *Pterostichus nigrita* *Agonum assimile*	Larvae often cannibalize other larvae (Heessen and Brunsting 1981; Brunsting and Heessen 1984; Thiele 1977)
Coccinellidae (ladybeetles)	
Harmonia axyridis	Larvae eat eggs (Osawa 1989)

Table 8.2 (*continued*)

Gyrinidae (whirligig beetles)	
Dineutes horni	Among larvae (Istock 1966)
Dineutes nigrior	
Dineutes assimilis	
Staphylinidae (rove beetles)	
Philonthus decorus	Among the larvae (Heessen and Brunsting 1981)

Relatedness is a key variable in the evolution of cannibalism, and consequently this chapter discusses cannibalism of kin and non-kin cannibalism in two separate sections. The costs and benefits of cannibalism are identified (e.g. nutrition, competition, reproductive potential) and are related to the ecology and evolution of cannibalism in beetles.

CANNIBALISM OF KIN

Female beetles often lay eggs in clutches. Sibling cannibalism occurs when early hatching larvae eat the eggs or late hatching larvae from the same clutch (Table 8.1). The time between hatching and obtaining a first meal can often be a period of high mortality for beetle larvae (Banks 1956; Dixon 1959; Kaddou 1960; Brown 1972; Kawai 1978) and nutritional benefits from sibling cannibalism provide the cannibal with an advantage in size and/or survivorship (Banks 1956; Kaddou 1960; Pienkowski 1965; Brown 1972; Kawai 1978; Breden and Wade 1989). When mortality rates are high, the benefit necessary for cannibalism to be favoured by natural selection can be small (Eickwort 1973). In the milkweed beetle, *Labidomera clivicollis* (Chrysomelidae), Eickwort (1973) estimated that when cannibalism is directed toward full sibs, a decrease in developmental time of two days will be enough to favour the evolution of cannibalism. If half-sibs are the victims, a decrease in development of less than one day is necessary. Eickwort (1973) concluded that because survivorship probabilities of larvae were low, only a small benefit from cannibalism was necessary to favour its evolution.

Studies of cannibalism in another chrysomelid, the imported willow leaf beetle *P. versicolora,* examined the selective forces governing cannibalism within and between kin groups (Breden and Wade 1985, 1987, 1989; Wade and Breden 1986). Females of *P. versicolora* lay eggs in groups, which hatch and form aggregate feeding groups that persist for one to two of the three larval instars. Clutches are almost exclusively laid by a single female, and as a result of multiple mating contain both full and half-sibs (McCauley and O'Donnell 1984; McCauley *et al.* 1988).

Cannibalism of kin occurs when early hatching larvae eat eggs: larvae are extremely cannibalistic for a period of 24 hours after hatching and subsequently are exclusively herbivorous. Cannibalism occurs in 41–77 per cent of the groups hatching in nature, leading to the loss of up to 45 per cent of the hatching individuals (Wade and Breden 1986). At the individual level, cannibals achieve an initial advantage in size over non-cannibals (Breden and Wade 1985). This initial advantage becomes greater with age through early instars (Breden and Wade 1989). However, at the group level, there is a positive correlation between larval group size (measured after cannibalism) and survivorship. Larval groups exhibit several social behaviours, including coordinated feeding, group defensive displays, trail following, and synchronous molting. Thus, there is a disadvantage to cannibalism in terms of reduced group size (Breden and Wade 1985, 1989; Wade and Breden 1986). Breden and Wade (1989) refer to the differential survival as a function of group size as group-level selection. An interaction exists between the advantage to cannibalism at the group and individual level, with a stronger positive correlation between group size and survivorship in the absence of cannibals than for mixed groups of cannibals and non-cannibals. Thus, group membership provides a survivorship and dry weight advantage to individuals, but these advantages may be diminished by the effects of cannibalism.

Within kin groups, cannibals have an advantage relative to non-cannibals, whereas between groups, non-cannibals have an advantage (Breden and Wade 1989). Over the several years during which the study took place, the two factors influencing the evolution of cannibalism were found to vary in space and time. First, the degree of relatedness within larval groups varies temporally and spatially across the range of the species (McCauley *et al.* 1988). Second, the effects of group membership on individual fitness varies between years and between localities. Groups exhibit coordinated defensive displays, and thus the relationship between group size and survivorship may be a function of the array of local predators, which Breden and Wade (1989) report varies considerably from year to year.

Filial cannibalism, the deliberate killing and consumption by parents of their own young, is extremely unusual in beetles, and has only been reported in the burying beetle *Nicrophorus vespilloides* (Bartlett 1987). Mating pairs raise their young on a limited food supply, usually a vertebrate corpse placed in a burial chamber or crypt. Eggs are laid in the soil, larvae migrate to the corpse and develop through three instars before returning to the soil to pupate. Both adults are normally present throughout larval development, feeding the young, repairing damage to the crypt, and driving off insects which trespass into the crypt. The number of eggs laid on 'small' mouse corpses (weighing 10–15 g) under laboratory conditions is such that complete hatching success will produce more larvae than the corpse can support without severe reduction in progeny weight. Parents reduce the brood by killing and eating almost half of the first instar larvae. Cannibalism occurs during the first 24 hours after the larvae reach the corpse, when food is least likely to be limiting, and does not occur on heavier (30 g) corpses. Excessive numbers of eggs are laid because not all larvae successfully reach the corpse. However, once a larvae reaches the corpse, survivorship is high, except for cannibalism. Adults cull the brood at the earliest stage, so that those which survive will be well fed and will presumably become larger adults. Larger adults may be more successful at defending breeding sites with the consequence that adult size may be important in reproductive success. Thus, brood reduction in *Nicrophorus* occurs in response to a shortfall of food relative to the requirements of the larvae arriving at the corpse (see also Baur, Haig, Stanback and Koenig, this volume).

Cannibalism of kin also occurs in several species of predacious ladybeetles (Coccinellidae) that feed on aphids. Clutches of eggs are often deposited on leaves that are not infested with aphid prey (Kawai 1978) and it is therefore difficult for first instar larvae to obtain their first meal (Banks 1956; Dixon 1959; Kaddou 1960; Brown 1972; Kawai 1978). Although there are a few non-cannibalistic species (Hawkes and Marriner 1927), in most species the larvae from the first-hatching eggs in a clutch often cannibalize unhatched eggs (Table 8.2). Many of these species lay 'trophic' (presumably unfertilized) eggs that are cannibalized, and are presumably provided as food for first instar larvae. For example, 15 per cent of *Harmonia axyridis* eggs collected in the field and maintained in the laboratory failed to hatch (Kawai 1978), and Pienkowski (1965) found that 8.4 per cent of the eggs of *Coleomegilla maculata*, collected in the field and maintained in the laboratory, were non-viable. Over 50 per cent of *C. maculata* eggs that did not hatch were yellow. Eggs of this species are bright yellow-orange until two to three hours prior to hatching when they turn grey following pigmentation of the embryo. Thus, yellow eggs are thought to be unfertilized. Laboratory studies of *Lioadalia flavomaculata* and *Cheilomenes lunata* revealed that 31 and 43 per cent, respectively, of the eggs were non-viable (Brown 1972). A large portion of these non-viable eggs (60 per cent) showed no development. Other laboratory studies estimate that *Adalia bipunctata*, *Harmonia quatturodecimpunctata*, and *Coccinella septempunctata* lay 5, 5 and 13 per cent non-fertilized (yellow) eggs respectively (Banks 1956). However, Mills (1982) suggested that the lack of fertilization in laboratory studies may be an artifact of the laboratory environment.

Cannibalism within the coccinellids generally occurs at the stage in an individual's life history when the nutritional benefits are high. For example, most, if not all of the non-viable eggs in a clutch are eaten, as well as some of the eggs that would have hatched into first instar larvae (Banks 1956; Brown 1972; Dixon 1959; Kaddou 1960; Kawai 1978; Pienkowski 1965). Cannibalism may also occur when earlier-hatching larvae consume larvae that hatch later (Dixon 1959; Kaddou 1960). First-instar larvae hatching from eggs laid in clutches must travel to find prey, and these small larvae have a low probability of successfully capturing their first meal. As the larvae feed and grow larger, their foraging success increases. Thus, cannibalism of eggs by first instar larvae provides them with an initial nutritional benefit thereby increasing their probability of foraging more successfully later (Banks 1956; Dixon 1959; Brown 1972; Kawai 1978).

CANNIBALISM OF KIN AND NON-KIN

Cannibalism of both kin and non-kin has been reported in a field study of the ladybeetle *Harmonia axyridis* (Osawa 1989). Ecological conditions apparently affect the frequency and occurrence of non-kin cannibalism, but do not affect cannibalism of kin. Osawa (1989) marked 80 egg batches in the field and censused cannibalism by kin and non-kin. Twenty-five per cent of the eggs were cannibalized by kin and 36 per cent by non-kin. This species oviposits 90 per cent of its eggs in late May and the oviposition period can be divided into an early, middle, and late phase. The frequency of egg batches cannibalized by kin in the early, middle, and late periods was similar. However, non-kin cannibalism was more frequent later in the oviposition period, being greatest in the late (90 per cent), intermediate in the middle (64 per cent), and lowest in the early period (12.5 per cent). The increase in non-kin cannibalism through the oviposition periods may be due to the higher density of coccinellid larvae relative to aphid density in the middle and late oviposition periods (Osawa 1989).

Some coccinellids lay clutches of eggs whereas others disperse their eggs singly (Hagen 1962; Brown 1972). Brown (1972) suggested that there is an association between egg dispersal and cannibalism, and this may depend on ecological conditions that influence non-sibling cannibalism (Osawa 1989). Those species which lay eggs singly often oviposit within prey groups, whereas species that lay eggs in batches often lay them far from aphid colonies. This may be because eggs are more likely to be cannnibalized by non-kin when they are near aphid colonies (Osawa 1989); non-sibling cannibalism of *H. axyridis* was significantly more intense for egg batches which were closer to the aphid colony, while distance from aphids did not have any effect on cannibalism by kin. Cannibalism of non-kin may be high near aphid colonies because a larva feeding within the aphid colony may find the egg batch and eat it. After feeding on aphids, larvae search intensely nearby, and do not discriminate between aphids and coccinellid eggs (Kawai 1976, 1978).

CANNIBALISM OF NON-KIN

Cannibalism of non-kin has been reported in several families of beetles, under a wide range of ecological conditions (Table 8.2). Non-kin cannibalism often removes a potential (or actual) intraspecific competitor and reduces population size. As a consequence of its effect on population size, cannibalism can affect food quality for the remaining individuals. However, it often reduces population size well below the environmental carrying capacity. For example, in the cannibalistic strains of *Tribolium confusum* the equilibrium total (larvae + pupae + adults) population sizes of the four cannibalistic strains are 467, 291, 82, and 40 (Park *et al.* 1961). These differences are mainly brought about by cannibalism (Park *et al.* 1965). Although cannibalism can have a large effect on population size, it does not finely tune population size to resource levels.

Initiation and control of cannibalistic behaviour may be a response primarily to the presence of vulnerable individuals (Polis 1981). Often the circumstances surrounding cannibalistic behaviour are benign; many species of beetles that are pests of stored products cannibalize at low densities in the presence of abundant food. Granivorous beetles, which can feed exclusively on grain, are often predatory (LeCato 1975*a*,*b*). Cannibalism may occur when such predators do not distinguish between conspecifics and other prey. The nutritional benefits from intra- and/or interspecific predation vary from no measured effect in nutritionally sufficient environments, to enabling individuals to survive in environments with insufficient nutritional resources (Dawson 1968).

Although cannibalism of non-kin may be viewed as indiscriminate or incidental predation, theoretical models predict that interference behaviour, such as cannibalism, will not have a selective advantage at the individual level if the perpetrators neither incur costs nor receive benefits other than the injury inflicted on their conspecific rivals (David Mertz, personal communiction). It seems likely that this is the case for the flour beetle *T. confusum*; however, in the closely related *T. castaneum*, individuals do benefit nutritionally from cannibalism (Ho and Dawson 1966). When such interference behaviours have a selective advantage, in the absence of these costs and benefits, it must be the result of selection at a level other than the individual. Kin selection and cohort selection can mold so-called indiscriminate cannibalism so that it is directed away from closely related individuals and consequently is no longer indiscriminate (Wade 1980; Mertz *et al.* 1984).

Recently, Mertz *et al.* (1984) predicted that a process called cohort selection can mold the age-specific pattern of cannibalism such that it will be directed against age classes in which the perpetrators' own progeny are underrepresented. Cohort selection is a form of natural selection that takes place because of interactions among the age groups in a population. Cohort selection is different from other models of selection in that it requires genetic differentiation among the age groups, which can be generated through random genetic drift. Selection then takes place when the age groups modify one another's survival and/or reproduction. In these Monte Carlo simulation models, the perpetrators incur no costs and receive no benefits other than the injury that they inflict on their conspecific rivals. Initial population sizes were over 500 individuals and there was no genetic differentiation among the age groups. Subsequent population sizes were usually over 200 and sometimes as high as 1000 individuals. Thus, the model was constructed to *minimize* the likelihood of cohort selection.

Another way in which cannibalism of non-kin can be directed away from closely related individuals is through spatial differences in egg laying and feeding: a female flour beetle tunnelling through the flour may consume eggs anteriorly while laying eggs posteriorly. In addition, variables such as predator preferences in egg size, pupation site, and age-specificity of predator and prey could be manipulated so that cannibalism is directed toward unrelated individuals.

Cannibalism of non-kin can be related to the level of resources available to the developing juveniles. For example, in granivorous beetles *Rhizopertha dominica*, early-instar larvae enter a kernel of grain in which they complete their development, and several studies suggest that cannibalism reduces intraspecific competition in granivorous species (Crombie 1944, 1946). Crombie (1944) reported that dead and partly eaten pupae of the granivorous beetle, *R. dominica*,were found in grains from which conspecifics had emerged as adults. Although not all parts of a kernel (germ, endosperm, seed coat, etc.) are likely to be nutritionally equal, resources are presumably sufficient to allow more than one beetle to survive. Nevertheless, *Rhizopertha* consumed on average only 23 per cent of a grain, and the average number of larvae surviving per grain (1.16) was independent of the initial number, which ranged from one to 20. In other species, the rate of population growth may be determined by cannibalism. For example, cannibalism in the saw-toothed grain beetle *Oryzaephilus surinamensis*s, probably occurs after random encounters between feeding stages and quiescent stages (Crombie 1946).

Non-kin cannibalism has been reported in several species of bark beetles (Scolytidae). There is anecdotal mention of cannibalism in two species of bark beetles that feed in elm, *Scolytus scolytus*, and *S. multistriatus*, and another species *Tomicus piniperda* that feed in pine (Beaver 1974). In these species, competition for food increases with population density. Both survivorship and adult weight decrease with increasing density, and mortality is probably not always due to starvation. Cannibalism is probably involved in some cases–a larger larva may simply eat straight through a smaller one as it elongates its gallery. As a result, larvae developing from later hatching eggs are more likely to be cannibalized. At high density, intraspecific competition is reduced as a result of cannibalism and possibly allows the survival of older, larger larvae (Beaver 1974).

Similar ecological factors are important in cannibalism of *Ips* spp. (Scolytidae). These bark beetles are major pests of southern pines. Cannibalism occurs when early hatching larvae eat later hatching larvae and increases with population size (Wagner *et al.* 1987, 1988). In *Ips calligraphus*, cannibalism of eggs occurs when parents inadvertently eat their own brood. This may occur when females excavate lateral turn arounds in the gallery; however only 0.6 per cent of eggs are eaten in this manner. Cannibalism of larvae by other larvae is high (35 per cent). About 5 per cent of pupae are cannibalized by larvae and teneral adults and 2.4 per cent of teneral adults are eaten by other teneral adults (Wagner *et al.* 1987). In a second species, *Ips avulsus*, cannibalism was less than 10 per cent for each life stage and was not influenced by oviposition time (Wagner *et al.* 1988). The differences between these two species in cannibalism rates is thought to be a function of gallery size and spacing. *I. avulsus* oviposits on alternate sides of the gallery with relatively large distances between eggs. Larval galleries are short, about 4–8 mm, and terminate in broad feeding cells (Wagner *et al.* 1988). In contrast, *I. calligraphus* oviposits on one side of the gallery with eggs closely spaced. The ecological factors governing cannibalism in *I. calligraphus* may have the result that larvae preferentially eat sibling larvae; normally, larvae do not cross parent galleries, so sibling larvae are confined to a single 'island' of phloem. Larval galleries are long, 100–120 mm. Resource use by *I. calligraphus* favours larvae originating in the proximal gallery which have more resources of better quality. These larvae develop faster, attain larger size as adults, and suffer less mortality from cannibalism (Wagner *et al.* 1987). Further experiments with these species explicitly examining cannibalism should provide insights into the ecological and evolutionary factors that influence cannibalism in this genus.

Cannibalism of non-kin also occurs in the herbivorous Colorado potato beetle *Leptinotarsa decemlineata*, and the false Colorado beetle, *L. juncta* (Tamaki *et al.* 1982; David McCauley, personal communication). In laboratory experiments, when adults of both sexes or older larvae of *L. juncta* encountered an egg clutch, they often preferentially consumed the eggs, leaving the leaf unscathed (David McCauley, personal communication).

The tenebrionid, *Zophobas atratus*, can have an elongated larval life stage; this unique behaviour apparently results from selective pressures generated by cannibalism. In this species there is intense cannibalism of pharate pupae by larvae and it is thought that pupation is inhibited by crowding. Pre-pupal dispersal may have evolved in response to cannibalism (Tschinkel 1981). If dispersal of the mature larva is prevented, metamorphosis is inhibited indefinitely, even until the larva dies of old age, after 18–24 months (Tschinkel 1978, 1981). The possible advantages of cannibalism have not yet been investigated in *Z. atratus,* although it may be that cannibalism of pharate pupae by larvae results in higher larval growth rates and higher fecundity (Tschinkel 1981). Larval cannibalism on pupae has also been shown to select for delayed pupation in *Tribolium castaneum* (McCauley 1978).

The effects of non-kin cannibalism among predacious beetles, such as *Pterostichus oblongopunctatus* (Carabidae) and *Philonothus decorus* (Staphylinidae), has been studied by Brunsting and Heessen (1984) and Heessen and Brunsting (1981). In these species, larvae eat other larvae. Cannibalism is reduced when alternative food is available, although it does not disappear entirely, and it was estimated to reduce population size by one-half when alternative food was available (Heessen and Brunsting 1981). In two other carabids, *Agonum assimile* and *Pterostichus nigrita*, the larvae are very aggressive towards members of their own species as well as larvae of other carabids. Cannibalism among these predatory beetles is viewed as a form of interference competition (Thiele 1977).

Larvae of whirligig beetles *Dineutes* spp. are cannibalistic and the intensity of individual larval aggression is inversely related to the amount of food received (Istock 1966). Single-species populations of larval *Dineutes* spp. adjust numerically to the existing food supply through cannibalism. Cannibalism rates decrease when additional prey are added, although the effect of food supplementation varies for the three cannibalistic species. Istock (1966) suggested regulation of population size by larval cannibalism in natural populations of this species is an adaptation for coping with a variable food supply. This conclusion suggests group selection may play a role in the evolution of cannibalism.

Non-kin cannibalism in *Tribolium*

Cannibalism in beetles has been most extensively examined in two species of flour beetle, *T. confusum* and *T. castaneum*. Studies of cannibalism in these species include: (1) its effect on the outcome of interspecific competition, (2) its effect on the population age-structure, (3) studies of the nutritional benefits of cannibalism, (4) kin selection and cannibalism, (5) the genetic basis of cannibalism, and (6) the evolution of cannibalism.

Cannibalism, competition and population age structure

Interspecific competition in *Tribolium* is influenced by inter- and intraspecific predation. Several studies suggest that predator preferences for small eggs rather than overall rates of mutual predation determine the winner of interspecific competition between two species of flour beetle (Teleky 1980; Craig 1986). When both *T. confusum* and *T. castaneum* are introduced into a vial containing flour medium (95 per cent by weight whole wheat flour and five per cent dried powdered brewer's yeast) one species always eliminates the other (Park 1948, 1954; Park *et al.* 1964; LeCato 1975*b*). Larvae of both species preferentially consume the eggs of *T. castaneum* and adults of both species prefer the eggs of *T. confusum* (Park *et al.* 1965). Eggs of *T. castaneum* are smaller than those of *T. confusum* (Craig 1986) and morphological differences mean that larvae are better able to eat smaller eggs whereas adults are more efficient at eating larger eggs. Experiments using a mutant strain of *T. castaneum* with eggs larger than *T. confusum* showed that egg size is a better predictor of cannibalism rates than any chemosensory preference (Craig 1986). Intraspecific predation rates among four cannibalism strains of *T. confusum* are positively correlated with interspecific predation rates (Park *et al.* 1965); however, these strains do not differ to any great extent in their ability to eliminate *T. castaneum* in experimental vials (Park *et al.* 1964). Preference of predators for various sizes of prey, rather than the predatory tendencies of the competing species seems to determine the outcome of interspecific competition in *Tribolium* (Teleky 1980).

Non-kin cannibalism of younger individuals, or defenceless life stages, can create dramatic asymmetries in the age and size distribution of a population (see also Quan and Polis, this volume). Mertz (1969) reported that the population size of strain cIV-a of *T. castaneum* was limited as a result of cannibalism, but was also due to a slight decrease in egg laying as a result of crowding. The observed age-distribution contained 15-40 per cent adults, while the expected stable age-distribution with no density-dependence was two per cent adults. The difference in the two age-distributions was caused mainly by the cannibalistic destruction of eggs and pupae in unmanipulated laboratory populations.

Both theoretical and empirical studies have attributed the phenomena of oscillations in immature numbers of *Tribolium* to cannibalism (Desharnais and Liu 1987; Hastings and Costantino 1987; Mertz 1969). In populations of strain cIV-a of *T. castaneum,* each immature stage fluctuates regularly in abundance

(Mertz 1969). These cycles can be attributed to changes in cannibalism rates associated with larval age. Newly-hatched larvae eat eggs at a rate of 0.001 per larva per egg per day. This increases to a maximum of 0.027 when the larvae are eight days old and decreases to 0.002 when the larvae are 18 days old and ready to pupate. The initial increase is due to an increase in larval size; however, as larvae approach pupation, they become less active and cannibalism declines. A population with 100 adults would lay eggs at a rate of about five per adult female per day and cannibalize at a rate of 0.0045 per adult per egg per day. In days 0–4 there are only adults and eggs in the population, and the percentage egg survival per day is 63.7 per cent. As small larvae begin to appear, egg survivorship decreases, and when the population is 11–22 days old (oldest larvae are 7–18 days of age) egg survivorship is less than 10 per cent. Thus, cannibalism creates a negative relationship between the number of larvae and the number of eggs in a population. As the larvae pupate, egg survivorship increases, resulting in a positive relationship between the numbers of eggs and pupae. The combined effects cause oscillations in the numbers of immatures in a population.

Theoretical studies focusing on cannibalistic egg-larva interactions in *Tribolium*, predict that cannibalism rates, fecundity, and the length of different life stages of *Tribolium* interact to determine abundance, and produce oscillations in the number of immatures (Hastings and Costantino 1987). Adult longevity is several times the duration of the developmental period and Hastings and Costantino (1987) modelled the dynamics within the immature age classes ignoring changes in adult numbers. An increase in egg-eating by larvae results in a tendency toward oscillations in the numbers of immatures. An increase in the length of the larval stage relative to the egg stage tends to move the population away from oscillatory behaviour. Models with significant larval mortality had reduced the magnitude of egg-larval oscillations (Hastings and Costantino 1987). Cannibalistic immature-adult models forecast no oscillations in adult numbers.

Hastings and Costantino's (1987) models assumed that larval cannibalism rates were independent of larval age. However, this is the very phenomenon which Mertz (1969) attributed to the cycling. Both models suggest that the critical factor for producing waves is a period during which cannibalism is rare, followed by a period of frequent cannibalism. This is generated in laboratory populations by age-dependent cannibalism rates, and in the models by a non-cannibalistic egg stage and a cannibalistic larval stage. Other models make similar predictions (Desharnais and Liu 1987): increases in larval cannibalism rates and fecundity are destabilizing. Increases in adult cannibalism and mortality are stabilizing. For both sets of models, the period of the oscillations is roughly equal to the developmental period and adult numbers show small oscillations while the immature age-classes can oscillate wildly.

Nutritional benefits of cannibalism in Tribolium

When raised on standard laboratory medium at 29°C and 70 per cent relative humidity, cannibalistic *T. castaneum* larvae had higher subsequent adult fecundity (i.e. laid one more egg/female/day) than adults that were not fed eggs as larvae (Ho and Dawson 1966). In contrast, *T. confusum* cannibalistic larvae did not have higher subsequent adult fecundity. Cannibalism was more beneficial in nutritionally-poor environments (S. Via, unpublished). Larvae raised in oat flour have a slower rate of development than larvae raised on standard wheat flour. When both populations were provided with conspecific eggs, larvae in oat flour showed higher cannibalism rates and exhibited a growth spurt. Thus, the nutritional benefits of cannibalism increase under conditions of physiological stress.

Kin selection and cannibalism in Tribolium

Cannibalism rates in *T. confusum* decreased rapidly though kin selection (Wade 1980), as predicted by Hamilton's theory (1964*a*,*b*). Populations were established with different degrees of relatedness, *r*, between predatory larvae and their egg victims: $r = 0$, 0.25 and 0.50. These groups of larvae were fed eggs, and those eggs that were not cannibalized matured to adulthood within the larval group. Populations with high cannibalism produced smaller groups of adults than those with low cannibalism. Subsequent generations were founded from adults selected from among all the individuals that matured in a population, such that the probability of a population founding a new population was positively related to population size. Lower cannibalism rates on related eggs increased population size and thus the chance of an individual's genes being present in the next generation. After only three generations, the rate of cannibalism among larvae that were fed related eggs was significantly lower than larvae that were fed unrelated eggs, and the degree of relatedness was inversely correlated with cannibalism rate. Thus the rate of cannibalism was decreased through kin selection as a function of the relatedness between cannibal and victim as predicted by Hamilton (1964*a*,*b*).

Genetics and the evolution of cannibalism

Four strains of *T. confusum*, bI, bII, bIII, and bIV, were derived during the

Table 8.3 Life history parameters of the four cannibalism strains of *Tribolium confusum*

Strain	Pop. size[1]	Eggs per 3 days[1]	% hatch	Developmental time (days)[1]			Days old at death[2]	
				Eggs	Larvae	Pupae	Males	Females
bI	467	28.8	70.5	5.37	18.96	6.98	306	321
bII	291	27.1	71.5	5.43	20.11	6.48	256	349
bIII	82	24.1	71.0	5.63	18.89	6.78	363	288
bIV	40	31.2	65.0	5.60	19.92	6.72	348	357

[1] From Park *et al.* (1961)
[2] From Mertz *et al.* (1965)

mid-1950s by three to four episodes of brother–sister mating of a 20-year laboratory strain known as b+ or b-Chicago (Park *et al.* 1961). Table 8.3 shows that the strains exhibit a tenfold difference in equilibrium population size and have similar birth rates, death rates, and developmental times (Mertz *et al.* 1965; Park *et al.* 1961). However, Table 8.4 shows that cannibalism rates are inversely correlated with population size (Park *et al.* 1965) and have been relatively stable for over 20 years (Stevens and Mertz 1985). Between 1960 and 1980, the strains were maintained as laboratory stocks by transferring about 600 premated adults to 200 g of fresh flour medium every four months. The stock-change interval approximates one generation; thus 20 years represents about 60 generations. During this period, the populations were subjected only to within-strain individual selection; there was no differential proliferation or extinction of populations or migration between the populations.

Crosses between the high and low cannibalism strains (bI and bIV) showed that the differences in Table 8.4 have a genetic basis (Stevens 1989). The cannibalism differences between the high and low cannibalism strains are due primarily to additive genetic effects; the traits are autosomally inherited, and environmental factors influence the expression of these behaviours. Larval cannibalism of eggs is controlled by many genes whereas the between strain differences in adult cannibalism of eggs may be due to as few as two genes. Larval and adult cannibalism of eggs have a high genetic correlation. The number of genes for adult cannibalism of pupae was not estimated.

Table 8.4 Cannibalism rates of the four cannibalism strains of *Tribolium confusum*[1]

Strain	Larvae eating eggs[2]				Adults eating eggs[3]				Adults eating pupae[4]			
	1962	1981	stable[5]	s^2_G[6]	1962	1981	stable[5]	s^2_G[6]	1962	1981	stable5	s^2_G[6]
bI	50.5	76.0	yes	no	22.3	9.5	no	yes	10.2	11.8	yes	yes
bII	142.5	141.0	yes	no	69.3	60.0	no	yes	10.9	9.8	yes	no
bIII	214.0	234.0	yes	yes	76.0	79.8	yes	yes	20.9	17.3	yes	no
bIV	256.3	288.5	no	yes	73.2	70.0	yes	yes	21.8	22.5	yes	yes

[1] Data for 1959 and 1962 from Park *et al.* (1965); data for 1981 from Stevens and Mertz (1985); values of s^2_G from Stevens (1989).
[2] Mean total number of eggs eaten by 50 larvae over the age interval 0–22 days
[3] Mean number of 'marked eggs' eaten in 48 hours by 25 female and 25 male adults approximately 20 days of age.
[4] Data are the mean number of pupae eaten by 10 female and 10 male adults during two one-week assays.
[5] Values in 1959 or 1962 and those in 1981 are not significantly different: $P < 0.05$.
[6] s^2_G detected by inbreeding experiment.

Does selection favour an optimum cannibalism rate in a given environment? Hamilton (1964*a*,*b*, 1970) predicted that interference behaviours such as cannibalism should be selected against in the absence of a benefit to the individual, but that when such a benefit exists, it should evolve to a single environmentally-dependent optimum. A single optimum is also predicted from quantitative genetic models and models with frequency-dependent selection (Aoki 1982; Crow and Aoki 1982; Harper 1989). In contrast to these predictions, empirical studies of the four cannibalism strains suggest that multiple adaptive peaks exist for cannibalism (Stevens 1989) and provide empirical evidence supporting Wright's shifting balance theory (Wright 1977).

The cannibalism rates of the four genetic strains were remarkably stable in the absence of artificial selection for over 60 generations of laboratory husbandry (Table 8.4; Stevens and Mertz 1985). Although three of the 12 cannibalism rates measured changed during the 20-year interval, there was no tendency for the strains to converge on a single 'optimum' cannibalism rate; the low cannibalism bI strain became even less so for adults eating eggs, and the high cannibalism bIV strain became even more cannibalistic for larvae eating eggs.

One explanation for the stability is that there was no genetic variation within the strains, and further experiments examined this possibility. Inbreeding showed that for the 12 cannibalism traits studied (three types of cannibalism x four strains = 12 traits) eight exhibited significant genetic variance, s^2_G (Table 8.4, columns 5, 9, and 13). A response to natural selection only occurs when there is additive genetic variance, s^2_A, and the inbreeding experiment detected overall genetic variance, a combination of additive and non-additive components (s^2_{NA}): $s^2_G = s^2_A + s^2_{NA}$. Thus, these results do not refute the possibility that there was no additive genetic variance. However, in combination with the additive genetic variance between strains described above, it appears that there was sufficient variation to produce a change if selection had been operating.

The stability of the between-strain differences, the additive nature of inheritance, and the demonstration of within strain genetic variation suggested that cannibalism may be selectively neutral, or under stabilizing selection with many adaptive peaks (Stevens 1989). For those strains that showed genetic variation in 1981 ('yes' in Table 8.4, columns 5, 9, and 13), an F-test that compares the observed change in variance between populations with that expected by genetic drift (Lande 1977; Wade 1984; Stevens 1989) was used to distinguish between these two possibilities. (Because strains that were not genetically variable in 1981 could have been in 1960, this is a conservative criteria and biases the results against invoking selection or neutrality.)

Strains bI and bII showed no genetic variation for larval cannibalism of eggs (Table 8.4, column 5). Strain bIV increased in cannibalism and had genetic variation; bIII also had genetic variation, but there was no change in cannibalism (Table 8.4, columns 4 and 5). The F-test was inconclusive; the observed pattern could be the result of either weak selection or genetic drift in a large population. Although strain bI exhibited no genetic variance in 1981 for larvae eating eggs, this cannibalistic trait decreased between 1959 and 1981. This suggests that bI was genetically variable in 1959 and that mass selection carried it towards a peak. In the process of approaching the peak, the genetic variance may have been depleted, as predicted by Fisher's Fundamental Theorem.

All four strains varied in adult cannibalism of eggs (Table 8.4, column 9). The two low cannibalistic strains had significantly lower cannibalism rates in 1981 than in 1962 (Table 8.4, column 8). The other two strains had not changed (Table 8.4, column 8). The F-test suggests that if the heritability of this trait is less than 0.75, then the observed changes result from directional selection. Heritability estimates are often between 0.05–0.75 (Falconer 1981) and are highest for morphological characters and lower for characters most closely connected with fitness. Cannibalistic behaviour is influenced by environmental factors (author's personal observation) and as a result may have low heritability. This suggests that the high cannibalistic strains were being maintained at adaptive peaks while the low strains were being selected to adaptive peaks. There were no changes in adult cannibalism of pupae although two of the strains exhibited genetic variation for this trait (Table 8.4, columns 12 and 13). This stability suggests that the strains are being maintained at adaptive peaks.

CONCLUSIONS

Cannibalism in many species benefits individuals directly through immediate nutritional gains. This is particularly true of those species that exhibit sibling cannibalism, such as the willow leaf beetle, and predaceous coccinellids. Cannibalistic willow leaf beetles gain a size advantage and may be more successful at initiating feeding sites on the tough willow leaves (Breden and Wade 1989). In species of predacious ladybeetles, cannibalism of eggs by first instar larvae provides them with an initial nutritional benefit and increases larval survivorship. However, not all cannibalism of kin provides a nutritional benefit to the cannibal. In the burying beetle, *Nicrophorus*, filial cannibalism occurs when adults eat as much as half of their brood (Bartlett 1987). It is hypothesized that cannibalism increases the reproductive success of the remaining larvae by allowing them to become larger in size, and thus is a result of kin or group selection.

Non-kin cannibalism removes potential competitors and predators. It reduces the number of beetle larvae within grain, under bark, in ponds, on vertebrate carcasses, or on the forest floor so that food supplies are adequate for the surviving larvae. However, the reduction in population size may be as much as an order of magnitude below the carrying capacity. Although there are often clear nutritional benefits to cannibalism of non-kin, this is not always the case (Ho and Dawson 1966).

If cannibals incur no costs and receive no benefits other than the injury that they inflict on their conspecific rivals, they will not have a selective advantage at the individual level (D. Mertz, unpublished results). When such behaviour has a selective advantage, it must be the result of selection at a level other than the individual. Cohort selection and kin selection can mold so-called indiscriminate cannibalism so that it is directed away from closely related individuals (Mertz *et al.* 1984; Wade 1980) and therefore is no longer indiscriminate.

Studies of cannibalism in beetles have shown that it responds rapidly to kin selection. These studies suggest that multilevel selection pressures are important in the evolution of cannibalism and show that cannibalism influences the benefit individuals may gain from foraging in a group. Studies of cannibalism in beetles also allow us to test predictions from single and multilocus theory, both of which predict that cannibalism rates should evolve to an optimum that is a function of the nutritional qualities of the environment. In the absence of group selection, this prediction does not hold, perhaps owing to the frequency and density dependence of cannibalistic behaviour. Instead it has been suggested that multiple stable adaptive peaks exist for cannibalism, providing strong evidence supporting the 'shifting balance theory' (Wright 1977). Empirical studies of cannibalism in beetles support such a scenario for the evolution of cannibalism and challenge us to consider multilevel selection (e.g. cohort selection, kin selection, and group selection) as well as frequency and density dependence in our attempts to understand the evolution of social behaviour.

ACKNOWLEDGEMENTS

I would like to thank D. McCauley, M. Whitlock, F. Breden, D. Craig, M. Wade, G. Polis, P. Goff, and C. Goodnight for input into this chapter. I am grateful to W. Campbell and D. LaPoint for assistance with preparing this chapter. This work was supported by grants from NSF (BSR-8906347) and the University of Vermont Committee on Research and Scholarship.

REFERENCES

Aoki, K. (1982). Additive polygenic formulation of Hamilton's model of kin selection. *Heredity*, **49**, 163–9.

Banks, C. J. (1956). Observations on the behaviour and mortality in Coccinellidae before dispersal from the egg shells. *Proceedings of the Royal Entomological Society of London, A*, **31**, 56–60.

Bartlett, J. (1987). Filial cannibalism in burying beetles. *Behavioral Ecology and Sociobiology*, **21**, 179–93.

Beaver, R. Z. (1974). Intraspecific competition among bark beetle larvae (Coleoptera:Scolytidae). *Journal of Animal Ecology*, **43**, 455–67.

Breden, F. and Wade, M. J. (1985). The effect of group size and cannibalism rate on larval growth and survivorship in *Plagiodera versicolora. Entomography*, **3**, 455–63.

Breden, F. and Wade, M. J. (1987). An experimental study of the effect of group size on larval growth and survivorship in the imported willow leaf beetle, *Plagiodera versicolora* (Coleoptera: Chrysomelidae). *Environmental Entomology*, **16**, 1082–6.

Breden, F. and Wade, M. J. (1989). Selection within and between kin groups of the imported willow leaf beetle. *American Naturalist*, **134**, 35–50.
Brown, H. D. (1972). The behavior of newly hatched coccinellid larvae (Coleoptera: Coccinellidae). *Journal of the Entomological Society of South Africa*, **35**, 149–57.
Brunsting, A. M. H. and Heessen, H. J. L. (1984). Density regulation in the carabid beetle *Pterostichus oblongopunctatus*. *Journal of Animal Ecology*, **53**, 751–60.
Chapman, R. N. (1928). Quantitative analysis of environmental factors. *Ecology*, **9**, 111–22.
Craig, D. M. (1986). Stimuli governing intraspecific egg predation in the flour beetles, *Tribolium confusum* and *T. castaneum*. *Researches on Population Ecology*, **28**, 173–83.
Crombie, A. C. (1944). On intraspecific and interspecific competition in larvae of graminivorous insects. *Journal of Experimental Biology*, **20**, 135–51.
Crombie, A. C. (1946). Further experiments on insect competition. *Proceedings of the Royal Society, London, B*, **133**, 76–109.
Crow, J. F. and Aoki, K. (1982). Group selection for a polygenic behavioral trait: a differential proliferation model. *Proceedings of the National Academy of Sciences, USA*, **79**, 2628–31.
Dawson, P. S. (1968). Xenocide, suicide, and cannibalism in flour beetles. *American Naturalist*, **102**, 97–105.
Desharnais, R. A. and Liu, L. (1987). Stable demographic limit cycles in laboratory populations of *Tribolium castaneum*. *Journal of Animal Ecology*, **56**, 885–906.
Dimetry, N. A. (1974). The consequences of egg cannibalism in *Adalia bipunctata* (Coleoptera: Coccinellidae). *Entomophaga*, **19**, 445-52.
Dixon, A. F. G. (1959). An experimental study of the searching behaviour of the predatory coccinellid beetle *Adalia decempunctata* (L.). *Journal of Animal Ecology*, **28**, 259–81.
Eickwort, K. R. (1973). Cannibalism and kin selection in *Labidomera clivicollis* (Coleoptera: Chrysomelidae). *American Naturalist*, **107**, 452–3.
Falconer, D. S. (1981). *Introduction to Quantitative Genetics*. (2nd edn). Longman, New York.
Fox, L. R. (1975). Cannibalism in natural populations. *Annual Review of Ecology and Systematics*, **6**, 87–106.
Hagen, K. S. (1962). Biology and ecology of predaceous Coccinellidae. *Annual Review of Entomology*, **7**, 289–326.
Hamilton, W. D. (1964*a*). The genetical evolution of social behaviour, I. *Journal of Theoretical Biology*, **7**, 1–16.
Hamilton, W. D. (1964*b*). The genetical evolution of social behaviour, II. *Journal of Theoretical Biology*, **7**, 17–52.
Hamilton, W. D. (1970). Selection of selfish and altruistic behavior in some extreme models. *Man and beast: comparative social behavior*. Smithsonian Institute, Washington, DC.
Harper, A. B. (1989). Evolutionary stability for interactions among kin under quantitative inheritance. *Genetics*, **121**, 877–89.
Hastings, A. and Costantino, R. F. (1987). Cannibalistic egg-larva interactions in *Tribolium*: an explanation for the oscillations in population numbers. *American Naturalist*, **130**, 36–52.
Hawkes, O. A. M. and Marriner, T. F. (1927). A preliminary account of the life-history of *Coccinella 11-punctata* (L.). *Transactions of the Entomological Society of London*, **75**, 47–52.
Heessen, H. J. L. and Brunsting, A. M. H. (1981). Mortality of larvae of *Pterostichus*

oblongopunctatus (Fabricius) (Col., Carabidae) and *Philonthus decorus* (Gravenhorst) (Col., Staphylinidae). *Netherlands Journal of Zoology*, **31**, 729–45.

Ho, F. K. and Dawson, P. S. (1966). Egg cannnibalism by *Tribolium* larvae. *Ecology*, **47**, 318–22.

Iperti, G. (1966). Some components of efficiency in aphidophagous coccinellids. In *Ecology of aphidophagous insects*, (ed. I. Hodek), p. 253. Academia, Prague.

Istock, C. A. (1966). Distribution, coexistence, and competition of whirligig beetles. *Evolution*, **20**, 211–34.

Jillson, D. and Costantino, R. F. (1980). Growth, distribution, and competition of *Tribolium castaneum* and *Tribolium brevicornis* in fine-grained habitats. *American Naturalist*, **116**, 206–19.

Kaddou, I. K. (1960). The feeding behavior of *Hippodamia quinquesignata* (Kirby) larvae. *University of California Publications in Entomology*, **16**, 181–228.

Kawai, A. (1976). Analysis of the aggregation behaviour in the larvae of *Harmonia axyridis* Pallas (Coleoptera: Coccinellidae) to prey colony. *Researches on Population Ecology*, **18**, 123–34.

Kawai, A. (1978). Sibling cannibalism in the first instar larvae of *Harmonia axyridis* Pallas (Coleoptera: Coccinellidae). *Kontyu, Tokyo*, **46**, 14–19.

King, C. E. and Dawson, P. S. (1972). Population biology and the *Tribolium* model. *Evolutionary Biology*, **5**, 133–227.

Lande, R. (1977). Statistical tests for natural selection on quantitative characters. *Evolution*, **31**, 442–44.

LeCato, G. L. (1975*a*). Interactions among four species of stored-product insects in corn: a multifactorial study. *Annals of the Entomological Society of America*, **68**, 677–9.

LeCato, G. L. (1975*b*). Predation by red flour beetle on sawtoothed grain beetle. *Environmental Entomology*, **4**, 504–6.

McCauley, D. E. (1978). Demographic and genetic responses of two strains of *Tribolium castaneum* to a novel environment. *Evolution*, **32**, 398–415.

McCauley, D. E. and O'Donnell, R. (1984). The effect of multiple mating on genetic relatedness in larval aggregations of the imported willow leaf beetle (*Plagiodera versicolora* Coleoptera: Chrysomelidae). *Behavioral Ecology and Sociobiology*, **15**, 287–91.

McCauley, D. E., Wade, M. J., Breden, F., and Wohltman, M. (1988). Spatial and temporal variation in group relatedness: evidence from the imported willow leaf beetle. *Evolution*, 184–92.

Mertz, D. B. (1969). Age-distribution and abundance in populations of flour beetles. I. Experimental studies. *Ecological Monographs*, **39**, 1–31.

Mertz, D. B. (1972). The *Tribolium* model and the mathematics of population growth. *Annual Review of Ecology and Systematics*, **3**, 51–78.

Mertz, D. B., Craig, D. M., Wade, M. J., and Boyer, J. F. (1984). Cohort Selection. *Evolution*, **38**, 560–70.

Mertz, D. B., Park, T., and Youden, W. J. (1965). Mortality patterns in eight strains of flour beetle. *Biometrics*, **21**, 99–114.

Mills, N. J. (1982). Voracity, cannibalism and coccinellid predation. *Annals of Applied Biology*, **101**, 144–8.

Osawa, N. (1989). Sibling and non-sibling cannibalism by larvae of a lady beetle *Harmonia axyridis* Pallas (Coleoptera: Coccinellidae) in the field. *Researches on Population Ecology*, **31**, 153–60.

Park, T. (1948). Experimental studies of interspecies competition. I. Competition between

populations of flour beetles, *Tribolium confusum* Duval and *Tribolium castaneum* Herbst. *Ecological Monographs*, **18**, 265–308.
Park, T. (1954). Experimental studies of interspecies competition. II. Temperature and humidity and competition in two species of *Tribolium*. *Physiological Zoology*, **27**, 177–238.
Park, T., Leslie, P. H., and Mertz, D. B. (1964). Genetic strains and competition in populations of *Tribolium*. *Physiological Zoology*, **37**, 97–162.
Park, T., Mertz, D. B., Grodzinski, W., and Prus, T. (1965). Cannibalistic predation in populations of flour beetles. *Physiological Zoology*, **38**, 289–321.
Park, T., Mertz, D. B., and Petrusewicz, K. (1961). Genetic strains of *Tribolium*: their primary characteristics. *Physiological Zoology*, **38**, 289–321.
Pienkowski, R. L. (1965). The incidence and effect of egg cannibalism in first instar *Coleomegilla maculata lengi* (Coleoptera: Coccinellidae). *Annals of the Entomological Society of America*, **58**, 150–3.
Polis, G. A. (1981). The evolution and dynamics of intraspecific predation. *Annual Review of Ecology and Systematics*, **12**, 225–51.
Polis, G. A. (1984). Intraspecific predation and 'infant killing' among invertebrates. In *Infanticide: comparative and evolutionary perspectives*, (ed. G. Hausfater and S.B. Hrdy), pp. 87–104. Aldine, New York.
Stevens, L. (1989). The genetics and evolution of cannibalism in flour beetles, genus *Tribolium*. *Evolution*, **43**, 169–79.
Stevens, L. and Mertz, D. B. (1985). Genetic stability of cannibalism in *Tribolium confusum*. *Behavior Genetics*, **15**, 549–59.
Tamaki, G., Chauvin, R. L., and Hsiao, T. (1982). Rearing *Doryphorophaga doryphorae*, a tachinid parasite of the Colorado potato beetle, *Leptinotarsa decemlineata*. U.S. Department of Agriculture, Advances in Agricultural Technology, Western Series, No. 21. 14 pp.
Teleky, S. B. (1980). Egg cannibalism in *Tribolium* as a model of interference competition. *Researches on Population Ecology*, **21**, 217–27.
Thiele, H.-U. (1977). *Carabid beetles in their environments: a study on habitat selection by adaptations in physiology and behaviour*. Springer-Verlag, Berlin.
Tschinkel, W. (1978). Dispersal behavior of the larval tenebrionid beetle *Zophobas rugipes*. *Physiological Zoology*, **51**, 300–13.
Tschinkel, W. R. (1981). Larval dispersal and cannibalism in a natural population of *Zophabas atratus* (Coloeptera: Tenebrionidae). *Animal Behavior*, **29**, 990–6.
Wade, M. J. (1980). An experimental study of kin selection. *Evolution*, **34**, 844–55.
Wade, M. J. (1984). Changes in group selected traits that occur when group selection is relaxed. *Evolution*, **38**, 1039–46.
Wade, M. J. and Breden, F. (1986). Life history of natural populations of the imported willow leaf beetle, *Plagiodera versicolora* (Coleoptera: Chrysomelidae). *Annals of the Entomological Society of America*, **79**, 73–9.
Wagner, T. L., Fargo, W. S., Flamm, R. O., Coulson, R. N., and Pulley, P. E. (1987). Development and mortality of *Ips calligraphus* (Coleoptera: Scolytidae) at constant temperatures. *Environmental Entomology*, **16**, 484–96.
Wagner, T. L., Hennier, P. B., Flamm, R. O. , and Coulson, R. N. (1988). Development and mortality of *Ips avulsus* (Coleoptera: Scolytidae) at constant temperatures. *Environmental Entomology*, **17**, 181–91.
Wright, S. (1977). *Evolution and the genetics of populations*, Vol 3. *Experimental results and evolutionary deductions*. University of Chicago Press.

9

Cannibalism and trophic eggs in subsocial and eusocial insects

Bernard J. Crespi

'What is to be thought of Nature, when she seems to give to insects the faculty of destroying their own species, when she permits hivebees to kill their males, and gives bumblebees the right and the desire to devour the newly-laid eggs?'

Huber (1802, p. 260, translated from the original French)

Eusocial insects are characterized by the presence of reproductive division of labour, such that some individuals forego or reduce personal reproduction to help rear the offspring of others, usually kin (Wilson 1971; Michener 1974). Among Hymenoptera (ants, bees, and wasps) and Isoptera (termites), eusociality also involves cooperative brood care and generation overlap, whereas among Homoptera (aphids) and polyembryonic wasps (family Encyrtidae) more or less sterile juvenile 'soldier' morphs defend their habitat and the individuals therein against predation or usurpation (Aoki 1977, 1987; Cruz 1981). Subsocial insects exhibit only brood care, defined here as behaviour beyond oviposition, primarily feeding, defense or both, that increases offspring survivorship.

Cannibalism is defined here as eating conspecifics, either dead or alive prior to the interaction, and 'trophic eggs' are defined as ovarian-derived structures or fluid, homologous to fertile eggs, that cannot develop into viable offspring and are normally eaten. Some species with trophic eggs are neither eusocial nor subsocial, and these will also be considered; after all, if eggs (with DNA) are considered as individuals, then trophic eggs can be considered as a sterile caste destined to be cannibalized!

The purpose of this chapter is to review the presence, form, and evolution of cannibalism and trophic eggs in eusocial and subsocial insects. In the process, I address four central questions: (1) how do trophic eggs originate and evolve and what are their functions? (2) is cannibalism an important selective pressure in the evolution of subsociality? (3) to what extent do kin selection, parental manipulation (Alexander 1974), and ecology explain the distribution and form of cannibalism and trophic eggs in eusocial insects? and (4) what are the ecological and evolutionary implications and effects of cannibalism and trophic eggs? Each of these questions requires the theoretical focus of analysing how conflicts of interest between parents, offspring, siblings, and unrelated individuals are resolved, and how these conflicts occasionally lead to cooperation.

THE EVOLUTION OF TROPHIC EGGS IN NON-EUSOCIAL TAXA

Theory

In non-eusocial insects, plausible starting points for the evolution of trophic eggs are cannibalism of viable eggs and cannibalism of eggs that failed to develop. Initially, eggs may be either clumped or scattered. All else being equal, larval cannibalism of eggs should be more prevalent when eggs are scattered, because victims are more likely to be non-relatives; indeed, in several butterfly families, larval cannibalism occurs in species in which females lay eggs individually, but not in species in which females oviposit in batches (Chew and Robbins 1984; Courtney 1986). Alternatively, in a species with scattered oviposition and cannibalism already present, egg clumping may be favoured where the survivorship and growth advantages to mothers of larvae having an assured food source outweigh the inclusive fitness costs (Eickwort 1973). Egg grouping may be favoured because it could result in 'incidentally' inviable eggs being put to good use, and more slowly-developing, and perhaps less fit, offspring would most likely be eaten. How, then, does adaptive egg inviability (i.e. trophic egg production) evolve from cannibalism of either viable or inviable eggs? Presumably it can only evolve in kin groups with clumped ovipositions, since otherwise a mother's or egg's loss in fecundity or viability, respectively, will benefit primarily non-relatives. Eating of incidentally-inviable eggs by offspring entails no inclusive fitness costs. However, if incidental egg inviability is due to constraints, mothers will not be able to control egg fertility so as to produce maximal brood survivorship. Thus, trophic eggs probably originated in the context of larvae eating otherwise viable sibs, and (later) mothers making otherwise viable eggs sterile or eggs themselves being selected for inviability.

By eating eggs, a larva increases its own fitness through better nutrition or reduced competition, but for an imaginary allele favouring some degree of sib cannibalism to increase or be maintained, this benefit must outweigh the cost involved in killing sibs (Hamilton 1964; Eickwort 1973). In groups of diploid full sibs all related by one-half, the optimal cannibalism rate is higher for larvae than for parents because offspring are more closely related to themselves than to their siblings or mother. If, moreover, mothers use sperm from more than one male, then average sibling relatedness falls below one-half, and offspring should cannibalize even more eggs than their parents would prefer. What, then, is a mother to do?

There are at least four possibilities: adjusting egg viability, egg size and composition, egg hatching synchrony, and egg number. Egg inviability may evolve if the aggregate fitness of a mother from a clutch increases from the success of survivors more than it is decreased by a lower clutch size. Moreover, if a mother makes some eggs inviable, then offspring should choose to cannibalize these over viable eggs, since there is no inclusive fitness cost to doing so. However, the interests of mothers and offspring may still conflict over the fate of viable eggs (see Parker and Mock 1987; also Alexander 1974). One recourse of mothers may be to make eggs that are not only inviable, but that cost less for them to produce than do viable eggs (perhaps because they are immature when laid), but still give cannibalistic offspring sufficient nutritional benefit that they are not favoured to eat viable kin when the trophic eggs are depleted. Thus, sibling cannibalism should be more common with lower within-brood relatedness, but trophic eggs should be found more commonly in monogamous species (because mothers thereby reduce parent–offspring conflict), and trophic eggs may be smaller or less costly to mothers individually than viable ones, or more specialized for larval needs. Moreover, if the production of trophic eggs is at an optimum level for both parents and offspring, then viable eggs should not be eaten (see Godfray 1987, for a discussion of how such an optimum may be predicted).

Expectations for sibling cannibalism and trophic eggs among haplodiploids follow similar, though more complex, inclusive fitness logic. For example, females should prefer to eat brothers over sisters, but brothers should show no sex preference (see Pickering, 1980, for analysis of a similar type of situation in a gregarious hymenopteran parasitoid).

Certain types and degrees of environmental variability may favour either hatching synchrony adjustment or increased clutch size. Hatching synchrony adjustment might be favoured where cannibalism is facultative and all offspring could survive under some reasonably frequent circumstances predictable by the mother. Eisenberg and Hurd (1977) suggest that high offspring emergence synchrony in a mantid is an adaptation of mothers to reduce cannibalism among a brood, which is presumably maladaptive for parents and offspring (perhaps caused by strong selection to catch any prey when young). Alternatively, where survivorship of offspring varies more unpredictably, females may gain by increasing clutch size, such that in good circumstances all offspring may prosper, but when times are poor, some offspring serve as food for others (i. e. Alexander's 1974 'icebox' effect).

Parker and Mock (1987) analysed parent–offspring conflict over clutch size, including situations where offspring cannibalize eggs, using Evolutionary Stable Strategy models. One of their main results was that, if offspring kill a fixed number of sibs, as might be expected in insects where cannibals may need primarily to gain only sufficient energy to moult to the next instar (or where immediate food intake may be limited by morphology), then females may be selected to produce larger clutches. Such larger clutches could represent a 'solution' to the mother's problem of sibling cannibalism in cases where trophic egg production is for some reason not favoured.

Data

Trophic eggs are known or suspected in at least five non-eusocial taxa: owlflies (Henry 1972), the cricket *Anurogryllus muticus* (West and Alexander 1967), spiders (see Valerio 1974, 1977), coccinellid beetles (see references below, and Stevens this volume), and a psocid (order Psocoptera) (Mockford 1957).

Owlflies lay their eggs in clumps, on stems of vegetation (Henry 1972). In addition to laying fertile eggs, females of these and other species produce structures called 'repagula', which are ovarian bodies placed next to the fertile eggs and between such eggs and the ground. The repagula of *Ascaloptynx furciger* look like miniature fertile eggs, and are pierced and eaten by newborn larvae. Moreover, the fertile eggs have hard shells that prevent ant predation, and larvae have large mandibles used in active group defence against predators. *Ululodes mexicana* repagula, which are club-shaped rather than round, are never eaten by larvae, but serve instead as a 'fence' that protects the vulnerable fertile eggs from predation by ants. Free from the threat of ant attacks, *U. mexicana* larvae show no defensive behaviour.

Henry (1972) suggested that trophic eggs were followed evolutionarily by ant fences, since coating an abortive egg with ant repellent converts one to the other. Assuming that a combination of trophic eggs with ant-vulnerable fertile eggs is impossible, and that the characters of *A. furciger* are primitive rather than derived, then the transition from trophic eggs to defensive eggs, followed by the loss of hard eggshells and defensive larval structures, may be the only plausible evolutionary route from one set of traits to the other. The presumed benefit causing the transition is conversion of resources from heavily-shelled eggs, larval defensive morphology, and a large, energy-rich repagula to increased clutch size and smaller eggs (Henry 1972). However, since the fertile egg clutch sizes of these two species are similar (Henry 1972), an advantage of ant fence repagula may also lie in more efficient protection from ants and other predators.

The function of trophic eggs in *A. furciger* is uncertain. Presumably, feeding on one or two eggs by each larva increases survivorship by reducing the chances of starvation or facilitating group defence. Any such survivorship increase must, however, more than compensate for a female's probable loss in fecundity caused by investing in trophic eggs. Experimental removal of trophic eggs, and comparisons of egg sizes and clutch sizes among owlfly species with and without trophic eggs and larval defence, would help to identify the selective pressures present. Moreover, if phylogenetic analysis revealed that ant fences were derived, then they would represent the only known case of evolutionary loss of the nutritive function of trophic eggs.

Females of the cricket *Anurogryllus muticus*, dig, inhabit, and oviposit in soil burrows, and vigorously defend their offspring against intruders such as conspecific males (West and Alexander 1967). Females lay both fertile eggs and much smaller trophic eggs, which are fought over and eaten by the nymphs. The one female observed by these authors died when her offspring were nymphs, and although ingestion of the mother was not observed, West and Alexander (1967) state that 'under usual circumstances there is little doubt that the female's own offspring would consume it'. They suggest that the combination of trophic eggs, maternal care, and the female's death and cannibalism in the burrow indicate the effects of selection to maximize 'the chance of success of individual hatchlings'. The social or environmental agent of this selection is unknown. However, the presence of trophic eggs and consumption of the female in *A. muticus* may be related to its low mobility, such that females may have difficulty providing well for offspring just by physically transporting food, rather than by also transforming food with their ovaries and providing food by their own death. The finding that females of the wingless, subsocial burrowing cricket *Schizodactylus monstrosus*, are also eaten by their offspring (Choudhuri and Bagh 1974), and the similar 'gerantophagy' of some spiders (Seibt and Wickler 1987) for which reliable food supplies may also be problematic, provides some support for this hypothesis.

Trophic eggs may occur among several other arthropod taxa, but because eggs are not externally dimorphic or because of insufficient information, their trophic function is uncertain. In subsocial spiders from several families, spiderlings in the first or first few instars feed on 'inviable' eggs, and such feeding increases juvenile survivorship until the time when young can find prey independently without undue difficulty (e.g. Valerio 1974, 1977). Indeed, Schick (1972) suggested that the sole 'function' of the first, semi-quiescent instar of some spiders is to feed on inviable eggs. Among many coccinellid beetles, first-instar larvae from the egg mass of a single female feed on both viable and inviable eggs within their natal egg mass (see Stevens this volume). This form of sibling cannibalism may be adaptive because it helps larvae survive until they find their aphid prey, which requires dispersal for uncertain distances (e.g. Boldyrev and Wilde 1969; Brown 1972; Takahashi 1989). In many such species, a high percentage of eggs normally do not develop and are eaten; Mills (1982) suggests that such high egg inviability is an artifact of ineffective mating under laboratory conditions, but several studies present data on high egg inviability from field-collected egg masses (e.g. Pienkowski 1965; Kawai 1978). Within-brood relatedness is apparently unknown for any coccinellid, but in the chrysomelid *Labidomera clivicollis*, which exhibits egg cannibalism (Eickwort 1973), some broods are multiply sired (Dickinson 1988).

The psocopteran *Archipsocus gurneyi*, which lives in colonies under silk, lays eggs in communal batches and the adults stay with the eggs, apparently guarding them (Mockford 1957). Virgin females of this species laid both normal-sized eggs and smaller, more spherical eggs; both egg types failed to develop, but females ate the smaller eggs. Small eggs were not reported in normal field colonies (but they may, or course, have been eaten immediately after oviposition if they were trophic!) Such eggs were not laid by two congeneric oviparous species, neither of which lays eggs in batches or shows egg attendance. The higher degree of social development in *A. gurneyi* (within which juveniles and adults contribute silk and care to the web) and the morphological differences between egg types, suggest that small eggs are trophic. However, further work is needed to exclude the alternative hypothesis of a developmental artifact of forced virginity.

Interpretation

The proximate cause of trophic eggs and egg cannibalism in owlflies, coccinellids, and spiders appears to be a separation of the oviposition site from the normal juvenile feeding sites, coupled in spiders with a need to pass through a stage when prey capture is difficult due to their small size, and in the owlfly *A. furciger* with a need for juveniles to defend themselves prior to dispersal. Thus, the primary selective pressure favouring the evolution of trophic eggs in subsocial insects is probably the assurance and regulation of offspring feeding and survival at and near the oviposition site, wherever offspring needs apparently cannot be better met through increased egg size, optimal hatching asynchrony, and cannibalism, or maternal care. Further studies of trophic eggs require:

(1) experimental removals and additions of trophic, inviable, and viable eggs, followed by measurement of offspring survivorship;
(2) measurement of relatedness in relation to trophic egg production, egg inviability, and cannibalism; and
(3) assessment of the cost to mothers of producing trophic versus viable eggs, and the incremental benefits to offspring of ingesting each.

CANNIBALISM UNDER SUBSOCIALITY

Theory

Wilson (1971) proposed that insect subsociality has evolved under four general circumstances: (1) localized, rich, and ephemeral resources, which must be defended against intruders, (2) harsh physical environments, where offspring care is required for survival, (3) high predation rates on juvenile stages, which can be reduced by care, and (4) stable or structured habitats, wherein care of altricial young improves their fitness expectations (see also Eickwort 1981; Tallamy 1984; Tallamy and Wood 1986). Cannibalism may be either a form of predation more or less important in causing the evolution of subsocial behaviour, or a direct effect of these peculiar environmental circumstances, or instead a consequence rather than a cause of subsociality. Distinguishing these alternatives requires knowledge of the evolutionary transitions, inferred from phylogeny, between solitary behaviour, subsociality, trophic egg production, and other manifestations of cannibalism. In the absence of such data, I focus on the functional design (i.e. the correspondence between specific aspects of phenotype and environment) of cannibalistic behaviour and trophic eggs in subsocial insects, and on any inter-specific comparisons that can be made.

Data

Since both trophic eggs and parental care are known only in the cricket *Anurogryllus muticus* (and possibly in the psocopteran *Archipsocus gurneyi*), a comparative approach to the coincidence of these two traits is problematic. In

Anurogryllus muticus, food transport to young may be enhanced through trophic egg production, and in this and other species, trophic eggs may be small relative to normal eggs in order to facilitate more equitable trophic egg distribution among offspring in a brood.

Cannibalism of viable eggs and parental care are known to coincide in five taxa: a thrips, a staphylinid beetle, burying beetles, bethylid wasps, and an opilionid spider. The haplodiploid thrips *Elaphrothrips tuberculatus* lives on clusters of hanging dead oak leaves and feeds on fungal spores (Crespi 1987, 1988, 1989). Females are reproductively dimorphic, such that in any given brood they either give birth to fully-formed, mobile male first-instar larvae that are not defended, or lay undeveloped female eggs onto an egg mass that they defend until the eggs have hatched. By each reproductive mode, offspring are produced one at a time over one to three weeks. Observations of marked females and experimental removal of guarding females, controlling for the presence of other females on a patch, demonstrate that egg cannibalism is the primary selective pressure favouring maternal care in this species (Crespi 1990). Most cannibalism occurs when females have temporarily left their eggs to feed. Overall, 15–20 per cent of eggs were cannibalized in the spring breeding period, and about 40 per cent were cannibalized in summer (Crespi 1990, unpublished data). Among leaf clusters, cannibalism increased with density (female per leaf), but it was unrelated to female size or the presence of viviparous females on the cluster. In *E. tuberculatus*, circumstances favouring cannibalism and subsociality include: (1) a localized resource (the leaf cluster), that compels group living often at high densities, (2) a simple habitat structure, such that females easily encounter each other and their eggs, and (3) the requirement that oviparous females leave their eggs to feed.

Female *Eumicrota corruscula* and *E. socia* beetles form chambers on the ventral surface of shelf fungi by modifying a natural cavity. They lay their eggs inside, and guard the eggs and larvae, apparently until the fungus becomes unsuitable as habitat (Ashe 1987). When the females abandoned their egg chambers (in this study, usually as a result of disturbance by the investigator), the eggs were usually eaten by conspecific larvae and adults, or destroyed by fungal mycelium. Although the frequency of 'natural' cannibalism is unknown, and experimental field studies are needed to assess the causes of maternal care, cannibalism appears to be a major cause of guarding in *Eumicrota* (Ashe 1987).

In *Nicrophorus* burying beetles, males and females bury a small animal carcass and roll it into a ball, whereupon the female oviposits and the female or both parents feed the developing larvae (Pukowski, 1933; Scott 1990). Scott (1990) and Scott and Traniello (1990) found that in *Nicrophorus tomentosus*, 22 per cent of broods were reared by females alone, and that females were capable of defending their brood against heterospecific insect predators; moreover, male help in provisioning larvae actually reduced female reproductive success (Scott 1989). The primary benefit of male assistance is the decreased likelihood of intraspecific usurpation and infanticide of larvae, through male help in repelling conspecific intruders. Scott (1990) does not discuss whether or not usurpers or their offspring cannibalize the original brood, but Trumbo (1990*a*) reports that cannibalism usually follows infanticide in *N. orbicollis*. In *N. vespilloides*, Bartlett (1987) discovered filial cannibalism, whereby parents normally engage in brood reduction by killing and may eat up to half of their first-instar larvae. Bartlett (1987) proposed that parents will benefit from the 'insurance' value of a large clutch, if predation on eggs and newly-hatched larvae is common. However, parents kill and eat offspring on the first day of larval emergence; egg predation has not been observed and may be rare due to male and female guarding, and predators would presumably eat the entire clutch. A plausible alternative explanation is that cannibalism is a simple, inexpensive mechanism for parents to optimally adjust clutch size (and possibly sex ratio) to corpse size, close in time to when this adjustment can best be made. Muller and Eggert (1990) and Trumbo (1990*b*) discuss furthur the mechanisms and effects of cannibalism in *Nicrophorus*.

Females of the bethylid wasp *Goniozus nephantides* sting, poison, and immobilize lepidopteran larvae, lay a mass of eggs on the host, and remain with their brood, defending it until it has consumed the entire host (Hardy and Blackburn 1991). The primary benefit of egg defence in this species is the prevention of usurpation and egg cannibalism by female conspecifics, which will then lay a clutch of their own. Larvicide does not occur in this species, possibly because it takes too much time (Strand and Godfray 1989), or because the host is already depleted (Hardy and Blackburn 1991). However, other species of parasitoids (e.g. *Bracon hebetor*) will destroy unguarded larvae (Hardy and Blackburn 1991). Protection of eggs from conspecifics (and, in some cases, hyperparasitoids) and protection of larvae from other species may favour maternal care in other species of bethylids. Egg-eating in bethylids may have another function: Wheeler (1928) reported that females of *Scleroderma macrogaster* would eat a proportion of their own eggs, which he suggested served to adjust the number of larvae in relation to the size of the host.

Cannibalism also occurs among larvae in parasitic Hymenoptera: Doutt (1973) found that in the subsocial bethylid *Prosierola bicarinata*, larvae apparently cannibalize their siblings within the host. Cannibalism among adult Hymenoptera has only been reported in the subsocial sphecid wasp *Philanthus basilaris*, within which females sometimes prey upon conspecific males (O'Neill and Evans 1981).

Males of the neotropical harvestman spider *Zygopachylus albomarginatus* build open mud nests on the rainforest floor, within which multiple females oviposit (Mora 1990). Males guard the eggs, and observations and experimental removals of guarding males show that 'the main source of egg predation was conspecific males and females', although protection from ant predation and fungal infection were also important. Mora (1990) found that when males usurped the nests of others they always ate the original eggs, and that guarding males never ate their own eggs.

In web-building spiders, the suppression of cannibalism among adults and spiderlings is considered an important prerequisite for the evolution of sociality (e.g. Rypstra 1986). Thus, high abundance of heterospecific prey may favour reduced cannibalism and allow individuals to benefit from group-living.

Nevertheless, some egg cannibalism still occurs among the more social web-building spiders.

Interpretation

In these five diverse taxa, *Nicrophorus* beetles, bethylid wasps, opilionid spiders, *Eumicrota* fungus beetles, and the thrips *Elaphrothrips tuberculatus*, one of the primary functions of egg and larva defence is to avoid cannibalism by conspecifics. In the former three taxa, usurpation of a valuable resource for young appears more important than nutrition for the cannibal, whereas in the latter two taxa, cannibalism may be simply a matter of utilizing a convenient, rich, food resource, which is possible under conditions of sufficiently high adult and offspring density. Both conditions are probably met in the opilionid *Z. albomarginatus*, which Mora (1990) reports as 'very common in patches of old forest'. Cannibalism and parental care may coincide where rich, ephemeral resources must be defended from conspecifics, and because cannibals are common predators that must be repelled.

The importance of cannibalism for parental care, relative to other selective pressures such as generalist predators, specialist parasitoids, and abiotic factors which are known to be important (e.g. Eberhard 1975; Edgerly 1987; reviewed in Eickwort 1981), is difficult to assess because few subsocial species have been studied in sufficient detail. However, the data indicate that the environments of parental care and cannibalism often correspond, and that cannibalism may often be sufficient to cause subsociality.

CANNIBALISM UNDER EUSOCIALITY

Theory

Because evolved sterility appears paradoxical to evolutionary theory, the study of the selective pressures involved in the origin, maintenance, and forms of eusociality has gained central importance for understanding the natural selection of behaviour. The analysis of cannibalism and trophic eggs in eusocial insects provides a window into the source and resolution of conflicts and confluences of social insect interests because the behaviours are simple, well-defined, and have obvious implications for the interpretation of behaviour in terms of inclusive fitness theory and the theory of parental manipulation (Alexander 1974). I consider cannibalism of viable eggs, whenever it clearly represents reproductive competition, as evidence against the importance of mutualism (Lin and Michener 1972).

Individuals should cannibalize one another whenever more distant relatives can be replaced by closer ones, or where resources can thereby be transferred to closer kin (Hamilton 1964; Eickwort 1973). Thus, under haplodiploidy and queen sperm use from one male (or two males, with unequal sperm mixing), mothers should eat their daughter's eggs, and virgin reproductive daughters should eat their sister's male and female eggs and their mother's male eggs, but not their mother's female eggs. By contrast, with queen sperm use by more than two males, daughters should prefer to rear the queen's male eggs over those of a randomly chosen sister (Starr 1984; Woyciechowski and Lomnicki 1987; Ratneiks 1988), unless they can recognize kinship and favour the eggs of full sisters over those of half-sisters (e.g. in honey bees, Page *et al.*, 1989; but not *Polistes* wasps, Queller *et al.* 1990). Hamilton (1972) and Trivers and Hare (1976) predict that workers in a eusocial colony should prefer a sex allocation ratio more female biased than that preferred by their mother. Thus, workers will prefer to cannibalize brothers and nephews as a means of diverting colony resources to daughters. Sex recognition, and perhaps kin recognition, are necessary for such predictions, based only on relatedness, to be realized. The presence of sex recognition is problematic both empirically and theoretically, since under worker control of sex allocation we might expect selection to favour male eggs or larvae to resemble female eggs and larvae, females to differ from males, mothers to make the sexes indistinguishable, and workers to be capable of sorting out the resulting confusion (Nonacs 1991).

The first part of this section discusses the presence and form of cannibalism in 'primitively-eusocial' insects – those in which female reproduction covaries with behaviour but not with distinct morphology (Peeters and Crozier 1988). In these societies, which are 'primitive' only in the sense that they probably resemble species in which eusociality originated, all females are capable of reproduction. The following part considers 'advanced-eusocial' insects – those with queens discrete in external morphology from workers. This distinction is necessary because the processes occurring in each are profoundly different: for example, no primitively-eusocial insects have trophic eggs! Kukuk (this volume) presents a more detailed discussion of cannibalism in bees, especially the Xylocopinae, Euglossini, and *Apis*, which are not considered in this chapter. The bee genera *Bombus* and *Lasioglossum* are included here to provide direct comparisons with other primitively-eusocial taxa (see Kukuk this volume, for futher descriptions). The distribution and forms of cannibalism and trophic eggs in stingless bees, wasps, ants and termites are reviewed in Sakagami (1982), Jeanne (1980), Hölldobler and Wilson (1990), and Wilson (1971), respectively. This section focuses on those groups that have not previously been considered in detail, and discusses the implications for theory.

Data

Halictine bees

Batra (1964, 1968) observed two cases of cannibalism in observation nests of the bee *Lasioglossum zephyrum*, wherein one female ate the egg of another and replaced it with her own. She saw no hostility among adults associated with the process, and suggested that such cannibalism occurs when the number of females ready to oviposit exceeds the number of provisioned cells. Michener and Brothers

(1974) observed three cases of cannibalism in this species, and in each case the queen ate a new worker-laid egg and immediately laid one of her own (see also Brothers and Michener 1974). Batra (1964) saw a two-day-old egg eaten – cannibalized eggs apparently need not always be freshly laid.

Bumblebees

Some *Bombus* have sufficiently large size differences between castes and large colonies to be considered more advanced than their 'primitively eusocial' classification warrants, but their behaviour and life cycle appear closer to other primitively-eusocial forms than to advanced ones such as *Apis* or the Meliponini. Cannibalism and associated behaviour has been studied in various *Bombus* species, and the circumstances appear similar in all (Plath 1923; Free 1955*a,b*; Free *et al.* 1969; Katayama 1974; Pomeroy 1979; Honk and Hogeweg 1981; Owen and Plowright 1982; Fisher 1987; Duchateau and Velthuis 1988; Duchateau 1989; Kukuk this volume). Before the initiation of rearing reproductives in the autumn, queens produce pheromones which have the effect of suppressing worker reproduction; workers usually resorb eggs, and if worker eggs are laid during this period the queen normally eats them and replaces them with her own. The queen begins to produce male and queen-destined eggs in mid to late summer (usually producing male eggs first), and reproductives are reared for several weeks. The social situation in colonies changes dramatically when the queen's pheromonally-mediated and behavioural dominance begins to fail in late summer. After this point, workers eat and attempt to eat queen-laid and sometimes worker-laid eggs, queens eat and attempt to eat worker-laid eggs, and all try to replace the eaten eggs with their own. Older workers are usually most aggressive, cannibalistic, and fecund (Duchateau 1989). Fighting occurs between queens and workers, and among workers as individuals try to defend cells containing their newly-laid eggs; although attacks usually appear 'restrained', queens are sometimes pushed off the comb if they do not stop interfering with worker reproduction. Only newly-laid eggs are cannibalized (Katayama 1974), and workers often readily rear the larvae from eggs that they attempted to eat a few days before. However, larval ejection (without cannibalism) occurs in some species; Röseler (1967) and Plowright and Laverty (1984) provide evidence that male larvae are preferentially ejected, but Pomeroy (1979) found no such sex bias.

Duchateau (1989) noted 'a high correlation . . . between the moment at which the production of young queens started and the moment at which workers started to lay eggs'. In some circumstances, aggression and cannibalism result in no new brood being reared, but where each individual defends its own cell (and therefore spends less time eating eggs), many males are reared to adulthood (see also Kukuk this volume).

Polistine wasps

Polistes colonies in the temperate zone are founded in spring by one or several females; the first workers emerge in late spring, and reproductives are reared in the autumn. Unlike halictine bees and bumble-bees, the *Polistes* larvae are fed directly, with macerated insect tissue. One of the females in cofoundress nests is behaviourally dominant over the others, attacking them and eating their eggs if they show signs of being reproductive (Pardi 1948; West-Eberhard 1969; Spradbery 1973; Jeanne 1980). In *P. fuscatus*, West-Eberhard (1969) found that only dominant egg-laying females ate eggs; egg-eating was always followed by oviposition, eaten eggs had always been freshly-laid by subordinate foundresses, and wasps never ate their own eggs. West-Eberhard also suggested that a female will not eat an egg unless she has one ready to lay. Gervet (1964) discovered that *P. gallicus* females antennate the first few eggs that they lay, and that they may eat one or two of these. Moreover, when females are prevented from engaging in such auto-cannibalism, they are apparently unable to discriminate between other female's eggs and their own. As in *P. fuscatus*, females will not eat eggs that are more than a few hours old (Gervet 1964). Females of *Mischocyttarus drewseni*, which has a similar life cycle to *Polistes*, were observed on four occasions to eat their own eggs just after laying – three times by the queen, and once by a subordinate female; in a fifth observed instance, a queen ate a subordinate-laid egg (Jeanne 1972). As in *P. fuscatus*, only egg-laying females ate eggs, and cannibalism was followed by oviposition, although sometimes in a different cell. In *M. drewseni*, however, some cannibalized eggs had not just been laid. Dominant foundresses of *P. annularis* eat eggs laid by subordinates until the first larvae appear (Hermann and Dirks 1975), and oophagy is always followed by oviposition by the cannibal. Similarly, in a species of *Belonogaster* wasp, large, mated, dominant females both laid most eggs and ate most eggs of smaller, unmated, subordinate females (Jeanne 1980).

In *Polistes* and the polybiine wasp genera *Protopolybia*, *Polybia*, *Synoeca*, and *Metapolybia*, females that have just oviposited guard their eggs against other colony members, and unguarded eggs are occasionally eaten by other females (West-Eberhard 1981). The duration of the guarding vigil increases as competition among developing queens in the polygynous colonies of *M. azteca* and *M. aztecoides* becomes more intense; by contrast, in monogynous colonies, the queen does not guard her eggs. In *Protopolybia sedula*, laying workers commonly eat their own eggs (Naumann 1970; Jeanne 1980), and in *P. exigua*, unmated workers apparently oviposit when young, but their ovaries regress when they begin to forage; Simoes (1977) states that in this species, egg-eating 'must occur', as few males developed in the colonies studied. Finally, Sakagami and Fukushima (1957) observed 'reciprocal thieving' of larvae and nests between two foundresses in two nests of *Polistes fedwigae*; the females, which were apparently highly food-stressed, ate each other's larvae, and ultimately both nests were abandoned.

Fletcher and Ross (1985) state that in some swarm-founding Polybiini with a high degree of caste dimorphism, 'worker-laid eggs seem to be trophic', presumably because in species such as *Protopolybia sedula* they are normally eaten.

Vespine wasps

Vespine wasps are not primitively-eusocial *per se*, because the queen and workers are morphologically distinct (reviewed in Jeanne 1980). However, they are considered here because they straddle the apparent threshold between primitive and advanced eusociality in behaviour and colony size, and they may therefore be illuminating about the causes of caste dimorphism, behavioural and pheromonal interactions between queens and workers, and cannibalism.

Vespine wasp colonies in the temperate zone are founded by single females in spring, and the colonies often develop to much larger sizes than those of halictine bees, *Bombus*, or *Polistes* (Spradbery 1973). Greene *et al.* (1976) reported cannibalism of worker-laid eggs by foundress queens of *Dolichovespula arenaria*; twice, queens ate worker-laid eggs and then laid their own. Greene (1979) stated that such oophagy, coupled with queen patrolling the comb, and chasing and 'mauling' workers, is common in both this species and *D. maculata.* The behavioural dominance by queens is probably associated with the presence of worker oviposition early in colony development, and with the small colony size and low caste dimorphism in *Dolichovespula* compared with other Vespinae (Greene 1979). In *Vespula*, which forms larger colonies and has more pronounced caste dimorphism, queens apparently do not engage in oophagy, and they express their reproductive dominance through pheromones; moreover, workers usually oviposit only in mature colonies (Greene 1979).

Archer (1981) reports that in *Paravespula vulgaris*, workers destroy pupae and late-stage larvae near the end of the season, ejecting them from the comb but not eating them. Workers apparently tend to eject queen-produced males, which are reared in large cells, and they replace them with males that they produce in small cells. Ishay (1964) observed similar ejection of larvae without cannibalism in queenless colonies of *Vespa orientalis*, in which workers defended their own small areas from other workers, and laid their own eggs there. Landolt *et al.* (1977) observed similar conflict among workers when they divided a colony of *Vespula atropilosa* in half and left the queen on one side; some workers on the queenless side developed ovaries, showed intense aggression ('mauling' each other), and destroyed each other's eggs (see also Jeanne 1980). Most aggression was perpetrated by three workers that became physogastric (highly gravid), fought over cells, and removed and replaced each other's eggs. Eggs were usually dropped rather than eaten and some of these were picked up and fed to larvae by other workers.

Interpretation

The similarities between *Lasioglossum*, *Polistes*, *Bombus*, and *Dolichovespula* in cannibalism and associated behaviour are remarkable. In all four taxa, queens eat the eggs, usually newly-laid, of workers or cofoundresses, and replace them with their own. In *Bombus*, *Polistes*, *Vespa orientalis* workers, and some polybiine wasps, females guard cells or areas of comb containing their newly-laid eggs. In *Bombus*, *Polistes*, and some vespines, workers will eject (or neglect) larvae and pupae, without eating them, toward the end of the season, and in *Bombus*, *Polistes*, and Vespinae (Spradbery 1973, pp. 244–5) colonies disintegrate in similar ways after most reproductives have been reared: workers fight among themselves (and sometimes with the principal egg-layer) to produce males of their own.

Given that queens of *Lasioglossum*, *Dolichovespula*, *Polistes*, and *Bombus* physically dominate workers, cannibalism in these taxa appears to be an expression of a queen–worker power asymmetry usually resolved in the queen's favour. Queens are apparently able to dominate workers because the only options to workers are: (1) to leave the nest and attempt to form a colony of their own (which probably has little or no chance of success, depending upon the taxon and time of year), or (2) to revolt against the queen, singly or en masse. Workers may be restrained from revolting against the queen because she is the sole producer of female workers and reproductives; thus, until very late in the colony cycle, workers have little to gain by defecting and much to lose. However, in primitively-eusocial taxa, the aggression of queens to workers is probably not so much aggression *per se*, but more a matter of the queen constantly monitoring the reproductive status of workers; workers are strongly selected to compete among themselves for queenship should the queen falter or die, and the queen is selected to maintain her status by suppressing any reproductive tendencies in workers (Alexander *et al.* 1991). Oophagy is probably an important form of reproductive suppression, because if worker-laid eggs are normally eaten by the queen, then workers will gain in inclusive fitness by directing their energies towards more productive activities.

In those taxa (e.g. *Polistes* and *Lasioglossum*) where workers can replace queens as producers of both sexes, colony members are competing primarily for queenship until the start of rearing reproductives. By contrast, in taxa where only males are produced if the queen dies (e.g. *Bombus* and *Dolichovespula*), all workers should strive to produce them should this disaster occur. Before this time, workers should not eat any of the queen's eggs (since these will develop into workers), unless they can replace them with their own male ones, which the queen will certainly contest. Near the start of rearing reproductives, the focus of competition begins to shift towards queen–worker conflict over the production of males. The timing and intensity of queen–worker conflict over male production should depend at least in part on relatednesses. Each worker would prefer to lay male eggs, but to help to rear her full (reproductive) sisters, with an investment ratio 1:3 female biased. By contrast, the queen prefers to have the workers rear females and (her) males in a 1:1 ratio (Trivers and Hare 1976). The link in *Bombus* between the start of queen production and the initiation of worker oviposition and oophagy may arise because once investment in female reproductives begins, the value of the queen to the workers declines, such that selfishness becomes a more viable option.

With queen sperm use from a single male, which is common in primitively eusocial insects (Starr 1984; Crozier *et al.* 1987; Kukuk *et al.* 1987; Ross and Matthews 1989; Strassmann *et al.* 1989), queens and workers should engage in reciprocal eating of male eggs, and workers should eat the eggs of the queen if these are replaced by either their own or those of another worker. In contrast, with queen sperm use by more than two males and no kin recognition, workers should eat all male eggs not their own, especially if one or a few workers can dominate male production. Thus, inclusive fitness theory predicts a tendency toward only queen–worker oophagy under single mating, and both queen–worker and worker–worker oophagy under multiple mating. Data from *Bombus* accord roughly with this prediction: queens of most, if not all *Bombus* species probably mate only once under natural conditions (see Morse 1982; also Kukuk this volume), and queen–worker oophagy has been reported in five species, *B. lapidarius* (Free *et al.* 1969), *B. terrestris* (Honk and Hogeweg 1981; Duchateau 1989), *B. atratus* (Michener 1974), *B. ruderatus* (Pomeroy 1979), and *B. affinis* (Fisher 1987), whereas worker–worker aggression and oophagy has been noted as common only in *B. hypocrita* (Katayama 1974). Queen–worker oophagy also occurs in *Lasioglossum zephyrum*; thus, the limited evidence from these species suggests that queen–worker oophagy is more common than worker–worker oophagy. However, this difference may in part reflect a tendency to observe queen behaviour rather than that of workers.

Oophagy among workers in queenless nests is predicted by inclusive fitness theory, since workers are always more closely related to their own male eggs than to those of nestmates. In nests with both full and half-sisters, workers might be more likely to bear the colony-level and individual-level costs associated with worker-worker aggression, whereas in nests with only full sisters, competition and social anarchy should be less extreme.

Two important questions concerning the presence and form of cannibalism and infanticide in primitively-eusocial insects remain. First, why, across diverse taxa, are only young eggs eaten, and why is oophagy nearly always followed by oviposition by the oophagist? One possibility is that potential oophagists eat only newly-laid eggs because only these are recognizable as laid by others. Thus, in *Lasioglossum*, *Polistes*, and small colonies of *Bombus*, where females do not guard newly-laid eggs, queen oviposition rates may be sufficiently low that queens could retain information about the timing and placement of their most recent oviposition. (If queens do not remember recent ovipositions, then oviposition may immediately follow oophagy because queens that have an egg ready to lay cannot have just laid one!) Gervet's (1964) studies suggest that females of *Polistes* can recognize their own eggs, and the auto-cannibalism of *Mischocyttarus*, *Protopolybia*, and a species of allodapine bee (Skaife 1953), could also be associated with learning to recognize one's own eggs, in a somewhat more costly way than bees and wasps learn to recognize nestmates (reviewed in Grafen 1990). An alternative possibility is that if eggs are laid by a relative of the oophagist, then the reproductive value of the offspring increases with time as less future investment is needed for rearing to adulthood. However, the only investment in eggs prior to hatching is their 'use' of a pollen ball or cell for that time, so this hypothesis does not strongly predict that only the youngest eggs should be eaten unless fecundities are quite low or eggs are large and costly to produce.

Second, why do no primitively-eusocial insects have trophic eggs, although such eggs are found in both subsocial and eusocial species? The cause may lie in a higher magnitude of reproductive differentials among individuals within colonies of these social forms, such that some females have opportunities to greatly out-reproduce others; as a result, all individuals are strongly selected to never forego any chance of producing viable (rather than trophic) eggs. In subsocial insects, the interests of parents and offspring usually overlap greatly, since parental fitness depends primarily on aggregate offspring success. By contrast, reproductive differentials among workers in advanced-eusocial insects are probably much less than those between workers and queens, or among workers, in primitively-eusocial species. Thus, in advanced-eusocial forms, caste dimorphism and resultant common worker interest in the queen's health and vigour as a producer of viable eggs favour situations where workers produce trophic eggs at least in part to increase colony efficiency in growth and reproduction (queen–worker reproductive differentials are usually moot in advanced eusocial forms, except as concerns male production). Partial communalities of interest, which trophic eggs apparently often reflect, are thus tempered in primitively-eusocial taxa by the chance that any individual would be queen or queenlike. This hypothesis of covariation between reproductive differentials and social systems could be investigated by determining relatedness of queens and workers to reproductive brood, preferably using precise genetic markers such as hypervariable fingerprint loci.

TROPHIC EGGS UNDER ADVANCED EUSOCIALITY

Theory

Queen–worker dimorphism reduces the reproductive options of workers: rather than strive for queenship, workers now can only help to rear relatives in cooperation with other workers, or try to produce males, in conflict with the queen and often other workers. Advanced eusocial taxa are characterized by a lack of behavioural dominance of workers by the queen; instead, queens produce pheromones that are usually considered to 'control' (i. e. prevent) worker reproduction. Such pheromones probably do not exert 'control' *per se*, but rather serve as signals to workers indicating the queen's status as a healthy prime reproductive (Alexander *et al.* 1991). The signals should be accurate indicators of vigour, and data from the ant *Solenopsis invicta* suggests that they are: in this species, there is a positive correlation between pheromone production and fecundity (Fletcher and Blum 1983). Although they are subject to their dependence upon the queen as an efficient egg-layer and usual sole producer of female eggs, and constrained by their conflicts with other workers, workers are otherwise necessarily in control of their own ability to respond to chemical signals by becoming more or less reproductive, or by being more or less inclined to lay trophic eggs or cannibalize nestmates.

Two hypotheses have thus far been proposed to explain the selective basis of trophic egg origin and maintenance among workers in ants and other advanced-eusocial forms. First, Trivers and Hare (1976) stated that 'the repeated evolution in ants of trophic eggs . . . suggests that in some groups of ants male production by workers was formerly more important than it is now, the queen having regained control of male production and forced a new function on worker-laid eggs'. This hypothesis assumes that trophic eggs serve primarily as food, and have no other adaptive significance for the egg-layer. Second, West-Eberhard (1981) suggested that trophic eggs are adaptive for their producers in part because they keep an individual's ovaries active, which would be beneficial to workers (and also to queens if the queen dies) should opportunities for producing viable eggs arise. These two hypotheses are not mutually exclusive, but the Trivers-Hare hypothesis stresses the importance of parental manipulation, whereas West-Eberhard's idea focuses on the adaptive significance of trophic eggs to workers.

The next section reviews the presence and form of trophic eggs in two advanced-eusocial insect groups, ants, and stingless bees, and evaluates these two hypotheses. The comparative and intraspecific patterns are then interpreted according to specific aspects of social systems and ecology.

Data

Ants

Trophic egg production is more taxonomically widespread and diverse in form among ants than among any other insect taxon, with the possible exception of stingless bees. Ant trophic eggs are produced in three circumstances (Table 9.1):

(1) by queens, soon after founding new colonies;
(2) by young, virgin queens before they leave the nest, and
(3) by workers in established colonies, usually with the queen present.

Colony-founding queens feed the eggs to their first larvae and workers, and where young virgin queens lay trophic eggs, they are eaten mostly by larvae, but also by workers. Worker-produced trophic eggs are eaten by larvae of all ages, workers, queens, or some combination of these stages, depending on the species. Trophic eggs appear to be important as food for larvae, especially young ones; indeed, these eggs form the main food for first-instar larvae of *Myrmica rubra* (Brian and Rigby 1978). Trophic eggs are an important source of food only for workers in *Eciton* spp., *Amonna* spp., and *Myrmecia gulosa*.

Wilson (1971) and Hölldobler and Wilson (1990) describe trophic egg morphology, which includes (a) fluid, ovarian-derived 'proctodeal nutriment', (b) eggs either smaller or larger than viable eggs, and (c) large 'omelettes' formed of multiple fused oocytes. The immediate function of trophic eggs is certainly nutrition of the fed individuals and efficient transfer of resources within the colony. For example, Brian and Rigby (1978) showed that in *Myrmica rubra*, eating of trophic eggs increases survival of larvae, and in some species trophic eggs form the main diet of the queen (e.g. in *Pogonomyrmex badius* and *Hypoclinea quadripunctata*, Wilson 1971). The proximate cause of workers producing trophic rather than reproductive eggs is the presence of either larvae (in *Temnothorax recedens*, Dejean and Passera

Table 9.1 Trophic eggs and their correlates in ants.

Taxon	Egg structure	Reprod. eggs?	Laid by?	Eaten by?	Trophallaxis?	Reference
Aphaenogaster cockerelli	Misshapen, flaccid	W	Q, large	L		Hölldobler and Wilson (1990)
Aphaenogaster subterranea		QA	W			Bruniquel (1972)
Atta laevigata	'Omelette'		FQ	L	No	Bazire-Bénazet (1970)
	Same		W	Q		
Atta cephalotes			FQ	L	No	Bazire-Bénazet (1970)
Atta rubopilosa	'Omelette'		FQ	L	No	Bazire-Bénazet (1957)
Atta sexdens						Bazire-Bénazet (1957)
Basiceros manni	Same		W	Q; L?	No	Wilson and Hölldobler (1986)
Dolichoderus quadripunctatus	Fluid	QA	W	WQLM		Torossian (1959, 1968, 1978)
Formica pergandei	Less elongated?	QA	W	QWL		Hung (1973)
Formica rufa						Weyer (1929)
Harpagoxenus americanus	Fluid	QP	slavemaker, W	slave W		Stuart (1981)
Hypoclinea quadripunctata			W	Q, large L		Torossian (1959)
Iridomyrmex humilis	Fluid		W	L	Yes	Torossian (1961, 1964); Keller (1988)
Lasius spp.	Thin shelled	QP	FQ, W	L		Taki (1987); Bourke (1988); Baroni Urbani (unpubl.)
Leptothorax recendens		QP	W			Dejean and Passera (1974)
Messor capitatus		QA	W			Delage-Darchen (1974); Delage (1968)
Messor semirufus			W	L		Baroni Urbani (unpubl.)
Myrmecia forceps	Same?	QA	W	W	No	Freeland (1958); Haskins and Haskins (1950)

Myrmecia gulosa	Same?	QA	W	QWL	No	Freeland (1958); Haskins and Haskins (1950)
Myrmecia simillima	Same?	QA	W	W	No	Crosland *et al.* (1988)
Myrmica rubra	Smaller, thin shell	QP	W	Q, small L	Yes	Brian and Rigby (1978)
Nothomyrmecia macrops						Hölldobler and Taylor (unpubl.)
Oecophylla longinoda	Same	QA	W	Q, W if QA		Hölldobler and Wilson (1983)
Oecophylla smaragdina	Same	QA	W	Q, W if QA		Hölldobler and Wilson (1983)
Paratrechina flavipes	Soft, ellipsoid		FQ	L		Taki (1987)
Pheidole pallidula		No	VQ	L, some W		Passera (1978)
Plagiolepis pygmaea	Smaller, flaccid	QA	W			Passera (1966); Passera *et al.* (1968)
Pogonomyrmex badius	Misshapen, flaccid		W	Q, large L	No	Hölldobler and Wilson (1990)
Prionopelta amabilis	Same		W	Q, L	No	Hölldobler and Wilson (1986)
Solenopsis invicta	Smaller	No	VG, FQ	W, small L		Glancy *et al.* (1973); Voss (1981); Vargo and Ross (1989)
Tapinoma erraticum	Fluid		W	mostly L		Torossian (1960)
Tetramorium caespitum						Poldi (1963)
Zacryptocerus varians	Fluid	QA	W	W	Yes	Wilson (1976)
	Smaller, soft	W	L			

QA = queen absent, QP = queen present, W = worker, Q = queen, VQ = virgin queen, FQ = founding queen, L = larvae, M = males. Reprod. eggs refers to the laying of viable eggs, and Trophallaxis refers to the presence of oral trophallaxis. If 'trophic' eggs have the same morphology as viable eggs, their viability status is probably uncertain.

1974), the egg cluster (in *M. rubra*, Brian 1979) or the queen herself (in *Plagiolepis pygmaea*, Passera 1969, and in *Oecophylla* spp., Hölldobler and Wilson 1990).

Trivers and Hare's (1976) idea for the evolution of trophic eggs in ants supposes an evolutionary transition, driven by the queen, from viable worker-produced male eggs to inviable eggs. Such a transition seems plausible in the early evolution of ants, since their remote ancestors were probably primitively-eusocial forms, within which (in extant forms) worker production of males is common. However, it is unclear how the queen could 'force' recalcitrant workers, presumably via pheromones or aggression and oophagy in small colonies, to sterilize their eggs against their wishes. Instead, trophic egg production may have been favoured in workers of advanced-eusocial ants as an efficient means of investing in sibs (mainly sisters) while the queen is alive. If one or a few workers cannot monopolize male production, then the transition from viable to trophic eggs may be easier under multiple mating than single mating, because under multiple mating workers always gain by preventing other workers from producing viable male offspring.

West-Eberhard's (1981) hypothesis for the adaptive significance of trophic eggs, that workers producing them thereby keep their ovaries active to speed the transition to viable eggs, makes two clear predictions. First, the hypothesis would be weakened by the discovery that workers (or one particular worker caste) of any species laid only trophic eggs and no viable ones. The data in Table 9.1 indicate that in all species of ant in which workers lay trophic eggs, except *Iridomyrmex humilis* (Keller 1988), workers also lay viable eggs with the queen either present (four species) or absent (10 species). However, this finding may in part reflect the paucity of ant species in which workers are fully sterile (Hölldobler and Wilson 1990). In *Solenopsis* and *Pheidole*, two genera in which workers apparently never lay viable (or trophic) eggs, trophic eggs are laid either only by queens during colony founding (in *Pheidole*), or both during founding and by virgin queens in polygynous colonies (in *Solenopsis*) (Vargo and Ross 1989).

Second, West-Eberhard's (1981) hypothesis would be supported if those individuals can switch quickly from producing viable to trophic eggs, and if individuals that laid more trophic eggs were able to switch more readily. In *Plagiolepis pygmaea*, workers lay only trophic eggs in queenright colonies, and a mixture of (primarily) reproductive and (some) trophic eggs after the queen is removed (Passera 1984). The production of any trophic eggs in queenless colonies is contrary to West-Eberhard's hypothesis, unless workers of these species were asynchronously switching to laying only male eggs, or unless trophic eggs are necessary or important to larval survival. Hung (1973) also noted that in *Formica pergandei*, workers can lay reproductive eggs immediately after laying trophic eggs, and a similar facility for switching between the two is shown by founding queens of *Atta laevigata*, (Bazire-Bnazet 1970), *Lasius* spp., and *Paratrechina flavipes* (Taki 1987), and virgin queens of *S. invicta* (Voss *et al.* 1988). Evidence against West-Eberhard's hypothesis comes from *Oecophylla* spp. and *M. rubra*, in which workers in queenright colonies produce trophic eggs, but reproductive male eggs are not produced until a substantial time has elapsed following queen death

(Hölldobler and Wilson 1983; Brian 1979). At least in *Oecophylla* spp., the diffuse and scattered structure of colonies in tree foliage (Hölldobler and Wilson 1983) may slow the transition from trophic to viable eggs in workers.

In some cases trophic eggs are simply 'immature' versions of viable eggs (see Sakagami 1982 on stingless bee trophic eggs), which is what one might expect if trophic eggs have an individually-selfish function as well as a kin-selected or mutualistic one. West-Eberhard's hypothesis could be investigated more directly by testing for a correlation, among workers in a colony, between rate of trophic egg production and duration of the lag between removal of the queen and production of the first male eggs, and by comparing the duration of the lag between species with and without worker-laid trophic eggs.

A third hypothesis for the significance of trophic eggs in ants is ecological rather than social: in some species, the habitat, food type, or social system may favour food transfer within a colony by either trophic eggs, direct transfer of prey items, or oral trophallaxis. Indeed, Hölldobler and Wilson (1990) noted an inverse relationship, among species, between the importance of trophic eggs and trophallaxis (and Michener 1974 independently noted the same relationship among bees!). Trophic eggs may be more prevalent than trophallaxis under four conditions:

(1) where workers cannot feed prey items directly to larvae (or larvae cannot eat them) easily;

(2) where high queen egg-laying capacity requires that she be fed large amounts of protein-rich food (e.g. in *Atta*);

(3) where prey items cannot be stored (as trophic eggs often can be, for more than six weeks in *Solenopsis invicta*, Voss and Blum 1988), and where colony food intake rates are highly variable; and

(4) where high brood-to-worker ratios favour specializations for feeding larvae.

Alternatively, trophallaxis may be more widespread among species in four other situations:

(1) where direct food transfer among all or most workers is important, perhaps as a result of morphologically-specialized foraging and intranidal castes;

(2) where a decentralized food economy is favoured, perhaps because of habitat structure;

(3) where it is important for the colony to transfer information about its overall nutritional status among colony members; and

(4) where a food exchange mechanism that is continuously-variable with regard

to amount of nutrient transferred is better than discrete, fixed food packages (eggs), perhaps because different types of individual require very different amounts of food (D. Gordon personal communication).

At least two of these claims and predictions are supported by the data in Table 9.1. First, as suggested by Hölldobler and Wilson (1990), trophallaxis is absent in many of the species in which trophic eggs are important, even though trophallaxis is otherwise nearly universal in ants. Second, some of the species with trophic eggs feed substantially on arthropods (e.g. *Basiceros*, *Myrmecia*, *Myrmica*, and *Prionopelta*), which are probably less easily stored than other food, and may be less predictable in abundance. Further study of the respective causes of ant trophallaxis and trophic eggs as food exchange should involve study of closely-related species that differ in the importance of each (e.g. in *Myrmecia* species), or species in which both occur and where their frequencies vary over time.

How might trophic eggs originate in ants? Imagine a situation where workers occasionally lay eggs, because these eggs have a small chance of developing to adulthood; most, however, are taken by other workers or, perhaps, the queen. The use of this food resource increases colony reproduction, perhaps via more efficient transfer of food to larvae. This benefit can be enhanced through evolutionary change in morphology and content of the eggs, such as the increase in soluble protein content in *Pheidole pallidula* (Passera 1984), the increase in lipid content in *M.rubra* (Brian 1979), or the flaccid, weak eggs shells found in many species (Hölldobler and Wilson 1990). Egg sterility may then arise as an incidental by-product of such egg specialization. This hypothesis could be investigated by feeding trophic and viable eggs to larvae and measuring growth rates (with appropriate controls for calorific content, etc.), by studying the mechanism of egg sterility, and by determining the fate of sporadically-produced viable eggs in queenright colonies.

Stingless bees

Stingless bees have large, long-lived colonies, mass provisioning of cells, pronounced queen–worker dimorphism, single queens (except in *Melipona nigra schencki*), and apparently, single mating by queens in most if not all species (reviewed in Sakagami 1982). Most species also have worker-produced eggs, some of them clearly trophic, which may be larger, smaller, or equal in size to those of the queen, depending upon the species. These eggs are laid by workers either near the margin of a provisioned cell, on the larval food, or away from the cell, and they are usually devoured by the queen. In some species, the queen derives most of her nutrition from eating trophic eggs and larval food, and oral trophallaxis appears to be rare among Meliponini (Sakagami 1982; see also Kukuk this volume). Moreover, in some species, workers also produce many (or all) viable male offspring.

The most fascinating aspect of behaviour in stingless bees, and perhaps in all social insects, is their 'oviposition rituals', in which worker egg production and oophagy play prominant roles. These rituals involve highly complex, stereotyped, species-typical queen-worker interactions in four main phases:

(1) attraction of the workers and queen to growing and completed cells (or attraction of the queen to the cell and workers),

(2) discharge of larval food by workers, which includes 'escape' of workers from the queen, cell or both after discharging food, and inspection of the cell by queens and workers,

(3) in many species, oviposition by a worker, oophagy by the queen, and queen oviposition, and

(4) cell closure (Sakagami *et al.*, 1964, 1965, 1973, 1977; Sakagami and Zucchi 1968, 1974).

Permutations of this general pattern, including cases of worker oviposition following that of the queen and worker oviposition followed by queen or worker oophagy apart from the oviposition process, are summarized by Sakagami (1982). The workers taking part in the oviposition ritual have been attracted by the cell and queen and are usually young bees with developed ovaries. Workers usually avoid the queen as much as possible in performing their acts; and in some species, the queen violently 'taps' workers attending or inspecting cells, and workers sometimes 'dart' sharply at the queen, although neither workers nor queens are apparently harmed by such contacts.

What might be the adaptive significance of stingless bee oviposition rituals? Sakagami *et al.* (1973) state that 'the core of the complicated oviposition process in this group is a conflict in workers between fear of the queen and defense of the cell against her'. This conflict may be produced by a combination of monogyny, large colonies, and mass provisioning, which causes queens to 'wait' at cells as they are provisioned, and compells queen–worker interaction when eggs are laid. Workers may 'fear' (and avoid) the queen because she is the source of reproductive suppression (apparently affecting workers as they age), which any worker would rather wish upon other workers than itself. Workers may seek to defend cells because of queen–worker conflict over male production, which is resolved differently across and within species of stingless bee; among colonies of the same species, and among species, males are produced by either workers or queens or both (Sakagami 1982; Engels and Imperatriz-Fonseca 1990). The oviposition rituals apparently represent either:

(1) multiple more or less evolutionarily stable points of behavioural interaction reflecting various outcomes of conflict between queen and workers, and among workers (West-Eberhard 1981; Engels and Imperatriz-Fonseca 1990);

(2) forms of communication between workers and the queen primarily with regard to bringing the queen to a completed cell ready for provisioning and having her oviposit after provisioning is complete (Sommeijer and de Bruijn 1984); or

(3) both queen-worker conflict and communication.

But what evidence supports these hypothesis, and what are the causes of the marked behavioural diversity?

I will consider behavioural variation at the level of genera and subgenera, with Michener's (1990) caveat that 'sometimes closely related forms differ considerably in behavior' tempered by the clear similarities among multiple species studied within *Plebeia*, *Melipona*, *Scaptotrigona*, and *Trigona* (*Tetragonula*) (Sakagami 1982; Sakagami and Inoue 1990), and by the tendency for among-genus and subgenus variation to exceed that among species (Sakagami personal communication). I use Michener's (1990) classification and terminology. In four groups, *Trigonisca* (*Leurotrigona*), *Trigona* (*Duckeola*), *T.* (*Frieseomelitta*), and *T.* (*Tetragonula*), workers apparently do not oviposit in queenright colonies, whereas in all other groups studied worker oviposition is routine and, in some species, it leads to a high frequency of worker-produced males (e.g. *Scaptotrigona postica*, Beig 1972).

The pattern of differences in behaviour between taxa with and without worker oviposition in queenright colonies should indicate the relationship between oviposition rituals and queen–worker conflict over male production, since there is no such expressed conflict in the latter. Little behavioural information is available for *Trigona* (*Frieseomelitta*). For *Trigonisca* (*Leurotrigona*), Sakagami and Zucchi (1974) report that 'this group is noteworthy by the simplicity of behavior and the absence of high excitement during [the] oviposition process. There [are] found no exaggerated behavior patterns'. In particular, *Trigonisca* (*Leurotrigona*) exhibits a scarcity of worker body insertions and queen inspections, a lack of queen 'beating' of workers at cells, and a lack of overt avoidance of the queen by workers (Sakagami and Zucchi 1974). Similarly, in *Trigona* (*Duckeola*), oviposition behaviour is 'less exaggerated than in the other species studied' (Sakagami and Zucchi 1968), perhaps in association with the 'generally sluggish disposition' of this subgenus (Sakagami personal communication). In *Trigonisca* (*Leurotrigona*), queen cell inspections (which could be associated with worker oviposition) are rare, relative to *Scaptotrigona*, *Melipona*, *Cephalotrigona*, and *Trigona* (*Tetragona*), and certain queen-worker interactions during non-oviposition periods, such as stereotyped worker 'turning' in response to the queen, are absent (Sakagami and Zucchi 1974). Moreover, in *Trigonisca* (*Leurotrigona*) and *Trigona* (*Duckeola*), in contrast to the four genera above, workers are 'not very excited' during the 'arousal' phase just prior to provisioning. Finally, four species of *Trigona* (*Tetragonula*) show 'simple queen–worker interactions'. . .'few cell attendants and low preprovisioning activities', and 'rare cell inspections by the queen' (Sakagami and

Inoue 1990). Although patterns of cause and effect and the functions of many behaviours are uncertain, the general simplicity of oviposition rituals in groups without worker oviposition in queenright colonies suggests that ritualized queen–worker interactions are associated with conflict over the production of male eggs. Observations contrary to this hypothesis include the presence of worker 'defence' of cells from the queen in *T.* (*Frieseomelitta*), and the nearly complete lack of cell inspection in *Plebeia* sensu strictu (Sakagami personal communication).

Direct evidence for queen-worker conflict comes from observation of *Melipona scutellaris trinitatis* by Sommeijer *et al.* (1984), who reported that once when the queen was absent from a cell during provisioning, a worker oviposited and began to close the cell. Just before closure, the queen arrived, opened the cell, ate the worker-produced eggs, and attempted to lay one of her own. In accordance with these observations, a plausible function for the frequent cell inspections after provisioning by both queen and workers is checking for the presence of eggs, and queen 'beating' of workers may strongly signal her presence and vigour, and thus help to behaviourally prevent worker oviposition. Clear aggression between queens and workers has been described for *Scaptotrigona postica*, and Engels and Imperatriz-Fonseca (1990, p. 200–1) interpret this behaviour as 'rudimentary defence of the worker's egg'. Ritualization and simplified 'intentionality' of these and other acts probably benefits both queens and workers through making the processes less demanding, but some degree of overt conflict is always expected, lest 'cheating' arise and begin to spread.

Trophic eggs certainly function to feed the queen, although they may also serve to keep worker ovaries active (West-Eberhard 1981). However, in species like *Scaptotrigona postica*, one of two forms of worker-laid egg may be produced to 'deceive' the queen. In this species workers lay two types of egg: large, inviable ones, laid on the upper part of the cells and eaten by the queen, and eggs similar in size to those of the queen, which are laid on the larval food after queen oviposition (Beig 1972; Engels and Imperatriz-Fonseca 1990). In cells with both queen and worker eggs, the worker-produced male larva kills its cell-mate within a few days of hatching (whether or not ingestion occurs is as yet unknown). Such egg dimorphism and killing may also occur in other genera where worker oviposition follows queen oviposition (Sakagami 1982; Kukuk this volume). Moreover, in *Melipona*, *Meliponula*, *Plebeia* (*Scaura*), *Plebeia* (*Schwarziana*), and *Scaptotrigona*, trophic eggs are laid after food provisioning and may therefore develop into (murderous) males (Sakagami 1982), the workers exhibit a much more pronounced 'darting' at the queen, with their mandibles open, than in other genera (Sakagami *et al.* 1977).

Sommeijer and de Bruijn (1984) believe that certain aspects of oviposition rituals function as communication from the workers to the queen, designed to prompt the queen to approach a completed cell with workers in attendance. This hypothesis is supported by the observation that 'court acts are initiated by courting workers', and that such acts are immediately followed by queen arrival at the cell. However, if queen prompting is the primary or only function of the oviposition rituals, then it is difficult to explain the marked diversity in behaviour among species, and in particular the differences between species with and without worker oviposition in queenright colonies. Thus, as Sommeijer and de Bruijn (1984) indeed contend (see also Sommeijer *et al.* 1984), oviposition rituals probably represent both communication, adaptive for both queens and workers, and ritualized queen-worker (and, perhaps, worker-worker) conflict, resolved in various ways in different species.

What might be some of the proximate causes of the pronounced among-genus and among-species variability in oviposition rituals and their apparent relation to male production? Sommeijer *et al.* (1984) suggest that the 'rhythmicity' of oviposition in queens of some stingless bees is an adaptation enhancing the queen's fitness. Thus, they suggest that queens group ovipositions temporally in order to avoid situations, such as sometimes happen in *Melipona* (see above), where workers would 'wait' over-long for the queen at a completed cell, provision it without her, and lay their own eggs. Ovipositions are facultatively or always grouped in 18 of 20 genera and subgenera described in Sakagami (1982); the only exceptions are *Melipona* and *Meliponula*, both of which show worker oviposition on the larval food after provisioning, which may indicate high levels of worker male production. *Trigona* (*Duckeola*) and *Trigonisca* (*Leurotrigona*) exhibit a combination of 'semisynchronous' cell construction (starting out partially synchronous, and becoming more synchronous near completion) with exclusively grouped ovipositions, whereas *T.* (*Frieseomelitta*) and *T.* (*Tetragonula*) have successive cell construction and facultatively grouped ovipositions (Sakagami 1982). However, one of these two combinations is found in all but three of the 17 groups with worker oviposition in queenright colonies. Thus, it appears that throughout the stingless bees, when workers have synchronized cell construction, queens always group ovipositions (seven groups), but that when cell construction is successive, queens group ovipositions facultatively (nine groups), or, rarely, always group ovipositions (*Scaptotrigona*), or oviposit singularly (*Melipona* and *Meliponula*). Interpretation of this variation is problematic without additional data on the prevalence and mechanisms of worker oviposition leading to male production.

How might oviposition rituals and trophic eggs have evolved in stingless bees? From a cladistic analysis of morphology and behaviour, Michener (1990) concluded that 'in all of the characters . . . related to oviposition, *Melipona* has the presumably plesiomorphic [ancestral] behavior'. Such ancestral behaviour includes:

(1) successive cell construction;

(2) singular queen ovipositions;

(3) worker oviposition directly onto the food mass after provisioning; and

(4) oophagy by the queen immediately after worker oviposition (Sakagami and Oniki 1963; Sakagami *et al.* 1965).

These character states are strikingly similar to the behaviour of many primitively-eusocial insects such as *Bombus* (see previous section), in which cells are provisioned one by one, workers often oviposit, and the queen eats newly-laid worker-produced eggs and replaces them with her own.

Mapping of two characters–(1) presence and absence of worker oviposition in queenright colonies and (2) the site of worker oviposition, on the larval food, on the cell margin, or on the comb, and separated from the oviposition ritual–onto Michener's (1990) cladogram leads to four inferences (Fig. 9.1). First, at least one, and probably more than two, transitions occurred from worker oviposition on the food to worker oviposition when confronted by the queen out of the oviposition process. In groups with this latter pattern (*Lestrimelitta*, *Hypotrigona*, and some *Plebeia*), worker oviposition may be an expression of queen dominance (and worker submission), that follows from worker dependence on the queen as the producer of female reproductives and workers coupled with conflict over male production. Second, assuming that worker oviposition on the larval food was retained in the main Meliponini branches leading to *Scaptotrigona* and *Trigona*, there have been three or more transitions from worker oviposition on the larval food to laying on the cell margin. Third, worker oviposition in queenright colonies has been lost in at least one clade, leading to the *Trigona* subgenera *Duckeola* and *Frieseomelitta*, and this transition may have been preceded by a change from worker oviposition on the larval food to oviposition on the cell margin. Fourth,

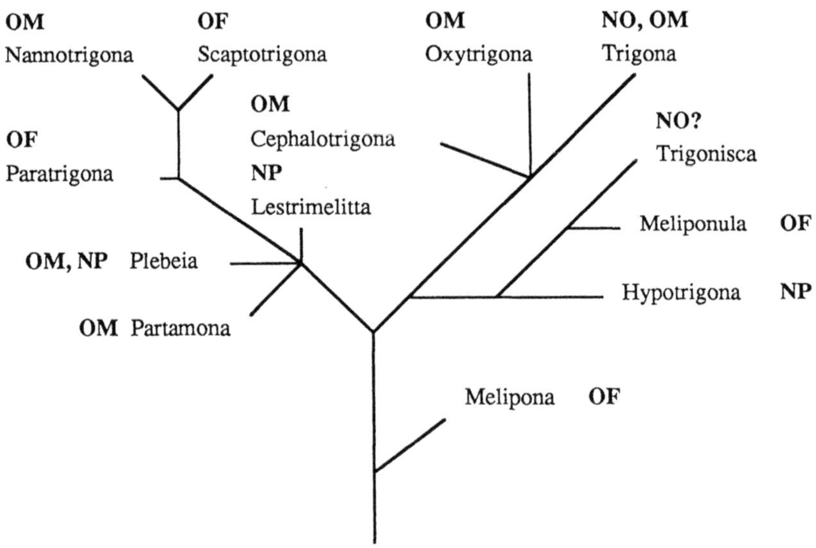

Fig. 9.1 Michener's (1990) phylogeny of stingless bee genera, showing only the groups for which data are available on the presence and site of worker oviposition (from Sakagami 1982). NO: no worker oviposition, NP: oviposition out of the queen oviposition process, OF: on larval food, OM: on cell margin.

worker egg dimorphism, and worker-produced male larvae killing of cell co-inhabitants, has apparently evolved in *Scaptotrigona* (and perhaps other genera) directly from the ancestral state of worker oviposition on the larval food.

The diversity and apparent lability of these inferred transitions support West-Eberhard's (1981) view that queen–worker relations in stingless bees involve strong social selection that promotes behavioural diversification by unending evolutionary 'races'. This hypothesis could be tested further by elucidation of the functions and effects of specific queen–worker interactions. The high prevalence of worker oviposition in Meliponini, relative to other advanced-eusocial forms, may be caused in part by queen monogamy, because it leads to worker genetic preference for rearing of other worker's male eggs over those of the queen, and tends to reduce 'policing' (Ratneiks 1988) by other than the queen, which she does primarily in the context of the oviposition ritual. Thus, once workers lay eggs frequently, the circumstances are set for queen oophagy that may sometimes benefit both the queen and other workers.

Perhaps the most bizarre form of cannibalism in animals occurs in the stingless bee *Lestrimelitta limao* (Sakagami 1982), in which workers that are confronted with the queen lay an egg and then eat it in front of her! This behaviour may represent one way in which submission can be tempered by selfishness, when the interests of mutually-dependent parties conflict.

Interpretation

In advanced eusocial taxa, inviable trophic eggs may have evolved from two states: worker oviposition of viable eggs in queenright colonies, or a general lack of worker oviposition. The former situation appears more likely, because it does not require the more or less simultaneous origin of both worker oviposition and egg inviability. Unlike subsocial insects, where eggs are likely to be under direct selection for inviability, inviability of worker eggs in advanced eusocial forms probably arises as an incidental byproduct of egg specialization, because workers should be under strong selection to retain viability of eggs unless there are some strong overriding advantages to trophic eggs in terms of their usefulness in the social economy. (Some ants may, in fact, now be 'trapped' into trophic egg production because these eggs are the sole food of young larvae.) The hypothesis above predicts that trophic eggs in advanced eusocial groups will be inviable because of some aspect of their nutritive function, but that in subsocial forms, inviability will come more from direct adjustment of mechanisms affecting egg viability. Indeed, trophism of eggs may evolve more readily under diploidy than haplodiploidy, because mothers in diploid taxa can simply not fertilize eggs and have them fail to develop.

For ants, Trivers and Hare's (1976) hypotheses for the origin of trophic eggs seems unlikely to be true, but West-Eberhard's (1981) hypothesis for their maintenance is in general supported by the data. In such a highly-derived group, comparative patterns and plausible evolutionary transitions are difficult to discern, because more primitive forms (and thus useful outgroups) are much more distant in behaviour and reproductive strategies than in other eusocial taxa. To study the evolution of ant trophic eggs, perhaps the most useful current approach is to compare the socio-ecologies of ant species that use varying degrees of trophallaxis and trophic eggs, to determine when and why each is favoured.

West-Eberhard's (1981) hypothesis of queen–worker conflict appears to explain, at least on a broad scale, the source of stingless bee oviposition rituals. Sommeijer and de Bruijn's (1984) hypothesis, that the oviposition ritual is a form of communication between the queen and workers, is compatible with West-Eberhard's idea but cannot explain the extent of behavioural diversity. The ultimate cause of among-species variation in oviposition rituals probably lies in the interactions of aspects of ecology, demography, and morphology with phylogeny. For example, in species with large colonies, which have high queen oviposition rates, 'ovipositions are always grouped in time' (Sommeijer *et al.* 1984). The causal links of ecology, morphology, and demography with behaviour are as yet unclear, and their elucidation requires comparison of closely-related species, with known phylogenies, in which the species differ in only a few of the interesting behavioural traits (e.g. Sakagami 1982 on *Plebeia* spp.). Further studies of worker oviposition in stingless bees must seek to connect behavioural variation within and between species with variation in colony genetic structure, to determine how the ritualized (and sometimes deadly) queen–worker conflicts translate into reproductive variation.

CONCLUSIONS

Since trophic eggs may usually or always derive from cannibalism of viable eggs, their evolution typifies the transitions from competition to varying degrees of cooperation. Under subsociality and advanced eusociality, dependencies of offspring on parents, and between queens, workers, and offspring, have probably facilitated trophic egg evolution. Thus, the sterility and trophic function evolves, either directly or indirectly, as a result of parental manipulation, inclusive fitness effects, or both; under only mutualism (reciprocity), 'voluntary' egg sterilization is unlikely to evolve because of the prevalence of potential 'cheaters' and constant strong selection on mothers for egg viability. In subsocial insects, trophic eggs are expected more under monogamy than multiple mating, but the reverse may be true for ants, since in these forms the workers (offspring) have more of an influence in determining egg fate and form than do their mothers. The hypothesis of multiple mating favouring trophic eggs in ants could be tested using the observed range of queen mating frequencies (e.g. Starr 1984). Among stingless bees, all of which apparently have single mating of queens, queen–worker interactions, rather than worker–worker interactions, are central to both oviposition and trophic egg production, so 'policing' effects may be relatively slight. The queen–worker conflict over queenship and male production that follows from inclusive fitness theory are the probable ultimate causes of most of the cannibalism and trophic eggs in eusocial taxa, but parental manipulation and behavioural peculiarities in any given taxon, such as mass provisioning in stingless bees, probably help determine the outcome of conflicts.

What are the evolutionary and ecological implications or effects of cannibalism and trophic eggs in subsocial and eusocial insects? Under subsociality, sibling cannibalism may provide strong selection for decreased egg developmental time, or it may favour monogamy (or high sperm precedence) in mothers. In many subsocial taxa, cannibalism has apparently led to parental care. Under primitive eusociality, cannibalism may reduce population size (e.g. in *Bombus* and other taxa with social breakdown in the fall), or affect sex allocation ratios, and in advanced eusocial taxa, cannibalism could also be used to adjust caste ratios (Flanders 1970), or sex allocation ratios (perhaps leading to single-sex broods in ants), and it has probably caused the remarkable rituals of stingless bee oviposition. Cannibalism may also favour segregation of eggs from larvae (which may also have ergonomic advantages) in ants (Hölldobler and Wilson 1990) and allodapine bees (Michener 1974), since the interests of larvae and workers probably conflict with regard to larval oophagy. Trophic eggs may also provide a more flexible food ecomony, which may help colonies survive unfavourable periods, since eggs can be stored.

ACKNOWLEDGEMENTS

I am grateful to R. D. Alexander, C. Baroni-Urbani, M. V. Brian, A. Bourke, R. H. Crozier, G. C. Eickwort, D. Gordon, A. Grafen, P. Jaisson, L. Keller, P. Kukuk, C. D. Michener, N. E. Pierce, M. J. Sommeijer, B. Wcislo, M. J. West-Eberhard, and E. O. Wilson for helpful comments and discussion. I am especially grateful to S. Sakagami, for his comments on stingless bee oviposition rituals.

REFERENCES

Alexander, R. D. (1974). The evolution of social behavior. *Annual Review of Ecology and Systematics*, **5**, 325–83.

Alexander, R. D., Noonan, K., and Crespi, B. J. (1991). The evolution of eusociality. In *The biology of the naked mole rat*, (ed. P. W. Sherman, J. Jarvis, and R. D. Alexander), pp. 3–44, Princeton University Press.

Aoki, S. (1977). *Colophina clematis* (Homoptera, Pemphigidae), an aphid species with 'soldiers'. *Kontyû*, **45**, 276–82.

Aoki, S. (1987). Evolution of sterile soldiers in aphids. In *Animal societies: theories and facts* (ed. Y. Itô, J. L. Brown, and J. Kikkawa), pp. 53–65. Japan Science Society Press, Tokyo.

Archer, M. E. (1981). Successful and unsuccessful development of colonies of *Vespula vulgaris* (Linn.) (Hymenoptera: Vespidae). *Ecological Entomology*, **6**, 1–10.

Ashe, J. S. (1987). Egg chamber production, egg protection and clutch size among fungivorous beetles of the genus *Eumicrota* (Coleoptera: Staphylinidae) and their evolutionary implications. *Zoological Journal of the Linnean Society*, **90**, 255–73.

Bartlett, J. (1987). Filial cannibalism in burying beetles. *Behavioral Ecology and Sociobiology*, **21**, 179–83.

Batra, S. W. T. (1964). Behavior of the social bee, *Lasioglossum zephyrum*, within the nest (Hymenoptera: Halictidae). *Insectes Sociaux*, **11**, 159–86.

Batra, S. W. T. (1968). Behavior of some social and solitary halictine bees within their nests: a comparative study. *Journal of the Kansas Entomological Society*, **41**, 120–33.

Bazire-Bénazet, M. (1957). Sur la formation de l'oeuf alimentaire chez *Atta sexdens rubropilosa*, Forel, 1908. (Hym. Formicidae). *Comptes Rendus de l'Académie des Sciences, Paris*, **244**, 1277–80.

Bazire-Bénazet, M. (1970). La ponte des ouvrières d'*Atta laevigata* Fred. Smith 1858 (Hym. Form.). *Comptes Rendus de l'Académie des Sciences, Paris*, **270**, 1614–15.

Beig, D. (1972). The production of males in queenright colonies of *Trigona* (*Scaptotrigona*) *postica*. *Journal of Apicultural Research*, **11**, 33–9.

Boldyrev, M. I. and Wilde, W. H. A. (1969). Food seeking and survival in predaceous coccinellid larvae. *Canadian Entomologist*, **101**, 1218–22.

Bourke, A. F. G. (1988). Worker reproduction in the higher eusocial Hymenoptera. *Quarterly Review of Biology*, **63**, 291–311.

Brian, M. V. and Rigby, C. (1978). The trophic eggs of *Myrmica rubra* L. (Hym., Formicidae). *Insectes Sociaux*, **25**, 89–110.

Brian, M. V. (1979). Caste differentiation and division of labor. In *Social insects*, Vol.1, (ed. H. R. Hermann), pp. 121–222. Academic Press, New York.

Brothers, D. J. and Michener, C. D. (1974). Interactions in colonies of primitively social bees. III. Ethometry of division of labor in *Lasioglossum zephyrum* (Hymenoptera: Halictidae). *Journal of Comparative Physiology*, **90**, 129–68.

Brown, H. D. (1972). The behaviour of newly hatched coccinellid larvae (Coleoptera: Coccinellidae). *Journal of the Entomological Society of South Africa*, **35**, 149–57.

Bruniquel, S. (1972). La ponte de la Fourmi *Aphaenogaster subterranea* (Latr.): oeufs reproducteurs–oeufs alimentaires. *Comptes Rendus de l'Académie des Sciences, Paris*, **275**, 397–9.

Chew, F. S. and Robbins, R. K. (1984). Egg-laying in butterflies. In *The biology of butterflies*, XI Symp. Royal Ent. Soc. Lond. (ed. R. I. Vane-Wright and P. R. Ackery), pp. 65–80. Academic Press, London.

Choudhuri, D. and Bagh, R. (1974). On the sub-social behavior and cannibalism in *Schizodactylus monstrosus* (Orthoptera: Schizodactylidae). *Revue d'Ecologie et de Biologie du Sol*, **1**, 569–73.

Courtney, S. P. (1986). The ecology of pierid butterflies: dynamics and interactions. *Advances in Ecology Research*, **15**, 51–131.

Crespi, B. J. (1987). Behavioral ecology of mycophagous Thysanoptera. Ph. D. Thesis. University of Michigan, Ann Arbor.

Crespi, B. J. (1988). Sex ratio selection in a bivoltine thrips. I. Conditional sex ratio manipulation and fitness variation. *Evolution*, **42**, 1199–211.

Crespi, B. J. (1989). Facultative viviparity in a thrips. *Nature*, **337**, 357–8.

Crespi, B. J. (1990). Subsociality and female reproductive success in a mycophagous thrips: an observational and experimental analysis. *Journal of Insect Behavior*, **3**, 61–74.

Crosland, M. W. J., Crozier, R. H., and Jefferson, E. (1988). Aspects of the biology of the primitive ant genus *Myrmecia* F. (Hymenoptera: Formicidae). *Journal of the Australian Entomological Society*, **27**, 305–9.

Crozier, R. H., Smith, B. H., and Crozier, Y.-C. (1987). Relatedness and population structure of the primitively eusocial bee *Lasioglossum zephyrum* (Hymenoptera: Halictidae) in Kansas. *Evolution*, **41**, 902–10.

Cruz, Y. P. (1981). A sterile defender morph in a polyembryonic hymenopterous parasite. *Nature*, **294**, 446–7.

Dejean, A. and Passera, L. (1974). Ponte des ouvrières et inhibition royale chez la fourmi *Temnothorax recedens* (Nyl.) (Formicidae, Myrmicinae). *Insectes Sociaux*, **21**, 343–56.

Delage, B. (1968). Recherches sur les fourmis moissonneuses du bassin aquitain: écologie et biologie. *Bulletin Biologique de la France et de la Belgique*, **102**, 315–67.

Delage-Darchen, B. (1974). Polymorphismus in der Ameisengattung *Messor* und ein Vergleich mit *Pheidole*. In *Sozialpolymorphismus bei Insekten*, (ed. G. H. Schmidt) pp. 590–603. Wiss. Verlagsges, Stuttgart.

Dickinson, J. L. (1988). Determinants of paternity in the milkweed leaf beetle. *Behavioral Ecology and Sociobiology*, **23**, 9–19.

Doutt, R. L. (1973). Maternal care of immature progeny in parasitoids. *Annals of the Entomological Society of America*, **66**, 486–7.

Duchateau, M. J. (1989). The regulation of colony development in the bumble-bee, *Bombus terrestris*. Ph. D. Thesis, Rijksuniversitiet te Utrecht, Netherlands.

Duchateau, M. J. and Velthuis, H. H. W. (1988). Development and reproductive strategies in *Bombus terrestris* colonies. *Behaviour*, **107**, 186–207.

Eberhard, W. G. (1975). The ecology and behaviour of a subsocial pentatomid bug and two scelionid wasps: strategy and counterstrategy in a host and its parasites. *Smithsonian Contributions to Zoology*, **205**, 1–39.

Edgerly, J. S. (1987). Maternal behaviour of a webspinner. *Ecological Entomology*, **12**, 1–11.

Eickwort, G. C. (1981). Presocial insects. In *Social Insects*, Vol. 2, (ed. H. R. Hermann), pp. 199–280. Academic Press, New York.

Eickwort, K. R. (1973). Cannibalism and kin selection in *Labidomera clivicollis* (Coleoptera: Chrysomelidae). *American Naturalist*, **107**, 452–3.

Eisenberg, R. M. and Hurd, L. E. (1977). An ecological study of the emergence characteristics for egg cases of the Chinese mantis (*Tenodera ardifolia sinensis* Suassure). *American Midland Naturalist*, **97**, 478–82.

Engels, W. and Imperatriz-Fonseca, V. L. (1990). Caste development, reproductive strategies, and control of fertility in honey bees and stingless bees. In *Social insects. An evolutionary approach to castes and reproduction*, (ed. W. Engels), pp. 167–230. Springer-Verlag, Berlin.

Fisher, R. M. (1987). Queen-worker conflict and social parasitism in bumble bees (Hymenoptera: Apidae). *Animal Behaviour*, **35**, 1026–36.

Flanders, S. E. (1970). Cannibalistic infanticide in social Hymenoptera related to adult caste ratios. *Canadian Entomologist*, **102**, 898–905.

Fletcher, D. J. C. and Blum, M. S. (1983). Regulation of queen number by workers in colonies of social insects. *Science*, **219**, 312–14.

Fletcher, D. J. C and Ross, K. G. (1985). Regulation of reproduction in eusocial Hymenoptera. *Annual Review of Entomology*, **30**, 319–43.

Free, J. B. (1955*a*). Queen production in colonies of bumble-bees. *Proceedings of the Royal Entomological Society of London*, **30**, 19–25.

Free, J. B. (1955*b*). The behaviour of egg-laying workers of bumble-bee colonies. *British Journal of Animal Behaviour*, **3**, 147–53.

Free, J. B, Weinberg, I., and Whiten, A. (1969). The egg-eating behaviour of *Bombus lapidarius* L. *Behaviour*, **24**, 313–17.

Freeland, J. (1958). Biological and social patterns in the Australian bulldog ants of the genus *Myrmecia*. *Australian Journal of Zoology*, **6**, 1–18.

Gervet, J. (1964). Le comportement d'oophagie différentielle chez *Polistes gallicus* L. (Hymen. Vesp.). *Insectes Sociaux*, **11**, 343–82.

Glancey, B. M., Stringer, C. E., and Bishop, P. M. (1973). Trophic egg production in the imported fire ant *Solenopsis invicta* (Hym., Formicidae). *Journal of the Georgia Entomological Society*, **8**, 217–20.

Godfray, H. J. C. (1987). The evolution of clutch size in invertebrates. *Oxford Surveys in Evolutionary Biology*, **4**, 117–54.

Grafen, A. (1990). Do animals really recognize kin? *Animal Behaviour*, **39**, 42–54.

Greene, A. (1979). Behavioral characters as indicators of yellowjacket phylogeny (Hymenoptera: Vespidae). *Annals of the Entomological Society of America*, **72**, 614–19.

Greene, A., Akre, R. D., and Landolt, P. (1976). The aerial yellowjacket *Dolichovespula arenaria* (Fab.): nesting biology, reproductive production, and behavior (Hymenoptera: Vespidae). *Melandaria*, **26**, 1–34.

Hamilton, W. D. (1964). The genetical evolution of social behaviour. I, II. *Journal of Theoretical Biology*, **7**, 1–52.

Hamilton, W. D. (1972). Altruism and related phenomena, mainly in social insects. *Annual Review of Ecology and Systematics*, **3**, 193–232.

Hardy, I. C. W. and Blackburn, T. M. (1991). Brood guarding in a bethylid wasp. *Ecological Entomology*, **16**, 55–62.

Haskins, C. P. and Haskins, E. F. (1950). Notes on the biology and social behavior of the archaic ponerine ants of the genera *Myrmecia* and *Promyrmecia*. *Annals of the Entomological Society of America*, **43**, 461–91.

Henry, C. S. (1972). Eggs and repagula of *Ululodes* and *Ascaloptynx* (Neuroptera: Ascalaphidae): a comparative study. *Psyche*, **79**, 1–22.

Hermann, H. R. and Dirks, T. F. (1975). Biology of *Polistes annularis* (Hymenoptera: Vespidae) I. Spring behavior. *Psyche*, **82**, 97–108.

Hölldobler, B. and Wilson, E. O. (1983). Queen control in colonies of weaver ants. *Annals of the Entomological Society of America*, **76**, 235–8.

Hölldobler, B. and Wilson, E. O. (1986). Ecology and behavior of the primitive cryptobiotic ant *Prionopelta amabilis* (Hymenoptera: Formicidae). *Insectes Sociaux*, **33**, 45–58.

Hölldobler, B. and Wilson, E. O. (1990). *The ants*. Belknap Press of Harvard University Press, Cambridge, MA.

Honk, C. van and Hogeweg, P. (1981). The ontogeny of the social structure in a captive *Bombus terrestris* colony. *Behavioral Ecology and Sociobiology*, **9**, 111–19.

Huber, P. (1802). Observations on several species of the genus *Apis*, known by the name of humble-bees, and called Bombinatrices by Linnaeus. *Transactions of the Linnean Society*, **6**, 214–98.

Hung, A. C.-F. (1973). Reproductive biology in dulotic ants: preliminary report (Hymenoptera: Formicidae). *Entomological News*, **84**, 253–9.

Ishay, J. (1964). Observations sur la biologie de la guêpe orientale *Vespa orientalis* F. *Insectes Sociaux*, **11**, 193–206.

Jeanne, R. L. (1972). Social biology of the Neotropical wasp *Mischocyttarus drewseni*. *Bulletin of the Harvard Museum of Comparative Zoology*, **144**, 63–150.

Jeanne, R. L. (1980). Evolution of social behavior in the Vespidae. *Annual Review of Entomology*, **25**, 371–96.

Katayama, E. (1974). Egg-laying habits and brood development in *Bombus hypocrita* (Hymenoptera, Apidae). Kontyû, **42**, 416–38.

Kawai, A. (1978). Sibling cannibalism in the first instar larvae of *Harmonia axyridis* Pallas (Coleoptera, Coccinellidae). *Kontyû*, **46**, 14–19.

Keller, L. (1988). Evolutionary implications of polygyny in the Argentine ant, *Iridomyrmex humilis* (Mayr) (Hymenoptera: Formicidae): an experimental study. *Animal Behaviour*, **36**, 159–65.

Kukuk, P. F., Eickwort, G. C., and May, B. (1987). Multiple maternity and multiple paternity in first generation brood from single foundress colonies of the sweat bee *Dialictus zephyrus* (Hymenoptera: Halictidae). *Insectes Sociaux*, **34**, 131–5.

Landolt, P. J., Akre, R. D., and Greene, A. (1977). Effects of colony division on *Vespula atropilosa* (Sladen) (Hymenoptera: Vespidae). *Journal of the Kansas Entomological Society*, **50**, 135–47.

Lin, N. and Michener , C. D. (1972). Evolution of sociality in insects. *Quarterly Review of Biology*, **47**, 131–59.

Michener, C. D. (1974). *The social behavior of the bees*. Harvard University Press, Cambridge, MA.

Michener, C. D. and Brothers, D. J. (1974). Were workers of eusocial Hymenoptera initially oppressed or altruistic? *Proceedings of the National Academy of Sciences, USA*, **71**, 671–4.

Michener, C. D. (1990). Classification of the Apidae. *University of Kansas Science Bulletin*, **54**, 75–164.

Mills, N. J. (1982). Voracity, cannibalism and coccinellid predation. *Annals of Applied Biology*, **101**, 144–8.

Mockford, E. L. (1957). Life history studies on some Florida insects of the genus *Archipsocus* (Psocoptera). *Bulletin of the Florida State Museum*, **1**, 253–74.

Mora, G. (1990). Paternal care in a neotropical harvestman, *Zygopachylus albomarginis* (Arachnida, Opiliones, Gonyleptidae). *Animal Behaviour*, **39**, 582–93.

Morse, D. H. (1982). Behavior and ecology of bumble-bees. In *Social Insects*, Vol 3, (ed. H. R. Hermann), pp. 246–322. Academic Press, New York.

Muller, J. K and Eggert , A.-K. (1990). Time-dependent shifts between infanticidal and parental behavior in female burying beetles: a mechanism of indirect mother-offspring recognition. *Behavioral Ecology and Sociobiology*, **27**, 11–16.

Naumann, M. G. (1970). The nesting behavior of *Protopolybia pumila* in Panama. Ph. D. thesis, University of Kansas, Lawrence.

Nonacs, P. (1991). Male parentage and sexual deception in the social Hymenoptera. In *Diversity of sex ratio in haplodiploid insects and mites*. (ed. D. Wrensch and M. Ebbert), Chapman and Hall, New York. (In press)

O'Neill, K. M. and Evans, H. E. (1981). Predation on conspecific males by females of the beewolf *Philanthus basilaris* Cresson (Hymenoptera: Sphecidae). *Journal of the Kansas Entomological Society*, **54**, 553–6.

Owen, R. E. and Plowright, R. C. (1982). Worker-queen conflict and male parentage in bumble-bees. *Behavioral Ecology and Sociobiology*, **11**, 91–9.

Page, R. E. Jr., Robinson, G. E., and Fondrk, M. K. (1989). Genetic specialists, kin recognition and nepotism in honey-bee colonies. *Nature*, **338**, 576–9.

Pardi, L. (1948). Dominance order in *Polistes* wasps. *Physiological Zoology*, **21**, 1–13.
Parker, G. A. and Mock, D. W. (1987). Parent-offspring conflict over clutch size. *Evolutionary Ecology*, **1**, 161–74.
Passera, L. (1966). La ponte des ouvrières de la fourmi *Plagiolepis pygmaea* Latr. (Hym. Formicidae): oeufs reproducteurs et oeufs alimentaires. *Comptes Rendus de l'Académie des Sciences, Paris*, **263**, 1095–8.
Passera, L. (1969). La rôle de la reine dans l'ovogénèse ouvrière chez la fourmi *Plagiolepis pygmaea* Latr. *Colloques internationaux du C. N. R. S.*, **189**, 129–46.
Passera, L. (1978). Une nouvelle catégorie d'oeufs alimentaires: les oeufs alimentaires mis par les reines vierges de *Pheidole pallidula* (Nyl.) (Formididae, Myrmicinae). *Insectes Sociaux*, **25**, 117–26.
Passera, L. (1984). *L'organisation Sociale des Fourmis*. Universite Paul Sabatier, Privat, Toulouse.
Passera, L., Bitsch, J., and Bressac, C. (1968). Observations histologiques sur la formation des oeufs reproducteurs chez les ouvrières de *Plagiolepis pygmaea* Latr. (Hymenoptera Formicidae). *Comptes Rendus de l'Académie des Sciences, Paris*, **266**, 2270–2.
Peeters, C. and Crozier, R. H. (1988). Caste and reproduction in ants: not all mated egg-layers are 'queens'. *Psyche*, **95**, 283–8.
Pickering, J. (1980). Larval competition and brood sex ratios in the gregarious parasitoid *Pachysomoides stupidus*. *Nature*, **283**, 291–2.
Pienkowski, R. L. (1965). The incidence and effect of egg cannibalism in first-instar *Coleomegilla maculata lengi* (Coleoptera: Coccinellidae). *Annals of the Entomological Society of America*, **58**, 150–2.
Plath, O. E. (1923). Notes on the egg-eating habit of bumblebees. *Psyche*, **30**, 193–202.
Plowright, R. C. and Laverty, T. M. (1984). The ecology and sociobiology of bumble-bees. *Annual Reviews of Entomology*, **29**, 175–99.
Poldi, B. (1963). Studi sulla Fondazione dei nidi formicidi. I. *Tetramorium caespitum*. *Symposia Genetica et Biologica Italica*, **12**, 132–99.
Pomeroy, N. (1979). Brood bionomics of *Bombus ruderatus* in New Zealand (Hymenoptera: Apidae). *Canadian Entomologist*, **111**, 865–74.
Pukowski, E. (1933). Okologische untersuchungen an *Necrophorus*. *Zeitschrift für Morphologie und Ökologie der Tiere*, **27**, 518–86.
Queller, D. C., Hughes, C. R., and Strassmann, J. E. (1990). Wasps fail to make distinctions. *Nature*, **344**, 388.
Ratneiks, F. L. W. (1988). Reproductive harmony via mutual policing by workers in eusocial Hymenoptera. *American Naturalist*, **132**, 217–36.
Röseler, P. -F. (1967). Untersuch ungen ber das Auftreten der 3 formen im Hummel staat. *Zoologische Jahrbücher*, **74**, 179–97.
Ross, K. G. and Matthews, R. W. (1989). Population genetic structure and social evolution in the sphecid wasp *Microstigmus comes*. *American Naturalist*, **134**, 574–98.
Rypstra, A. L. (1986). High prey abundance and a reduction in cannibalism: the first step to sociality in spiders (Arachnida). *Journal of Arachnology*, **14**, 193–200.
Sakagami, S. F. (1982) Stingless bees. In *Social Insects*, Vol. 3, (ed H. R. Hermann) pp. 361–423, Academic Press, New York.
Sakagami, S. F. and Fukushima, K. (1957). Reciprocal thieving found in *Polistes fedwigae* Dalla Torre. (Hymenoptera, Vespidae). *Journal of the Kansas Entomological Society*, **30**, 140.
Sakagami, S. F. and Inoue, T. (1990). Oviposition behavior of two Sumatran stingless bees,

Trigona (Tetragonula) laeviceps and *T. (T.) fuscobalteata*. In *Natural history of social wasps and bees in equatorial Sumatra*, (ed. S. F. Sakagami, R. Ohgushi, and D. W. Roubik), pp. 201–17, Hokkaido University Press, Sapporo.

Sakagami, S. F. and Oniki, Y. (1963). Behavior studies of the stingless bees, with special reference to the oviposition process. I. *Melipona compressipes manaosensis* Schwarz. *Journal of the Faculty of Science of Hokkaido University Series VI, Zoology*, **15**, 300–18.

Sakagami, S. F., Beig, D., and Akahira, Y. (1964). Behavior studies of the stingless bees, with special reference to the oviposition process III. Appearance of laying workers in an orphan colony of *Partamona (Partamona) testacea testacea* (Klug). *Japanese Journal of Ecology*, **14**, 50–7.

Sakagami, S. F., Montenegro, M. J., and Kerr, W. E. (1965). Behavior studies of the stingless bees, with special reference to the oviposition process. V. *Melipona quadrifasciata anthidioides* Lepeletier. *Journal of the Faculty of Science of Hokkaido University Series VI, Zoology*, **15**, 578–607.

Sakagami, S. F., Camilo, C., and Zucchi, R. (1973). Oviposition behavior of a Brazilian stingless bee, *Plebeia (Friesella) schrottkyi*, with some remarks on the behavioral evolution in stingless bees. *Journal of the Faculty of Science of Hokkaido University Series VI, Zoology*, **19**, 163–89.

Sakagami, S. F. and Zucchi, R. (1968). Oviposition behavior of an Amazonic stingless bee, *Trigona (Duckeola) ghilianii*. *Journal of the Faculty of Science of Hokkaido University Series VI, Zoology*, **16**, 564–81.

Sakagami, S. F. and Zucchi, R. (1974). Oviposition behavior of two dwarf stingless bees, *Hypotrigona (Leurotrigona) muelleri* and *H. (Trigonisca) duckei*, with notes on the temporal articulation of oviposition process in stingless bees. *Journal of the Faculty of Science of Hokkaido University Series VI, Zoology*, **19**, 361–421.

Sakagami, S. F., Zucchi, R., and Araujo, V. de P. (1977). Oviposition behavior of an aberrant African Stingless bee, *Meliponula bocandei*, with notes on the mechanism and evolution of oviposition behavior in stingless bees. *Journal of the Faculty of Science of Hokkaido University Series VI, Zoology*, **20**, 647–90.

Schick, R. X. (1972). The early instars, larval feeding and the significance of larval feeding in the spider genus *Misumenops* (Araneida: Thomisidae). *Notes of the Southwest Arachnological Society*, **3**, 12–19.

Scott, M. P. (1989). Male parental care and reproductive success in the burying beetle, *Nicrophorus orbicollis*. *Journal of Insect Behavior*, **2**, 133–7.

Scott, M. P. (1990). Brood guarding and the evolution of male parental care in burying beetles. *Behavioral Ecology and Sociobiology*, **26**, 31–9.

Scott, M. P. and Traniello, J. F. A. (1990). Behavioural and ecological correlates of male and female parental care and reproductive success in burying beetles. *Animal Behaviour*, **39**, 274–83.

Siebt, U. and Wickler, W. (1987). Gerontophagy versus cannibalism in the social spiders *Stegodyphus mimosarum* Pavesi and *Stegodyphus dumicola* Pocock. *Animal Behaviour*, **35**, 1903–4.

Simoes, D. (1977). Etologia e deferenciacao de casta em algumas vespas socialis. Ph. D. Thesis, Univ, Sao Paulo, Ribeirao Preto, Sao Paulo.

Skaife, S. H. (1953). Subsocial bees of the genus *Allodape* Lep. and Serv. *Journal of the Entomological Society of South Africa*, **16**, 3–16.

Sommeijer, M. J. and de Bruijn, L. L. M. (1984). Social behaviour of stingless bees:

'bee-dances' by workers of the royal court and the rhythmicity of brood cell provisioning and oviposition behaviour. *Behaviour*, **89**, 299–315.

Sommeijer, M. J., Houtekamer, J. L., and Bos, W. (1984). Cell construction and egg-laying in *Trigona nigra paupera* with a note on the adaptive significance of oviposition behaviour of stingless bees. *Insectes Sociaux*, **31**, 199–217.

Spradbery, J. P. (1973). *Wasps. An account of the biology and natural history of solitary and social wasps*. University of Washington Press, Seattle.

Starr, C. K. (1984). Sperm competition, kinship, and sociality in the aculeate Hymenoptera. In *Sperm competition and the evolution of animal mating systems* (ed. R. L. Smith), pp. 427–466. Academic Press, New York.

Strand, M. R. and Godfray, H. J. C. (1989). Superparasitism and ovicide in parasitic Hymenoptera: a case study of the ectoparasitoid *Bracon hebetor*. *Behavioral Ecology and Sociobiology*, **24**, 421–32.

Strassmann, J. E., Highes, C. R., Queller, D. C., Turillazzi, S., Cervo, R., Davis, S. K., and Goodnight, K. E. (1989). Relatedness in primitively eusocial wasps. *Nature*, **342**, 268–9.

Stuart, R. J. (1981). Abdominal trophallaxis in the slave making ant, *Harpagoxenus americanus* (Hymenoptera: Formicidae). *Psyche*, **88**, 331–4.

Takahashi, K. (1989). Intra- and inter-specific predations of lady beetles in spring alfalfa fields. *Japanese Journal of Entomology*, **57**, 199–203.

Taki, A. (1987). The trophic eggs of colony founding ant queens. In *Chemistry and Biology of Social Insects*, (ed. J. Eder and H. Rembold), p. 268. Verlag J. Peperny, Munich.

Tallamy, D. W. (1984). Insect parental care. *Bioscience*, **34**, 20–4.

Tallamy, D. W. and Wood, T. K. (1986). Convergence patterns in subsocial insects. *Annual Reviews of Ecology and Systematics*, **31**, 369–90.

Torossian, C. (1959). Les échanges trophallactiques proctodeaux chez la fourmi *Dolichoderus quadripunctatus* (Hymenoptere-Formicoidea). *Insectes Sociaux*, **6**, 369–74.

Torossian, C. (1960). Les échanges trophallactiques proctodeaux chez la fourmi *Tapinoma erraticum*. *Insectes Sociaux*, **7**, 171–5.

Torossian, C. (1964). Les échanges trophallactiques proctodeaux chez la fourmi d'Argentine: *Iridomyrmex humilis*. (Hym. Form. Dolichoderinae). *Insectes Sociaux*, **8**, 189–91.

Torossian, C. (1968). Recherches sur la biologie et l'ethologie de *Dolichoderus quadripunctatus* (Hym. Form. Dolichoderinae). *Insectes Sociaux*, **15**, 51–72.

Torossian, C. (1978). Formation des oeufs abortifs trophiques (oeufs 'alimentaires'), chez la fourmi *Dolichoderus quadripunctatus*. *Comptes Rendus de l'Académie des Sciences, Paris*, **286**, 1889–92.

Trivers, R. L. and Hare, H. (1976). Haplodiploidy and the evolution of the social insects. *Science*, **191**, 249–63.

Trumbo, S. T. (1990*a*). Reproductive benefits of infanticide in a biparental burying beetle *Nicrophorus orbicollis*. *Behavioral Ecology and Sociobiology*, **27**, 269–73.

Trumbo, S. T. (1990*b*). Regulation of brood size in a burying beetle *Nicrophorus tomentosus* (Silphidae). *Journal of Insect Behavior*, **3**, 491–500.

Valerio, C. E. (1974). Feeding on eggs by spiderlings of *Achaearanea tepidariorum* (Araneae, Theridiidae), and the significance of the quiescent instar in spiders. *Journal of Arachnology*, **2**, 57–63.

Valerio, C. E. (1977). Population structure in the spider *Achaearanea tepidariorum* (Araneae, Theridiidae). *Journal of Arachnology*, **3**, 185–90.

Vargo, E. L and Ross, K. G. (1989). Differential viability of eggs laid by queens in polygyne colonies of the fire ant, *Solenopsis invicta*. *Journal of Insect Physiology*, **35**, 587–93.

Voss, S. H. (1981). Trophic egg production by virgin fire ant queens. *Journal of the Georgia Entomological Society*, **16**, 437–40.

Voss, S. H. and Blum, M. S. (1988). Trophic and embryonated egg production in founding colonies of the fire ant *Solenopsis invicta* (Hymenoptera: Formicidae). *Sociobiology*, **13**, 271–8.

Voss, S. H., McDonald, J. F., and Keith, C. H. (1988). Production and abortive development of fire ant trophic eggs. In *Advances in Mymecology*, (ed. J. C. Trager), pp. 517–534. E. J. Brill, New York.

West, M. J. and Alexander, R. D. (1967). Subsocial behavior in a burrowing cricket *Anurogryllus muticus* (De Geer). *Ohio Journal of Science*, **63**, 19–24.

West-Eberhard, M. J. (1969). The social biology of polistine wasps. *Miscellaneous Publications of the Museum of Zoology, University of Michigan*, **140**, 1–101.

West-Eberhard, M. J. (1981). Intragroup selection and the evolution of insect societies. In *Natural selection and social behavior*, (ed. R. D. Alexander and D. W. Tinkle), pp. 3–17. Chiron Press, New York.

Weyer, F. von. (1929). Die eiablage bei *Formica rufa*-arbeiterinnen. *Zoologischer Anzeiger*, **84**, 253–6.

Wheeler, W. M. (1928). *The social insects, their origin and evolution*. Harcourt Brace, NY.

Wilson, E. O. (1971). *The insect societies*. Belknap Press of Harvard University Press, Cambridge, MA.

Wilson, E. O. (1976). A social ethogram of the neotropical arboreal ant *Zacrytocerus varians* (Fr. Smith). *Animal Behaviour*, **24**, 354–63.

Wilson, E. O. and Hölldobler, B. (1986). Ecology and behavior of the Neotropical cryptobiotic ant *Basiceros manni* (Hymenoptera: Formicidae: Basicerotini). *Insectes Sociaux*, **33**, 70–84.

Woyciechowski, M. and Lomnicki, A. (1987). Multiple mating of queens and the sterility of workers among eusocial Hymenoptera. *Journal of Theoretical Biology*, **128**, 317–27.

10

Cannibalism in social bees

Penelope F. Kukuk

Opportunities for both cooperation and competition are abundant in group-living species and the fitness consequences depend in complex ways on the social organization and genealogical relationships among group members. In simple aggregations of unrelated individuals, cannibalism may be beneficial through elimination of competitors and through direct nutritional gains (see Dong and Polis this volume). If members of a group are relatives or cooperative in some way, then the benefits to the cannibal are more complex. For example, these benefits may be counterbalanced by the costs of lost opportunities for cooperation or reduction of inclusive fitness through cannibalism of relatives.

The most prevalent form of cannibalism in social bees is oophagy, where eggs are eaten by adult females. This has been observed in all social bees that have been studied in sufficient detail (Table 10.1). Oophagy occurs within a shared domicile, the nest, where females compete for the stored pollen and nectar, or glandular secretions, that constitute larval food. Although the cannibal may gain some nutritional benefit, more importantly, she obtains the resources necessary to rear her own offspring. Because the context for cannibalism in social bees is competition for stored resources in a nest, oophagy may be influenced by one or all of several relevant factors that are also important in cooperative interactions among colony members: sociality, provisioning method, and colony longevity.

Genealogical relationships within colonies will also influence the fitness consequences of oophagy through kin selection. Whereas social organization and genealogical relationships among group members provide the context for oophagy in group-living bees, the effects of oophagy on the reproductive success of individual group members may have important implications for the evolution of social organization and even sex allocation.

Social organization in group-living bees is quite variable. All the forms of sociality described by Michener (1974) and Wilson (1971) are found in bees. These include small, apparently cooperative, egalitarian societies in which all females are reproductively active (e.g. Abrams and Eickwort 1980; Eickwort 1981; Michener 1974) as well as hierarchical societies consisting of a long-lived, morphologically distinct queen and a large number of sterile workers (Michener 1974). Group-living bees are found principally in three families: two closely related, 'long-tongued' bee families– the Apidae and the Anthophoridae–and one 'short-tongued' bee family–the

Table 10.1 The social context, pattern and proximate outcomes of oophagy in social bees*

Taxon	Context	Pattern	Outcomes
Anthophoridae			
Xylocopini and Ceratinini	Facultative group-living Small, short-lived colonies Mass provisioning No morphological castes	Early in colony cycle Among adult females	Reproductive dominance of forager
Apidae			
Euglossini	Facultative group-living Small, short-lived colonies Mass provisioning No morphological castes	Early in colony cycle Among adult females	Reproductive dominance of sedentary guard
Bombini	Obligate group-living Small to moderate size Usually short-lived, tropical perennial species Progressive provisioning No morphological castes Singly mated queens	Late in colony cycle Queen versus workers Among workers	Reduction of colony productivity? Apparent male bias in sex allocation? Worker male production or reinforcement of queen's dominance
Meliponini**	Obligate group-living Large, long-lived colonies Mass provisioning Morphological castes Singly mated queens	Throughout colony cycle Queens and workers Larvae	Trophic eggs Worker male production[§] Queen production[§]

Apini	Obligate group-living Large, long-lived colonies Mass provisioning Morphological castes Multiply mated queens	Throughout colony cycle Workers	Regulation of worker and drone production? Worker Policing
Halictidae			
Halictinae[¶]	Obligately group-living Small, short-lived colonies Mass provisioning No morphological castes	Timing unknown Queens and workers	Reinforcement of sedentary queen's dominance?

* This summary is based on the following references: Anthophoridae (Sakagami and Maeta 1982, 1985, 1987*a*,*b*, Stark *et al*. 1990); Euglossini (Garófalo 1985), Bombini (Alford 1975, Blom 1986, van Doorn and Heringa 1986, Duchateau and Velthuis 1988 1989, Free and Butler 1959, Free *et al*. 1969, Hobbs 1967, van Honk and Hogeweg 1981, van Honk *et al*. 1981, Katayama 1971, 1974, Plath 1923, Plowright and Laverty 1984, Pomeroy 1979, Richards 1977, Sladen 1912, Zucchi 1973); Meliponini (Sakagami 1982, Sakagami and Yamane 1987, Sakagami *et al*. 1983); Apini (Morse 1982, Myser 1952, Nolan 1925, Robinson *et al*. 1990, Woyke 1963, 1967, 1976, 1977, 1980, Visscher 1989); Halictidae (Batra 1964, 1968, Packer 1987, Brothers and Michener 1974).

** Michener 1974, also given subfamily rank as Meliponinae (Winston and Michener 1977, Michener 1990*c*).

§ Due to cannibalism by larvae.

¶ Only the sociality of species for which oophagy is known is included. This does not reflect the wide variation in sociality found in the Halictinae.

Halictidae. The social Anthophoridae and the Apidae will be considered first because they are more closely related to one another than either is to the Halictidae. The higher level classification of the bees is currently in flux; for this review, the classifications of Michener (1974) will be used for taxa at the tribal level and above. The aims of this chapter are threefold: to bring together the information on cannibalism in the bees, to provide an appraisal of its evolutionary context and implications, and to call attention to the potential importance of cannibalism as a topic for studies of the balance between cooperation and competition in group-living Hymenoptera. A brief description of the social organization and provisioning method will be presented for each taxon in which information on oophagy is available in the literature. This will provide the context for a more detailed treatment of oophagy itself that will lead to consideration of its implications. In addition, the meagre information on cannibalism by larvae will be reviewed.

XYLOCOPINAE

This subfamily of the Anthophoridae (carpenter bees and allodapine bees) is closely related to the Apidae (euglossine bees, bumble-bees, stingless bees, and honey-bees; Sakagami and Michener 1987). All of the four tribes in this taxon have group-living species (Michener 1990*b*) and there is evidence of oophagy for two of them, the Xylocopini and the Ceratinini.

Xylocopini

Context

Social organization is quite variable in the Xylocopini–the large carpenter bees. The nest of these species consists of a simple or branched burrow in wood excavated by the bees. As noted in Michener's (1990*b*) recent review of caste in xylocopine bees, records of group-living *Xylocopa* include both 'pre-reproductive assemblages' and functional colonies. The former are groups of non-reproductive females sharing a nest prior to brood rearing. Functional colonies, which are associations of a small number of females within a single nest during brood rearing, may take a variety of forms that differ both within and between species.

X. pubescens nests either solitarily or in groups within which most females remain fully reproductive; each occupying her own branch of a complex nest (Ben Mordechi *et al.* 1978; Blom 1988; Blom and Velthuis 1988; Gerling *et al.* 1981, 1983; Velthuis and Gerling 1983). In *X. sulcatipes* and *X. virginica*, some females nest singly but hierarchical colonies also occur. These small groups (usually only two females) consist of a mother and daughter, sisters, or unrelated females. Usually, one of the females is reproductively active and the other is not; however, a small proportion of nests contain two females with enlarged ovaries (Gerling and Hefetz 1981; Gerling and Herman 1978; Gerling *et al.* 1983; Velthuis 1987). In these species, and in *X. sonorina*, the forager is the primary reproductive and the functionally sterile individual guards the nest (Gerling 1982; Gerling *et al.* 1989; Stark 1990; Michener 1990*b*).

Large carpenter bees are mass provisioners. They bring the pollen and nectar needed for rearing brood into a specially prepared chamber within the nest and mould it into a loaf. The females place an egg on this 'bee bread' and seal the chamber containing the egg and sufficient food to rear a single individual. Thereafter each cell normally remains undisturbed as the brood develops.

Stark *et al.* (1990) showed that oophagy plays a prominent role in the contest that establishes a reproductive dominance hierarchy among adult females in nests of *X. sulcatipes*. Many females nest solitarily. However, multi-female colonies are also established in the breeding season in three ways, as a result of

(1) incomplete dispersal from old nests;
(2) dispersal into old nests at the onset of the breeding season; or
(3) new associations of females in empty nests throughout the breeding season.

In addition, two or more sisters may stay in a nest later in the season, following maturation and emergence of the first brood. Alternatively, if the original, solitary foundress survives, one or more young daughters may stay in a nest producing a mother–daughter association.

Detailed observation of 17 colonies of *X. sulcatipes*, established by two, apparently unrelated, females that independently colonized a single nest during the reproductive season, revealed a cycle of repeated oophagy (Stark *et al.* 1990). In these colonies, one female became the forager and the second remained in the nest as a guard. If either individual completed a provision mass, oviposited on it, and sealed the brood chamber, her nestmate would then open the chamber, eat the egg, oviposit, and reconstruct the cell closure. The initial brood cells were thus destroyed and rebuilt repeatedly. In most cases, the surviving egg was produced by the forager. After the first three cells were finally completed oophagy ceased and a stable division of labour became established with the forager as the reproductive and her cofoundress as a guard.

In contrast to the ubiquitous oophagy observed in colonies of apparently unrelated females of *X. sulcatipes*, oophagy was observed in only two of 30 associations between mother and daughter (in which the mother was the reproductive) and in none of eight associations involving sisters (Stark *et al.* 1990). These observations suggest that genealogy may be an important factor in intracolony competition and oophagy in *X. sulcatipes*. This view is compatible with the expected limitations on cannibalism predicted by kin selection theory (Hamilton 1964*a*,*b*; Hamilton 1972). It may not benefit a female to cannibalize close relatives, and relatives should be more cooperative than non-relatives. Moreover, the lack of oophagy in colonies consisting of relatives suggests that females behave as if they can assess whether or not they are related to their nestmates. Such an assessment does not necessarily imply kin recognition but could be based on a behavioural strategy contingent upon nesting conditions. For example, competition and oophagy could be a strategy adopted by a dispersing female colonizing a new nest that is already occupied or if she is later joined by another female, and cooperation could be adopted if a female remains in her natal nest.

Implications

Oophagy occurs early in the colony cycle in the context of competition between adult females inhabiting the same nest for sequestered resources. Secondary nutritional benefits are also gained by the cannibal, especially because xylocopine bees are known to have very large eggs; some of these bees have the largest egg to body size ratios known among insects (Iwata and Sakagami 1966). The reproductive success of the cannibal is increased while that of her nestmate is decreased. These differences in the reproductive success of colony members will shape the social system evolving from any association. In the large carpenter bees, the most reproductively successful female is the individual that does most of the foraging and the sedentary female that guards the nest is functionally sterile. This contrasts with the Apidae and Halictidae, where the primary reproductive is a female that remains within the nest and the less reproductively active or sterile group members forage (Michener 1974).

Michener (1985, 1990*b*) suggested that the lack of obligate sociality in the Xylocopinae is due to weaker selection for group life. This is a consequence of the three dimensional placement of nests, making them more difficult for predators and parasites to find than those distributed on a two dimensional surface such as in aggregations of ground burrowing bees. The general pattern of reproductive success may be an important additional factor. The greater reproductive success for the primary forager found in the large carpenter bees is qualitatively different from that found in apid and halictid societies where non-foragers are the primary egg producers and obligate eusociality has evolved many times (Michener 1974). The evolution of queen-like morphological characteristics such as physogastry (abdominal enlargement as a consequence of increased egg production), would be less likely if the principal reproductive foraged. Evolution of the behavioural characteristics of a sedentary queen might be similarly inhibited.

An additional effect on the evolution of sociality could stem from the foraging of the principal reproductive. If foraging is more risky than guarding, then the survivorship of the reproductive would be on average less than that of the non-reproductive guard. Upon the death of the forager, the guard would take over the role of forager-queen. Thus, the evolution of long lived, sedentary queens would be unlikely. The opportunity to inherit a nest is an important benefit for non-reproducing females (Blom 1988).

Ceratinini

Context

Oophagy is also a common phenomenon in the second xylocopine tribe with group living species–the small carpenter bees or Ceratinini. Small carpenter bees nest in pithy stems, excavating a simple tunnel. Like the large carpenter bees, they are mass provisioners and lay a single egg on a mass of pollen and nectar sufficient to rear a single individual to adulthood. The following summarizes the studies of Sakagami and Maeta (1982, 1985, 1987*a*,*b*). Two species (*Ceratina japonica* and *C. flavipes*) can be induced to form small, multifemale colonies in the laboratory if insufficient nesting sites are supplied. A division of labour developed in these laboratory colonies, and it was thought that colonies consisted of a sedentary reproductive and a sterile forager.

The most recent of these studies (Sakagami and Maeta 1987*a*,*b*) include careful observations of oophagy and its role in differential reproduction in multifemale nests. A division of labour occurs in *C. flavipes*, with one female staying in the nest (usually the larger of the two) and the other doing most of the foraging, even though the proportion of laboratory induced multifemale nests is low (5 of 48 attempts). However, most of the eggs of the sedentary individual were eaten by the forager who then replaced them with her own. In *C. flavipes*, as in *X. sulcatipes*, the female that foraged for pollen and nectar was reproductively more successful than the female that remained in the nest as a guard.

Sakagami and Maeta (1982, 1985, 1987*a*,*b*) conducted similar studies with *C. japonica*, where induction of multifemale nests in the laboratory is easier than in *C. flavipes*. In six laboratory colonies (with one sedentary female and one forager) of 60 surviving eggs, 17 were laid by the sedentary female and 43 were by the forager. Again the forager became the principal reproductive.

Implications

As with the large carpenter bees, oophagy plays a strong role in the reproductive success of colony members. The evolution of social behaviour in small carpenter bees may be constrained in the same way as that for *Xylocopa*. The contrast between the results of oophagy in small social groups of xylocopine bees versus a primitively social apid is instructive. In the orchid bee *Euglossa cordata*, oophagy results in enhanced reproductive success for a female that remains in the nest (Garfalo 1985, see below). These alternative resolutions of reproductive competition through cannibalism may have produced a different evolutionary trajectory for sociality of the Xylocopinae compared with the Apidae.

Further work is required to investigate the possibility that the relatively greater reproductive success of foragers in the Xylocopinae is in part responsible for the lack of obligate sociality in this group. Additional behavioural observations focusing on oophagy combined with the use of genetic markers to elucidate the behaviour of females that are the most reproductively successful would be of great interest in the carpenter bees. Moreover, studies of oophagy in the allodapine bees, included in the Ceratinini by Michener (1974) may prove of interest. Unlike most bees, allodapines do not partition their nests into cells but rear the brood in a simple burrow, usually in a pithy stem, where it is provided with pollen and nectar throughout development. Thus, eggs laid in the nest chamber may be more susceptible to oophagy.

APIDAE

This family of bees has perhaps the greatest variation in social organization, and includes those species with the most elaborate and hierarchical social structures. It will be discussed in four sections: the Euglossini (orchid bees); the Bombini (bumble-bees); the Meliponini (stingless bees); and the Apini (honey-bees). Although the classification of the Apidae is currently in flux (see Cameron 1990; Kimsey 1984; Plant and Paulus 1987; Winston and Michener 1977) this order of presentation roughly reflects both phylogeny and increasingly complex sociality.

Euglossini

Context

Euglossine bees, known as 'orchid bees' because males of some species are known to pollinate orchids, are usually solitary and nest in pre-existing cavities such as tree hollows, crevices in cliffs and buildings, or on vegetation. However, many species also form small groups of between two and 25 individuals, that form as they re-use nests (Michener 1974). All the adult females occupying the same nest have well developed ovaries and appear reproductively active. It was thought that they comprised a communal, cooperative social group in which all females reproduced (Michener 1974). Orchid bees are mass provisioners; constructing vertically oriented brood cells, filling them with provisions (a mixture of pollen and nectar), ovipositing, and sealing the cell within which a single individual develops.

Like carpenter bees, oophagy by euglossine bees also appears to be an important component of adult-adult competition over sequestered resources required for brood rearing. Females consume the eggs of their nestmates and replace them with their own. This process was examined in *Euglossa cordata* by Garfalo (1985). Newly occupied nests contain single females. Nests that are re-used often contain two or more females; either a female and her daughters or a group of sisters. Even though all females have active ovaries, one becomes behaviourally dominant over the others. The dominant female rarely leaves the nest and becomes the major guard bee. She consumes the eggs of her nestmates and deposits her own eggs on the provisions they gather. The cannibal gains nutritional benefits but, more importantly, she co-opts the resources for rearing her own offspring.

Implications

Reproductive dominance in *E. cordata* is established through oophagy. In contrast to the xylocopine bees, however, the principal reproductive does not forage. If the social behaviour found in euglossine bees approximates the ancestral condition for the remainder of the Apidae, then the evolution of queen and worker castes could occur more readily in the Apidae than in the carpenter bees where the principal reproductive is also the principal forager. It is of interest to note again that there are no obligately-social xylocopine bees, whereas eusociality has arisen at least once and perhaps twice in the Apidae (Winston and Michener 1976; Michener personal communication). In addition, oophagy in *E. cordata* could represent a plausible ancestral condition preceding the evolution of oophagy in the Bombini, or the evolution of inviable (trophic) eggs found in many stingless bees.

Bombini

Context

Virtually all of the non-parasitic species in this tribe are group-living. With rare exceptions, bumble-bees are obligately social. Although there are no species of *Bombus* with morphologically distinct reproductive castes (*sensu* Peeters and Crozier 1988), they all have functional castes, where morphologically similar females function as either queen or workers in a colony. Generally, colonies are initiated by one, singly-mated female in the spring and she alone rears one brood of daughters. These young females become workers and help the foundress (now the queen in a eusocial colony) to rear additional workers or reproductives. At the 'switch' point queens begin to produce haploid (male) eggs.

The switch can occur early or late in the season in colonies of *Bombus terrestris*, the most intensively studied bumble-bee (Duchateau and Velthuis 1988). This switch coincides with either early or late low values in the fluctuating larvae/worker ratio. Eventually, the 'competition' point is reached, when the colony begins to rear gynes which are females destined to become queens in the next season. Following the competition point a small number of 'elite' workers (van Doorn and Heringa 1986, van Honk and Hogeweg 1981) begin to oviposit, and considerable oophagy by both the queen and these workers follows. Bumble-bees are progressive provisioners. Young are fed as they develop with pollen and possibly with some glandular secretions (Michener 1974; Röseler 1967). The fact that *Bombus* queens apparently mate only once (Röseler 1973; Alford 1975) may be an important element in the context of oophagy in bumble-bees.

Reproductive competition involving oophagy is quite conspicuous and may be very common during the final stages of the annual colony cycle for most species in the genus *Bombus*. In some cases, only about a third of a queen's eggs will be reared to adulthood, with most brood mortality occurring during the egg or early larval stages (Alford 1975). Most of the eggs laid by workers are consumed (see below). Some oophagy may function to prevent the production of small reproductives by decreasing the number reared and thereby increasing the resources available for each. However, oophagy in bumble-bees is consistent with queen–worker conflict theory (Trivers and Hare 1976), which predicts that workers should not only rear the queen's daughters but also attempt to rear their own sons, and differs from the competition among adult females described above for the xylocopine and euglossine bees in two important ways. First, it involves competition between reproductive castes; that is both between a queen and her workers (her daughters), as well as among workers. Second, it occurs after the establishment of the dominance hierarchy, and not as a mechanism by which dominance is established.

Laboratory observations by Duchateau and Velthuis (1988, 1989) were able to exclude several proximate causes of the onset of competition and oophagy in *B. terrestris*. These include: the size of the worker population (but see van Doorn and Heringa 1986); the age structure of the worker population; senescence of the queen; the queen's switch to laying haploid (male) eggs; and changes in behavioural traits of queens or workers. They also found that a low larval/worker ratio is a necessary, but not sufficient, condition for young queen production. The timing of the competition point is a characteristic of the queen and occurs from 23–45 days after the maturation of the first brood, when larval/worker ratios are sufficiently low. Duchateau and Velthuis (1988, 1989) suggest that the changes in queen pheromone production, necessary for cessation of queen rearing inhibition (Röseler 1970; van Honk *et al.* 1981), also removes an inhibitory factor that prevents workers from laying eggs.

Crespi (this volume) suggests that the value of the queen to her workers decreases once rearing of female reproductives is underway. The queen has provided the workers with sisters to rear as reproductives, and it is not to the worker's advantage to rear the queen's male eggs. This arises because of the hymenopteran haplodiploid system of sex determination whereby workers are more closely related to their own sons and to the sons of other workers than they are to the sons of the queen, if their mother (the queen) is singly mated. Perhaps unavoidable pheromone changes, necessary for queen-rearing, provide a cue to the workers indicating that queen-laid female gynes are being reared. This may signal a decrease in the queen's value and trigger male egg production by workers, resulting in intense reproductive competition.

This idea is in agreement with experimental data, which indicates that a reduction in queen pheromones increases the rate and duration of larval feeding (Röseler 1970). Increased larval feeding results in gyne production (Free 1987). It would be useful to know the precise timing of the competition point relative to the onset of young queen-rearing, in order to assess whether or not the production of gynes is causally related to the onset of oophagy. An alternative approach could be an experimental manipulation of queen pheromones.

In *B. terrestris* about half of the colonies appear to rear males long before rearing queens; that is, they have early 'switch' points. The queen's switch to laying male eggs does not result in attempted reproduction on the part of workers. The switch to laying male eggs may be cryptic and therefore not perceived by the workers because it can be made without pheromone changes. Perhaps in this way, early switching queens avoid competition with workers over parentage of males. Duchateau and Velthuis (1988) show that there is no difference in the average biomass of reproductives produced by early versus late switching colonies. However, early switching colonies produce many more males.

In many other species of *Bombus*, competition between a queen and her workers also occurs at the time that the colony begins to rear new queens (e.g. Alford 1975;

van Doorn and Heringa 1986; Richards 1977; Sladen 1912). Like *B. terrestris*, queens and workers of these species both consume eggs, and considerable disarray and apparent inefficiency results.

The fitness consequences of this cycle of oophagy and competition vary both within and between species. In *B. terrestris* some observations indicate that direct reproduction on the part of workers is minimal because most worker eggs and larvae are destroyed by the queen and other workers (Duchateau and Velthuis 1988, 1989; van Doorn and Heringa 1986). However, this is in contrast to the observations by van Honk *et al.* (1981), which suggest that a substantial proportion of males in *B. terrestris* are the sons of workers. The workers of some species within the genus *Bombus* produce few or no offspring (Alford 1975; Duchateau and Velthuis 1988, 1989; Morse 1982), but worker-produced males do occur in substantial numbers for other species (e.g. Katayama 1971, 1974; Owen and Plowright 1982; Owen *et al.* 1980). For example, worker laid eggs account for 90 per cent of males in *B. atratus* (Zucchi 1966). The sources of this variation both within and between species are not known, but further investigations will provide considerable insights into the process of cannibalism and competition in *Bombus*.

Another interesting aspect of oophagy in bumble-bees, first noted by Huber (1802; cited in Alford 1975) and subsequently confirmed by other observers (Alford 1975), is that eggs are only consumed if they are less than one-day-old. For example, Katayama (1971) observed that worker-produced eggs were often inspected by workers and only sometimes devoured. Those eggs that were consumed were always younger than 24 hours. This suggests the intriguing possibility of discriminate oophagy in *Bombus* and raises the question of which cues may be used to discriminate between eggs.

Implications

The differences between the genetic interests of a singly-mated queen versus her daughter workers provides the evolutionary context for the reproductive competition and oophagy in *Bombus*. This conflict of interests may influence the timing of male production and differential investment in the sexes. Many species of *Bombus* have highly male-biased numerical sex ratios and investment ratios (Duchateau and Velthuis 1988; Owen and Plowright 1982; Owen *et al.* 1980). This is remarkable because bumble-bee queens generally mate only once (Röseler 1973; Alford 1975). Consequently, they are expected to have an unbiased or female-biased investment in the sexes, depending on whether the queen or the workers 'control' sex allocation (Hamilton 1972; Trivers and Hare 1976). However, the competition point is tied to the production of young queens, and not to the production of males. Thus, the measurements of the investment in female reproductives might more appropriately include the time and energy expended in reproductive competition, even though this competition is largely over worker reproduction of males.

The cost of rearing males does not necessarily involve competition because, as Duchateau and Velthuis (1988) show, queens may switch to laying haploid eggs (rearing males) well in advance of the competition point. However, virtually all female reproductives are progressively fed after the competition point. Therefore, the relative investment in rearing females versus rearing males is greater than can be measured by differences in pupal weight and numbers of females and males. This higher investment cost is due to the additional cost of rearing brood after the competition point and with the subsequent advent of social disorganization. In addition, rearing a queen requires more food than rearing a male (Cumber 1949; Plowright and Jay 1968; Röseler 1970). An appropriate currency with which to evaluate the investment in the sexes in bumble-bees would be the time and energy expended (for example measured in worker-hours or days) to rear male versus female reproductives. Perhaps such an estimate would show that investment in male and female reproductives is more in accord with theoretical predictions than previously thought.

Mating biology has an important effect on the genetic interests of workers (see Ratnieks 1988). If a queen is singly mated, then the average genetic relationship of a worker to the possible males reared in the colony indicates that she should prefer to rear her own sons, then those produced by a randomly-chosen worker and finally those of the queen. If the queen mates multiply, this order of preference changes. A worker will still prefer to rear her own sons, but should then prefer rearing the son of a queen over that of a randomly-chosen worker. This distinction is of interest when comparing the prevalence of laying-workers in the singly mating Bombini and Meliponini, with the rarity of laying-workers in the Apini in which queens mate multiply. Ratnieks (1988) suggests that 'worker policing' exists to prevent worker-rearing of other workers' brood in situations where queens mate multiply.

Although most species of *Bombus* are thought to mate only once, studies of captive colonies indicate that multiple matings are common in *B. hypnorum* (Röseler 1973) and *B. huntii* (Hobbs 1967). If queens in nature use sperm from two or more males, examination of competition and oophagy late in the colony cycle of these species would provide a useful test of the 'worker policing' hypothesis, because competition and oophagy should be less evident.

Meliponini

Context

The Meliponini, otherwise known as stingless bees, have large, long-lived colonies. There is discrete morphological differentiation between queens and workers (morphological castes). These bees are obligately social, and neither queens nor workers can survive and reproduce alone. They are committed to group life with a well defined reproductive division of labour. Stingless bees rear their brood in individual cells and are mass provisioners. Each cell is filled with sufficient glandular secretions to provide the materials necessary for rearing a single individual, an egg is placed on these provisions and the cell is then sealed. Stingless bee queens are singly mated. The generic and subgeneric names used here follow Michener (1990*c*).

With some notable exceptions, trophic eggs are produced by workers throughout the year. Trophic eggs are unlike viable eggs in size (they are usually larger but may be smaller) and in composition, they are usually placed on the brood-cell margin or in the cell, and are usually consumed by the queen just prior to her oviposition. Trophic eggs may provide the main source of protein for the queen, along with the glandular secretions that make up larval food (Sakagami 1982; Sakagami and Yamane 1987; Sakagami *et al.* 1983). They may also function to keep the ovaries of the workers active (West-Eberhard 1981). The exceptions are those genera in which worker oviposition is either absent or confined to queenless colonies, including *Trigona* (subgenera *Trigona, Frieseomelitta,* and *Dukeola*) and perhaps also *Trigonisca* (subgenera *Trigonisca* and *Leurotrigona*). The evolution of trophic eggs is discussed by Crespi (this volume).

Workers in some species also produce viable male eggs. These slender eggs are laid by young workers after the queen oviposits and before the cell is closed. This has been observed in *Plebeia* (subgenera *Plebeia, Friesella,* and *Scaura*), *Scaptotrigona*, and *Melipona* (Sakagami 1982). The fate of these eggs is discussed below in the section on larval cannibalism.

Implications

In the Meliponini, an important issue is the evolution of trophic eggs. The initial evolutionary transition from worker production of viable eggs to worker production of trophic eggs is problematical since a worker gains little by producing inviable eggs unless they require less energy (see Crespi this volume). Moreover, the maintenance of this trait is also of interest because a single individual can produce both types within a short time (Sakagami 1982) suggesting that there is selection maintaining the production of trophic eggs in individuals that can produce both trophic and viable eggs.

Competitive oophagy, similar to that found in euglossine bees, could have been an early step in the evolution of trophic eggs. Stark *et al.*'s (1990) suggestion that oophagy, similar to that found in *Xylocopa*, has led to the production of trophic eggs seems less likely. This is because oophagy in the Xylocopinae results in enhanced reproductive success of the principal forager rather than a sedentary female. Considerable worker oviposition may be a precondition for the evolution of trophic eggs. If the 'worker policing' hypothesis is correct (Ratnieks 1988), worker oviposition will only be found if queens mate singly. In the Apidae, both the Bombini and Meliponini have singly mated queens and both have considerable egg production by workers. In *Bombus,* worker oviposition involves considerable outright competition and oophagy of viable eggs. In contrast, overt competition in stingless bees is ritualized into stereotyped behaviours that occur during queen oviposition and trophic eggs (see Crespi this volume). It would be of interest to know the mating biology of those genera in which worker oviposition is lacking.

The differences in oophagy between *Bombus* and stingless bees could be related to the difference in the colony cycle between these two groups. The colony cycle of bumble-bees is annual and that of stingless bees is perennial, which leads to two important consequences. First, the genealogy in *Bombus* colonies is likely to be more simple, if the original foundress remains the queen. In the stingless bees, longevity of the colony and the resulting greater likelihood of queen replacement may alter the asymmetries due to haplodiploidy and change the fitness consequences of worker reproductive strategies. For example, if the queen is a singly-mated sister of the workers then, on average, workers should still prefer to rear their own sons but should be indifferent about rearing the queen's eggs versus other worker's eggs. Thus a form of serial polygyny (Kukuk 1989) may alter the genetic context of oophagy in stingless bees.

The second consequence is that the value of the queen will fluctuate more widely in annual colonies. This will only occur if a decreasing value of the queen is important in the evolution of overt competition in *Bombus* colonies. In annual colonies, where reproductives are reared at the end of the colony cycle, and young queens are developing, the queen is of decreased value to the workers because she is no longer needed to supply female eggs. In perennial colonies, the value of a queen to her workers, who are also her daughters, may fluctuate, but will remain high because she is still needed to provide eggs for rearing female reproductives at some time in the future.

Apini

Context

Honey-bees of the genus *Apis* are probably the most extensively studied of all bee groups. The social structure of *Apis* is in some ways similar to that of meliponine bees. Colonies are large and long lived, have morphologically distinct reproductive castes, and are obligately social. There are many very striking differences, however (see Michener 1974). For this review, the most important differences are that *Apis* queens mate multiply (Page 1986), trophic eggs are absent, and worker production of males in queenright colonies is rare (Visscher 1989).

Oophagy has only recently received attention in *Apis*, even though it is known that daily egg production of *Apis mellifera* queens exceeds the amount of sealed brood produced from them (see review by Woyke 1977). Indirect evidence suggests that workers eat the brood in normal colonies and that a larger number of drone (male) brood than worker brood disappears (Nolan 1925; Woyke 1977). The disappearance of 103 out of 117 eggs in 12 hours was reported by Myser (1952). *Apis cerrana indica* workers also consume brood if pollen supplies are low (Woyke 1976). Robinson *et al.* (1990) observed behaviours associated with oviposition, nursing, and oophagy in queenless colonies of *A. mellifera*, and examined electrophoretically workers and the drone brood they produced. There was considerable intercolonial variation in the observed behaviours that may be genetically based.

Allozyme analysis suggested that the genotypic distributions found in drone larvae are due to genetic variation among workers in egg production, and in treatment of eggs. Both the production of eggs and their subsequent treatment contributed to the genetic composition of drone brood in queenless colonies.

Workers also eat diploid males (Woyke 1963). In *Apis*, and some other aculeate Hymenoptera, diploid males of zero fitness can occur if there is homozygosity at a sex determining locus (see Kukuk and May 1990). Such diploid males in *A. mellifera* are consumed within six hours of hatching. Woyke (1967) demonstrated the presence of a 'cannibalism substance' produced by young larvae that caused them to be eaten by workers. This substance may have evolved by kin selection. Thus, males with no possibility of direct reproduction can only pass on their genes through increasing the reproduction of relatives. This they can accomplish by sacrificing themselves and avoiding the waste of resources that would be used to rear them. Similarly, in *A. cerana indica*, few diploid males are reared to adulthood (Woyke 1980). Cannibalism of diploid drones would be indistinguishable from the disappearance of worker brood, because fertilized eggs are laid in worker comb. Therefore, the bias in favour of consumption of drone brood noted by observers is not due to removal of unfit diploid males.

Implications

Oophagy and eating of young larvae in *Apis* may involve both colony level selection and kin selection. First, it may function to regulate colony commitment to rearing of worker and drone brood. This can only be true if colonies are not egg limited, an assumption that is supported by the honey-bee queen's enormous capacity to produce eggs (Ribbands 1953). The fact that more drone brood than worker brood is eaten could be due to either worker consumption of eggs laid by other workers or a regulatory mechanism for adjusting drone production to ecological or seasonal conditions.

Regulation and reduction of the colony's investment in unfit, diploid males may be an important function in *Apis* for eating of young larvae. However, worker consumption of normal, haploid male brood (more frequent than cannibalism of worker brood) could also be based on kin selection. Honey-bee workers are more closely related to the sons of the queen than to the sons of other workers because *Apis* queens commonly mate multiply. Therefore, workers are expected to 'police' other workers and reduce worker oviposition or eat worker-produced male eggs (Ratnieks 1988). The results of Visscher (1989) are compatible with this idea. He found that only 0.12 per cent of males are the sons of workers in queenright colonies of *A. mellifera.*

HALICTIDAE

Most group-living halictids are found in the subfamily Halictinae and they exhibit a wide range of social organization.

Halictinae

Context

Social organization in halictine bees ranges widely from solitary to obligately group-living species with a reproductive division of labour. Although there is very little data on oophagy, the information available is for species that are 'primitively-eusocial' and the colony cycle is similar to that found in *Bombus*. Colonies are founded by one or more females in the spring. One of the foundress females then becomes the queen following the emergence of the mostly female first brood. One or more worker generations follow, and young queens and males are reared at the end of the active season. Workers and queens do not differ morphologically and ovarian development as well as mating of workers are not uncommon (see review by Michener 1990*a*). Fifteen per cent of males produced in laboratory colonies of *Lasioglossum zephyrum* can be attributed to workers and up to 50 per cent in some colonies (Kukuk and May 1991). Queens were thought to be singly mated, but recent electrophoretic evidence from natural, single queen colonies indicates that nearly a quarter of females may use the sperm of two males (Kukuk *et al.* 1987).

The instances of oophagy reported in the literature are surprisingly few, given the hundreds of hours that the halictine *L. zephyrum* has been observed in laboratory nests. Three are reported by Michener and Brothers (1974; see also Brothers and Michener 1974) in which a *L. zephyrum* queen consumed the egg of a worker and replaced it with her own, and Batra (1964) noted two instances of oophagy and 15 ovipositions of females of unknown caste. Three instances of oophagy carried out by unmarked females were reported in *L. versatum* (Batra 1968) and nine ovipositions. All cannibals entered the brood cell, ate the egg, manipulated the pollen mass, and replaced the consumed egg with their own. Cannibalism appears to occur in the context of competition for resources necessary to rear offspring and is not primarily nutritional. Packer (1987) also reports apparent oophagy in natural colonies of *Halictus ligatus*.

Implications

Oophagy in halictine bees occurs under circumstances similar to that found in bumble-bees, that is, competition for resources already in the nest that are necessary for rearing offspring. The direct nutritional benefit to the cannibal from oophagy appears to be only a secondary benefit. Although the data are meagre, in the instances where caste was known in *L. zephyrum*, oophagy appears to reinforce the reproductive dominance of the queen, who remains in the nest and does not forage. This pattern is apparently similar to that found in the euglossine bees. However, the data are currently insufficient for drawing any conclusions, and more research is required to ellucidate the effects of oophagy on reproductive success.

CANNIBALISM BY LARVAE

Context

In addition to cannibalism of eggs and young larvae by adults, there are several instances of cannibalism by larvae. These larval bees kill and presumably consume eggs or conspecific larvae. In each case, the usual condition of sequestering each developing individual in its own chamber or brood cell is absent or is circumvented by the cannibal.

In contrast to most other bees, Allodapines (anthophorid bees that are closely related to carpenter bees) rear brood together in a single chamber. In *Exoneura bicolor*, larvae feed on a large, shared provision mass of moistened pollen. Schwarz (1986) noted portions of dead *Exoneura* larvae in the pollen masses of two nests. Further, he noted a single nest in which one larva was consuming another, still living, larva. The frequency of this phenomenon in unknown.

Larval cannibalism may also occur in the stingless bee *Scaptotrigona postica* (Apidae). Some brood cells contain two viable eggs; one laid by the queen and the other by a laying worker. Beig (1972) demonstrated that the worker-produced, male larva is mobile; it kills (and presumably consumes) the other larva, most likely a worker female. In this case, the cannibalism is a side effect of the male larva's successful attempt to obtain all the provisions in the cell and is based on competition for those resources. This form of cannibalism may also occur in *Plebeia* (subgenera *Plebeia*, *Friesella*, and *Scaura*), and *Melipona* where workers produce viable eggs and oviposit in a manner similar to that of *S. postica* (Sakagami 1982).

Another type of larval cannibalism, or at least siblicide, is involved in queen production in *Trigonisca* (subgenus *Leurotrigona*) and *Trigona* (subgenus *Freseiomelitta*). Queens result when a larva perforates the wall of her own cell and breaks into an adjacent, younger cell, consumes the provisions contained in it and presumably also kills and consumes her sister (Sakagami 1982).

Implications

Selection for larval cannibalism in allodaopine bees may be competition among larvae for the resources available in the nest; it may also provide a direct, nutritional benefit for the cannibal. Intracolony relatedness among adults in *E. bicolor* is about 0.5 (Schwarz 1987). There are two types of colonies in these bees; in one type all females reproduce, and in the other from one to three principal reproductives occur (Schwarz 1987, 1988). Relatedness among larvae in the same nest will decrease with the number of reproductives in the nest, particularly if each female is mated to a different, unrelated male. If cannibalism is held in check by kin selection, the frequency of colonies where larval cannibalism occurs should be positively correlated with the number of reproductives in each colony. This might be expected if local environmental conditions are poor, leading to low levels of resources in the nest, but not if conditions are disastrous, leading to insufficient resources for rearing brood.

In the Meliponini, larval cannibalism appears to be primarily a case of competition between two larvae over resources sequestered in a brood cell and the direct nutritional benefits of cannibalism itself are secondary.

SYNTHESIS

Oophagy and eating of newly hatched larvae are ubiquitous in group-living bees (Table 10.1). In facultatively social species, the cannibal wins in competition for stored resources necessary for rearing brood. These resources already sequestered in the nest are much more valuable, even if there is an abundance of pollen and nectar available in the environment, because they can be obtained without risk and with little effort by cannibalism and egg replacement. A secondary nutritional benefit is also gained by the cannibal. In contrast, oophagy also appears to have cooperative elements in species with complex, obligate sociality.

The least complex social organization for which information on oophagy is available occurs in facultatively social species, where some individuals are solitary and others live in small groups that lack morphological castes and have annual colony cycles. These are found in carpenter bees (*Xylocopa* and *Ceratina*), along with the Euglossine bees (*Euglossa*). Oophagy in these groups occurs early in the colony cycle and leads to the reproductive dominance of the cannibal. While the social context for oophagy is similar for carpenter and euglossine bees, the outcomes are very different. In the carpenter bees the reproductively successful female is also the principal forager while in *E. cordata* the principal reproductive is a sedentary, guard bee that stays within the nest. The reasons for these dramatically different outcomes are unclear but their investigation could lead to new insights in the evolution of obligate sociality.

Obligately social species with annual colony cycles and lacking morphological reproductive castes (such as *Bombus* and the halictine bees for which data on oophagy are available) provide a different context for oophagy. In *Bombus*, oophagy is part of the competition between a queen and her workers for rearing male brood toward the end of the colony cycle. The results of oophagy and competition are variable both within and between species. In some cases, the dominance of the queen is reinforced by her consumption of worker eggs and in other cases a large proportion of the male reproductives are derived from worker-laid eggs. An additional consequence may be the apparently male biased investment in the sexes seen in *Bombus*. The data on halictine bees are too meagre for drawing any conclusions other than that it occurs and may influence the reproductive dominance of queens. Nevertheless, oophagy in halictine bees is not a conspicuous feature of within-nest behaviour. The reasons for this are not clear. It could be simply that studies of halictine bees have not focused sufficiently on oophagy to provide the necessary data. Studies of halictine sociality during the late phases of the colony cycle would be very useful in this regard.

Although both honey-bees and stingless bees are obligately social, with long-lived colonies and morphologically distinct castes, their patterns of oophagy are quite different. In *Apis*, oophagy of young worker and drone brood may be a method of regulating the colony's investment in brood rearing. In contrast to *Apis*, worker egg production in stingless bees is conspicuous, and trophic eggs may be the principal source of protein for the queen. Moreover, worker oviposition of viable, but unfertilized, eggs coupled with larval cannibalism leads to worker parentage of some male brood in some species. These differences suggest that for highly social species, the pattern of oophagy may also be influenced, in addition to the social context, by the method of provisioning and genealogical relationships within colonies.

The method of provisioning is unlikely to be a major influence on patterns of oophagy. Mass provisioning, where workers provide enough material for the maturation of a single individual in a chamber, occurs in large and small carpenter bees, euglossine bees, stingless bees and also halictine bees. However, the patterns of oophagy found within these groups are quite variable. Similarly, the patterns of oophagy differ between taxa such as *Apis* and *Bombus*, in which workers are progressive provisioners, feeding the larva as it matures.

Genealogical relationships may have a more important influence on the patterns of oophagy than provisioning method. These relationships are quite variable among colony members in facultatively social species and appear to be associated with predictable patterns of oophagy. Genealogical relationships among colony members are influenced by the number of reproductives and also by the mating frequency of these reproductive individuals. In *Xylocopa*, for example, a colony can consist of unrelated individuals, sisters, or a mother and her daughters. Stark *et al.* (1990) observed that oophagy in *Xylocopa* was common in colonies of unrelated females and rare in colonies made up of relatives. In species where queens mate only once, such as *Bombus* and stingless bees, worker oviposition and oophagy are more conspicuous than in those with multiply mated queens, such as *Apis*. The general pattern of worker oviposition and oophagy is therefore compatible with hypotheses based on kin selection and 'worker policing' (Hamilton 1964*a*,*b*; Ratneiks 1988). Nevertheless, oophagy may also have evolved via colony-level selection in the highly social groups as a mechanism of regulating brood production and eliminating unfit individuals such as diploid drones. Oophagy in the meliponine bees has taken on a role in the cooperative interactions within the colony and functions to provide nourishment for the queen.

In summary, sociality and mating biology are influential factors in providing the cooperative and genealogical context for patterns of oophagy. In contrast, there are no clearly predictable associations between provisioning method and patterns of oophagy. Clearly, the evolutionary implications of oophagy in social bees arise from the obvious effects it has on the reproductive success of colony members. Thus oophagy has important implications for both the evolution of social behaviour and also the patterns of sex allocation. Our understanding of the precise nature of these implications for individual species, and particularly the interspecific associations between oophagy, sociality, and sex allocation are currently limited by a lack of quantitative data. It seems likely that empirical studies of this form of cannibalism will provide interesting insights into the conflicts of interests and degrees of cooperation in social insects.

ACKNOWLEDGEMENTS

I wish to thank B. J. Crespi, M. A. Elgar, C. D. Michener, R. E. Stark, L. Stevens, H. H. W. Velthuis, and W. Wcislo for their helpful comments on this manuscript. Partial funding was provided by NSF grants BSR-8818461 and BSR-890374.

REFERENCES

Abrams, J. and Eickwort, G. C. (1980). Biology of the communal sweat bee *Agapostemon virescens* (Hymenoptera: Halictidae) in New York State. *Search* (Cornell Univ. Agr. Exp. Sta.), **1**, 1–20.

Alford, D. V. (1975). *Bumblebees*. Davis-Poynter Ltd., London.

Batra, S. W. T. (1964). Behavior of the social bee *Lasioglossum zephyrum* within the nest. *Insectes Sociaux*, **11**, 159–86.

Batra, S. W. T. (1968). Behavior of some social and solitary halictine bees within their nests: a comparative study. *Journal of the Kansas Entomological Society*, **41**, 120–33.

Beig, D. (1972). The production of males in queenright colonies of *Trigona (Scaptotrigona) postica*. *Journal of Apicultural Research*, **11**, 33–9.

Ben Mordechi, Y. R., Cohen, D., Gerling, D., and Moscovitz, E. (1978). The biology of *Xylocopa pubescens* Spinola (Hymenoptera: Anthophoridae) in Israel. *Israel Journal of Entomology*, **12**, 107–21.

Blom, J. van der (1988). Mutual benefit for reproducing and non-reproducing females in nests of the carpenter bee *Xylocopa pubescens* Spinola. *Actes Colloques Insectes Sociaux*, **5**, 131–6.

Blom, J. van der and Velthuis, H. H. (1988). Social behaviour of the carpenter bee *Xylocopa pubescens* (Spinola). *Ethology*, **79**, 281–4.

Brothers, D. J. and Michener, C. D. (1974). Interactions in colonies of primitively social bees. III. Ethometry and division of labor in *Lasioglossum zephyrum* (Hymenoptera: Halictaidae). *Journal of Comparative Physiology*, **90**, 129–68.

Cameron, S. A. (1990). Molecular perspectives on the phylogeny of the apidae. In *Social insects and the environment*, (ed. G. K. Veeresh, B. Mallik, and C. A. Vidraktamath), pp. 122–3. Mohan Primilani, New Delhi.

Cumber, R. A. (1949) The biology of humble-bees, with special reference to the production of the worker caste. *Transactions of the Royal Entomological Society (London)*, **100**, 1–45.

Doorn, A. van and Heringa, J. (1986). The ontogeny of a dominance hierarchy in colonies of the bumblebee *Bombus terrestris* (Hymenoptera, Apidae). *Insectes Sociaux*, **33**, 3–25.

Duchateau, M. J. and Velthuis, H. H. W. (1988). Development and reproductive strategies in *Bombus terrestris* colonies. *Behaviour*, **107**, 186–207.

Duchateau, M. J. and Velthuis, H. H. W. (1989) Ovarian development and egg laying in workers of *Bombus terrestris* (L). *Entomologia Experientia and Applicata*, **51**, 199–214.

Eickwort, G. C. (1981). Presocial insects. In *Social insects,* Vol 2., (ed. H. R. Hermann), pp. 199–280. Academic Press, New York.

Free, J. B. (1987). *Pheromones of social insects.* Comstock Publishing Associates, Ithaca.

Free, J. B. and Butler, C. G. (1959). *Bumblebees,* Collins, London.

Free, J. B. Weinberg, I., and Whiten, A. (1969). The egg-eating behaviour of *Bombus lapidarius* L. *Behaviour,* **35**, 313–17.

Garófalo, C. A. (1985). Social structure of *Euglossa cordata* nests (Hymenoptera: Apidae: Euglossini). *Entomologia Generalis,* **11**, 77–83.

Gerling, D. (1982). Nesting biology and flower relationships of *Xylocopa sonorina* Smith in Hawaii. *Pan-Pacific Entomolgist,* **58**, 336–51.

Gerling, D. and Hefetz, A. (1981). The ecology of the carpenter bee *Xylocopa sulcatipes* Maa in Israel. In *Developments in arid zone ecology and environmental quality,* (ed. H.Shuval), pp. 71–4. Balaban ISS, Philadelphia.

Gerling, D. and Herman, H. R. (1978). Biology and mating behavior of *Xylocopa virginica* L. *Behavioral Ecology and Sociobiology,* **3**, 99–111.

Gerling, D., Hurd, P. D., and Hefetz, A. (1981). In-nest behavior of the carpenter bee *Xylocopa pubescens* Spinola (Hymenoptera: Anthophoridae). *Journal of the Kansas Entomological Society,* **54**, 209–18.

Gerling, D., Hurd, P. D., and Hefetz, A. (1983). Comparative behavioral biology of two middle east species of carpenter bees (*Xylocopa* Latreille). *Smithsonian Contributions to Zoology,* **369**, 1–33.

Gerling, D., Velthuis, H. H. and Hefetz, A. (1989). Bionomics of the large carpenter bees of the genus *Xylocopa. Annual Review of Entomology,* **34**, 163-90.

Hamilton, W. D. (1964*a*). The genetical evolution of social behaviour. I. *Journal of Theoretical Biology,* **7**, 1–16.

Hamilton, W. D. (1964*b*). The genetical evolution of social behaviour. II. *Journal of Theoretical Biology,* **7**, 17–52.

Hamilton, W. D. (1972). Altruism and related phenomena, mainly in social insects. *Annual Review of Ecology and Systematics,* **3**, 193–232.

Hobbs, G. A. (1967). Ecology of species of *Bombus* (Hymenoptera: Apidae) in southern Alberta VI. Subgenus *Pyrobombus. Canadian Entomologist,* **99**, 1271–92.

Honk, C. G. J. van and Hogeweg, P. (1981). The ontogeny of the social structure in a captive *Bombus terrestris* colony. *Behavioral Ecology and Sociobiology,* **9**,111–19.

Honk, C. G. J. van, Velthuis, H. H. W., and Hoogeveen, J. C. (1981). Factors influencing the egg laying of workers in a captive *Bombus terrestris* colony. *Behavioral Ecology and Sociobiology,* **9**, 9–14.

Iwata, K. and Sakagami, S. F. (1965). Gigantism and dwarfism in bee eggs in relation to the modes of life, with notes on the number of ovarioles. *Japanese Journal of Ecology,* **16**, 4–16.

Katayama, E. (1971). Observations on the brood development in *Bombus ignitus* (Hymenoptera: Apidae). 1. Egg-laying of queens and workers. *Kontyû,* **39**, 189–203.

Katayama, E. (1974). Observations on the brood development in *Bombus ignitus* (Hymenoptera: Apidae). *Kontyû,* **42**, 416–38.

Kimsey, L. S. (1984). A re-evaluation of the phylogenetic relationships in the Apidae (Hymenoptera). *Systematic Entomology,* **9**, 435–41.

Kukuk, P. F. (1989). Evolutionary genetics of a primitively eusocial bee *Dialictus zephyrus.* In *Evolutionary genetics of social insects,* (ed. R. E. Page, Jr. and M. D. Breed), pp. 183–202. Westview Press, Boulder.

Kukuk, P. F. and May, B. (1990). Diploid males in a primitively eusocial bee *Lasioglossum (Dialictus) zephyrum* (Hymenoptera; Halictidae). *Evolution*, **44**, 1522–8.

Kukuk, P. F. and May, B. (1991). Colony dynamics in a primitively eusocial halictine bee, *Lasioglossum zephyrum* (Hymenoptera: Halictidae). *Insectes Sociaux*, **38**, 171–89.

Kukuk, P. F., Eickwort, G. C., and May, B. (1987). Multiple maternity and multiple paternity in first generation brood from single foundress colonies of the sweat bee *Dialictus zephyrus* (Hymenoptera: Halictidae). *Insectes Sociaux*, **34**, 131–5.

Michener, C. D. (1974). *The social behavior of the bees*. Harvard University Press, Cambridge, MA.

Michener, C. D. (1985). From solitary to eusocial: need there be a series of intervening species? In *Experimental behavioral ecology*, (ed. B. Hölldobler and M. Lindauer), pp. 293–305. B. Fischer Verlag, New York.

Michener, C. D. (1990*a*). Reproduction and castes in social halictine bees. In *Social insects: an evolutionary approach to castes and reproduction*, (ed. W. Engles), pp. 77–121. Springer-Verlag, Berlin.

Michener, C. D. (1990*b*). Caste in xylocopine bees. In *Social insects: an evolutionary approach to castes and reproduction*, (ed. W. Engles), pp. 123–46. Springer-Verlag, Berlin.

Michener, C. D. (1990*c*). Classification of the Apidae (Hymenoptera). *University of Kansas Science Bulletin*, **54**, 75–174.

Michener, C. D. and Brothers, D. H. (1974). Were workers of eusocial Hymenoptera initially oppressed or altruistic? *Proceedings of the National Academy of Science, USA*, **71**, 671–4.

Morse, D. H. (1982). Behavior and ecology of bumble bees. In *Social insects*, Vol. III, (ed. H. R. Herman). Academic Press, New York.

Myser, W. C. (1952). Ingestion of eggs by honeybee workers. *American Bee Journal*, **92**, 67.

Nolan, W. J. (1925). The brood rearing cycle of the honeybee. *Bulletin of the United States Department of Agriculture*, **1349**, 1–56.

Owen, R. E. and Plowright, R. C. (1982). Worker-queen conflict and male parentage in bumble bees. *Behavioral Ecology Sociobiology*, **11**, 91–9.

Owen, R. E., Rodd, F. M., and Plowright, R. C. (1980). Sex ratio in bumble-bee colonies: complications due to orphaning? *Behavioral Ecology and Sociobiology*, **7**, 287–91.

Packer, L. (1987). Competition over oviposition in subtropical and temperate populations of the social sweat bee *Halictus ligatus*. In *Chemistry and biology of social insects*, (ed. J. Eder and H. Rembold), pp. 553–4. Verlag J. Peperny, Münich.

Page, R. E. (1986). Sperm utilization in social insects. *Annual Review of Entomology*, **31**, 297–320.

Peeters, C. and Crozier, R. H. (1988). Caste and reproduction in ants: not all mated egg-layers are 'queens'. *Psyche*, **95**, 283–8.

Plant, J. D. and Paulus, H. F. (1987). Comparative morphology of the postmentum of bees (Hymenoptera: Apoidea) with special remarks on the evolution of the lorum. *Zeitschrift für Zoologische Systematic und Evolutionsforschung*, **25**, 81–103.

Plath, O. E. (1923). Notes on the egg-eating habits of bumblebees. *Psyche*, **30**, 193–202.

Plowright, R. C. and Laverty T. M. (1984). The ecology and sociobiology of bumblebees. *Annual Review of Entomology*, **29**, 191–8.

Plowright, R. C. and Jay, S. C. (1968). Caste differentiation in bumblebees (*Bombus* Latr.: Hym) I. The determination of female size. *Insectes Sociaux*, **15**, 171–92.

Pomeroy, N. (1979). Brood bionomics of *Bombus ruderatus* in New Zealand (Hymenoptera: Apidae). *Canadian Entomologist*, **111**, 865–74.

Ratnieks, F. L. W. (1988). Reproductive harmony via mutual policing by workers in eusocial Hymenoptera. *American Naturalist*, **132**, 217–36.

Ribbands, C. R. (1953). *The behavior and social life of honeybees*. Bee Research Association, London.

Richards, K. W. (1977). Ovarian development of queen and worker bumble-bees (Hymenoptera: Apidae) in southern Alberta. *Canadian Entomologist*, **109**, 109–16.

Robinson, G. E., Page, R. E., Jr., and Fondrk, M. K. (1990). Intracolonial behavioral variation in worker oviposition, oophagy, and larval care in queenless honey bee colonies. *Behavioral Ecology and Sociobiology*, **26**, 315–23.

Röseler, P. F. (1967). Arbeitsteilung und Drüsenzustände in Humelvökern. *Naturwissenschaften*, **54**, 146–7.

Röseler, P. F. (1970). Unterschiede in der Kastendetermination zwischen den Hummelarten *Bombus hypnorum* and *Bombus terrestris*. *Zeitschrift für Naturwissenschaften*, **25**b, 543–8.

Röseler, P. F. (1973). Die Anzahl der Spermien im Receptaculum Seminis von Hummelkniginnen (Hym. Apoidea, Bombinae). *Apidologie*, **4**, 267–74.

Sakagami, S. F. (1982). Stingless bees. In *Social insects*, Vol III, (ed. H. R. Herman). Academic Press, New York.

Sakagami, S. F. and Maeta, Y. (1982) Further experiments on the artificial induction of multifemale associations in the principally solitary bee genus *Ceratina*. In *The biology of social insects. Proceedings of the 9th Congress of IUSSI*, (ed. M. D. Breed, C. D. Michener, and H. E. Evans), pp. 171–4. Westview Press, Boulder.

Sakagami, S. F. and Maeta, Y. (1985) Multifemale nests and rudimentary castes in the normally solitary bee *Ceratina japonica*. *Journal of the Kansas Entomological Society*, **57**, 639–56.

Sakagami, S. F. and Maeta, Y. (1987*a*). Sociality, induced and/or natural, in the small basically solitary carpenter bees (*Ceratina*). In *Animal societies: theories and facts*, (ed. Y. Itô, J. L. Brown, and J. Kikkawa), pp. 1–16. Japan Science Society Press, Tokyo.

Sakagami, S. F. and Maeta, Y. (1987*b*). Multifemale nests and rudimentary castes of an 'almost' solitary bee *Ceratina flavipes*, with some additional observations on multifemale nests of *C. japonica*. *Kontyû*, **55**, 391–409.

Sakagami, S. F. and Michener, C. D. (1987). Tribes of the Xylocopinae and origin of the Apidae. *Annals of the Entomological Society of America*, **80**, 439–50.

Sakagami, S. F. and Yamane, S. (1987). Oviposition behavior and related notes of the Taiwanese stingless bee *Trigona (Lepidotrigona) ventralis hoozana*. *Journal of Ethology*, **5**, 17–27.

Sakagami, S. F., Yamane, S. and Inoue, T. (1983). Intranidal behaviors of *Trigona (Tetragonula) laeviceps* and *T. (Trigonella) moorei*. In *Ecological study on social insects in central Sumatra with special reference to wasps and bees*. Report on grant-in-aid for scientific research, Sumatra nature study (entomology), (ed. R. -I. Ohgushi), pp. 46–50. Kanaqawa University, Japan.

Schwarz, M. P. (1986). Persistent multi-female nests in an Australian allodapine bee, *Exoneura bicolor* (Hymenoptera, Anthophoridae). *Insectes Sociaux*, **33**, 258–77.

Schwarz, M. P. (1987). Intra-colony relatedness and sociality in the allodapine bee *Exoneura bicolor*. *Behavioral Ecology and Sociobiology*, **21**, 387–92.

Schwarz, M. P. (1988). Notes on cofounded nests in three species of social bees in the genus *Exoneura* (Hymenoptera; Anthophoridae). *Victorian Naturalist*, **105**, 212–15.

Sladen, F. W. L. (1912). *The humble-bee, its life history and how to domesticate it.* Macmillan, London.

Stark, R. E. (1990). Untersuchungen zur brutbiologie und zum Sozialverhalten der Groen Holzbiene *Xylocopa sulcatipes* Maa. PhD. dissertation, Albert-Ludwigs-Universität, Freiburg. pp. 182.

Stark, R. E., Hefetz, A., Velthuis, H. H. W., and Gerling, D. (1990). Reproductive competition by oophagy in the socially nesting bee *Xylocopa sulcatipes*. *Naturwissenschaften*, **77**, 38–40.

Trivers, R. L. and Hare, H. (1976). Haplodiploidy and the evolution of the social insects. *Science*, **191**, 249–63.

Velthuis, H. H. W. (1987). The evolution of sociality: ultimate and proximate factors leading to primitive social behavior in carpenter bees. *Experientia*, **54**, 405–30.

Velthuis, H. H. W. and Gerling, D. (1983). At the brink of sociality: interactions between adults of the carpenter bee *Xylocopa pubescens* Spinola. *Behavioral Ecology and Sociobiology*, **12**, 209–14.

Visscher, P. K. (1989). A quantitative study of worker reproduction in honey bee colonies. *Behavioral Ecology and Sociobiology*, **25**, 247–54.

West-Eberhard, M. J. (1981). Intragroup selection and the evolution of insect societies. In *Natural selection and social behavior*, (ed. R. D. Alexander and D. W. Tinkle), pp. 3–17. Chiron Press, New York.

Wilson, E. O. (1971). *The insect societies*. Harvard University Press, Boston.

Winston, M. L. and Michener, C. D. (1977). Dual origin of highly social behavior among bees. *Proceedings of the National Academy of Science, USA*, **74**, 1135–7.

Woyke, M. (1963). What happens to diploid drone larvae in a honeybee colony. *Journal of apicultural Research*, **2**, 73–6.

Woyke, M. (1967). Diploid drone substance–cannibalism substance. In *XXIth International Beekeepers Congress*, Maryland, pp. 471–2.

Woyke, J. (1976). Brood-rearing efficiency and absconding in Indian honeybees. *Journal of apicultural Research*, **15**, 133–43.

Woyke, J. (1977). Cannibalism and brood-rearing efficiency in the honeybee. *Journal of apicultural Research*, **16**, 84–94.

Woyke, J. (1980). Evidence and action of cannibalism substance in *Apis cerana indica*. *Journal of apicultural Research*, **19**, 6–16.

Zucchi, R. (1973). Aspectos bionomicos de *Examolopsis aureopilosa* e *Bombus atratus* incluindo consideraces sober e evoluco do compartments social (Hymenoptera: Apidae). PhD. Dissertation, Cincias e letras de Ribeirâo Preto, S. P. Brasil.

11

Empirical studies of cannibalism in fish

Gerard J. FitzGerald and Fredrick G. Whoriskey

Cannibalism can be classified as heterocannibalism, the killing and consumption of unrelated conspecifics, and filial cannibalism, the eating of kin. How distantly-related individuals must be, in order to be considered as non-kin by a fish is not known. However adult guppies prefer to eat their neighbour's offspring rather than their own (Loekle *et al.* 1982). Cannibalism is widespread and can have profound consequences on an individual's inclusive fitness. Behavioural ecologists are concerned with quantifying cannibalism in natural populations and identifying what selective advantage, if any, the behaviour confers to the cannibal. The existing theoretical framework explaining cannibalism in mammals can be attributed to Hrdy (1979) and for fishes to Dominey and Blumer (1984). Suggested benefits to cannibals include: (1) nutrition; conspecifics are a high quality food, (2) reduced resource competition; the number of potential competitors for food and space are reduced, (3) parental manipulation; the cannibal terminates parental care in order to improve its own survivorship or those of its other offspring, and (4) sexual selection; reduced competition for mates. Glass *et al.* (1985) provide an explanation for the adaptive significance of infanticide (relevant to cannibalism in fishes) which emphasizes a cost-benefit approach. They used game theory to reject a 'maladaptive' explanation for infanticide. Infanticide confers a relative advantage to the perpetrator when other population members are non-infanticidal. For additional discussion of the theoretical aspects of cannibalism, see Sargent (this volume).

This chapter will evaluate the recent theoretical and empirical research on cannibalism in fish with a focus on its effects on inclusive fitness; earlier studies are reviewed by Dominey and Blumer (1984). First, the various types of cannibalism in fishes are described. Second, two evolutionary models of cannibalism, developed to examine the effects of cannibalism upon inclusive fitness, are evaluated in the light of current empirical evidence. These are, Rohwer's (1978) model of male filial cannibalism and Vickery *et al.*'s (1988) model of female heterocannibalism.

Cannibalism also has important population level effects (Fox 1975; Polis 1981). An example of a population level effect would be if there is selective predation upon the smaller members of a species which are predominantly female (e.g. Marconato and Bisazza 1988). In this case cannibalism could generate a male-biased sex ratio resulting in destabilized population dynamics. In contrast, cannibalism could be an important stablizing factor of populations when food is scarce. At times large fish could sustain growth or survive by eating smaller conspecifics which feed on different levels of the trophic system. Dominey and Blumer (1984), Fox (1975) and Polis (1981) (and references therein) provide valuable information on some demographic aspects of cannibalism. Other recent studies that consider the potential importance of cannibalism for population dynamics include Alheit (1987), Valdes *et al.* (1987) and Hyatt and Ringler (1989*a*).

FACTORS PROMOTING CANNIBALISM IN FISH

Dominey and Blumer (1984) list several aspects of fish biology that contribute to the high prevalance of cannibalism in this taxon. First, eggs and fry are quite small compared with adults, which makes them easy to consume. Other factors favouring high rates of cannibalism in fish with external fertilization are the laying of undeveloped zygotes, which are often concentrated in one area; the presence of adults and older young together with smaller conspecifics on the breeding sites; and the high energetic value of eggs (Gross and Shine 1981).

Fish usually cannibalize the eggs, larvae, and fry, but adults can also cannibalize each other. For example, Marconato and Bisazza (1988) report that large male river bullhead *Cottus gobio* preyed upon adult conspecifics including gravid females. During courtship the male bites the head of the female and this behaviour is sufficiently dangerous to cause small females to avoid relatively large males. Although larger fish generally eat smaller ones, cases of sibling cannibalism occur where larvae attack similar sized individuals (Loadman *et al.* 1986; Valerio and Barlow 1986).

NUTRITIONAL BENEFITS

Eating conspecifics may provide additional nutritional benefits beyond the normal diet. Dionne (1985) investigated the relationships between food availability, cannibalism, and reproductive output in the live-bearing mosquitofish *Gambusia affinis*. Dionne (1985) tested the hypothesis that cannibalism allows the animal to maintain a commitment to reproduction during a reduction from a period of high food intake to one of low intake. She compared the amount of cannibalism and the reproductive performance of females fed on a low, a variable (high food level for 15 days followed by low food level for 10 days), and a high food ration. Half of the females were then given 10 juvenile conspecifics as potential prey. The mean number of juveniles consumed per female at the low and the variable food levels were both greater than at the high food levels, but there was no difference in cannibalism between the former two groups. The proportion of the total original juveniles cannibalized per day within each food treatment level showed an early peak, but then continued at a low level in her low and variable food treatment groups. Dionne (1985) concluded that mosquitofish regulate their cannibalistic behaviour according to the degree of food availability. Dionne's measure of reproductive output was the mean total weight of eggs and embryos per female. The reproductive output was similar for the low and variable food groups. However, reproductive output for the high food group was much *lower* when juveniles were present. She attributed this effect to exploitation competition for food between juveniles and adults in the juveniles-present treatment.

Meffe (1986) agreed with Dionne's (1985) conclusion that the frequency of cannibalism in mosquitofish is a function of food availability, but disagreed with her conclusion that cannibalism influenced reproductive output. He argued that the variable food treatment was inappropriate to test the hypothesis that cannibalism allows the animal to maintain a commitment to reproduction during a high food period, even after the food level is reduced. This is because mosquitofish are ovoviviparous and produce large, yolky ova containing all the energy necessary for complete development of the embryo at the beginning of the ovarian cycle. Maternal energetic input beyond vitellogenesis is minimal or non-existent, and any food treatment after vitellogenesis is irrelevant to the number or size of developing embryos because offspring weight is determined at yolking, even before fertilization (Meffe 1986).

Meffe and Crump (1987) also evaluated the nutritional benefits of cannibalism in mosquitofish. The fish were fed one of four diets, a control diet of commercially prepared dry food and three other diets that contained different levels of supplements of conspecific and heterospecific fish, shrimp, and mosquito larvae. The subsequent growth and reproduction of individuals were measured. Fish on the cannibal diets showed greatest wet weight gains and had the highest fertility (mature ova plus embryos) of the four experimental groups. The average developmental stage, either reflecting the speed of embryonic development or the speed with which the brood was initiated, or both, was most advanced in the cannibalistic group and significantly greater than in the controls. These results were impressive because the ingestion of conspecifics during only seven weeks of the organism's one to two year lifespan had a large effect. Presumably a life-long cannibalistic habit could result in a more dramatic growth and reproduction advantage. Belles-Isles (1991) compared the total seasonal egg production of female threespine sticklebacks *Gasterosteus aculeatus,* that had been fed a mixture of conspecific eggs and commercially prepared dry food, with those that had only been fed the dry food. The two groups were fed the same weight of food. Although Belles-Isles (1991) did not measure the energetic content of his diets, the fish were fed *ad libitum* so presumably all groups obtained enough food to meet their energetic needs. He showed that the cannibals produced a greater mass of eggs per gram of food eaten than the non-cannibals. Cannibalistic females produced about 1.5 times as many eggs as the non-cannibals over the two month breeding season.

Furthermore the cannibalistic females averaged higher somatic wet and dry weights than the other groups at the end of the experiment.

Meffe and Crump (1987) suggested that much more information is required on the nutritional benefits of cannibalism accruing over the entire lifespan of individuals, and its effects on interbrood intervals and gestation periods. Data are also needed on the actual nutritional content and value of conspecifics, the conversion efficiences of conspecifics versus other foodstuffs, and the presence of essential elements found in conspecifics, but lacking elsewhere.

Although the above studies showed clear nutritional benefits of cannibalism under certain laboratory conditions, other tests have not found such clear benefit. For example, under conditions of unlimited food, Loadman *et al.* (1986) were unable to demonstrate any increase in growth or survival of cannibalistic juvenile walleye compared with non-cannibals. However, the authors suggest that nutritional advantages of cannibalism are possible in this species because fish cannibalizing for the first time were significantly larger than the control fish. Juveniles switched from the laboratory diet to feed on conspecifics as soon as they were large enough. Loadman *et al.* (1986) suggested that their failure to detect any nutritional effect of cannibalism may have been due to their small sample sizes. Size differentiation within a cohort of growing fish (growth depensation) is a well-known phenomenon and larger samples sizes may have been required to detect any nutritional benefit of cannibalism. Paller and Lewis (1987) showed a strong link between growth depensation and cannibalism with striped bass. Fish were fed several diets to investigate the relationship between diet, growth, and cannibalism. Larvae fed large amounts of brine shrimp and then converted to a combination of brine shrimp and formulated feed, exhibited high growth rates without growth depensation or cannibalism. Larvae fed smaller amounts of brine shrimp showed growth depensation and were cannibalistic.

Despite its possible nutritional advantages, cannibalism may still occur for other reasons. For example, female threespine sticklebacks in the laboratory prefer to attack the eggs of conspecifics rather than those of the sympatric blackspotted stickleback. In tide pools, female threespine sticklebacks frequently raid the eggs of conspecifics but generally ignore those of the blackspotted stickleback. The nests of the two species are often no farther apart than 15 cm so it is unlikely that female threespine sticklebacks cannot find the blackspotted stickleback eggs. To test the hypothesis that eating conspecific eggs confers a nutritional advantage not obtained by eating heterospecific eggs, FitzGerald (unpublished) fed conspecific and blackspotted stickleback eggs to groups of female threespine sticklebacks and measured the effect of egg consumption upon the reproductive performance (number of spawnings per season, clutch sizes, total seasonal egg production, egg size, wet and dry gonad weight). There was no significant differences for any of these measures between the two groups of females, indicating that female threespine sticklebacks raid conspecific nests for reasons other than purely nutrition.

CANNIBALISM AND COMPETITION

Competition for limiting resources

Competition for food

An individual that eats a conspecific has removed a potential competitor from the population, if some food resource, which is in short supply, is consequently made available to the cannibal or its kin. Dominey and Blumer (1984) concluded that there is little evidence for resource competition in fish and thought it unlikely that heterocannibalism could evolve in fish as a primary response to competition for food. The validity of this conclusion is uncertain because most authors have not measured the availability of alternative prey in aquatic habitats.

Competition for mates

In some circumstances, mates may be considered as a limiting resource (Ward and FitzGerald 1988). Female threespine sticklebacks form shoals to raid the nests of males (FitzGerald 1983). Eggs provide the raiders with food, but it is unlikely that this is the only function of raiding. In tide pools where the study was done, there are abundant supplies of apparently suitable alternative food, such as zooplankton (Worgan and FitzGerald 1981) and benthic invertebrates (Ward and FitzGerald 1983). Cannibalism during the nest-raiding behaviour in threespine sticklebacks may be due to the lack of suitable mates. Vickery *et al.* (1988) argued that female threespine sticklebacks raid the nests of conspecific males and eat their eggs, not for the nutritional advantages, but rather to increase their mating success when they mate subsequently with the male whose nests they have destroyed. Males court less when they move into the parental phase, particularly if they have more clutches. Males from a Dutch population showed little or no sexual behaviour four to five days after their first fertilization, and males that had experienced five fertilizations and clutches showed little sexual behaviour (van Iersel 1953).

The model of Vickery *et al.* (1988) makes the implicit assumption that high quality males or nest sites are limiting resources for which females compete. Three predictions arise from their model. First, females prefer to spawn in nests containing small numbers of eggs rather than large numbers of eggs. This hypothesis is based on the finding that females spawning in nests with many eggs will have reduced hatching success because of egg crowding (Reebs *et al.* 1984). Reebs *et al.* (1984) found a significant negative correlation between the rate of oxygen consumption by eggs and the numbers of eggs per stickleback nest in a field population. At their site, low levels of dissolved oxygen are correlated with long incubation times and low hatching success (Poulin and FitzGerald 1989). Thus a female that spawns in a nest already containing many eggs may have a lower proportion of her eggs hatch than a female who spawns in a nest containing fewer eggs. By consuming the male's eggs and forcing him to rebuild his nest, a female raider might mate with him. In support of this prediction, Belles-Isles *et al.* (1990)

found that females were more likely to spawn in nests containing fewer eggs than in nests with many eggs. Second, females are more likely to raid the nests with the most eggs. Contrary to this prediction, males defending one clutch of eggs were as likely to have their nests raided by groups of females as males defending several clutches (Belles-Isles *et al.*1990). Third, females should avoid eating their own eggs and spawn in nests they had just previously raided. In support of this prediction, FitzGerald and van Havre (1987) found that females were less likely to eat their own eggs than those of their shoal mates and that those females who had initiated raids on nests resulting in cannibalism, were the first ones to spawn in the reconstructed nest.

The female cannibalism model (Vickery *et al.* 1988) ignored factors such as nutrition, differential predation upon non-cannibals, group selection pressures (which may operate if cannibalism becomes so widespread as to threaten the extinction of the whole population), or 'spiteful' behaviour, which decreases the reproductive output of other members of the population without increasing one's own. Thus, the evidence that female heterocannibalism increases reproductive potential through male competition in threespine sticklebacks is generally supportive since two of the model's three predictions were confirmed.

Cannibalism and egg recognition

In some populations of sticklebacks, females run a high risk of encountering and mistakenly eating their own eggs, which may explain why they have the ability to recognize their spawn (FitzGerald and van Havre 1987; Smith and Whoriskey 1988). Male sticklebacks apparently lack the ability to recognize the eggs they have fertilized (van den Assem 1967; FitzGerald and van Havre 1987). In contrast, male but not female pupfish, can recognize their own eggs, even though both sexes are filial cannibals (Loiselle 1983). Loiselle (1983) explained this asymmetry as the result of differences in movements between the two sexes during the reproductive season. Whereas males are mostly confined to their territories, females are free to roam among territories. Only a small proportion of the eggs present in a given male's territory will probably have been spawned by a particular female. Her access to the eggs is further limited by the aggressive behaviour of the territorial males. Presumably, the likelihood that a female could cannibalize her own progeny under such conditions is too small for selection to have favoured the ability to discriminate. Loiselle (1983) also argued that there is no selective advantage for males to tolerate unrelated eggs that will hatch competitors for his progeny. Unrelated eggs may occur in the male's nests through fertilizations by 'sneaker' males (e.g. van den Assem 1967). Selective predation on unrelated eggs allows him to prolong occupancy of his territory without engaging in extensive filial cannibalism. It is surprising that males of other species, such as sticklebacks and cichlids, do not have the ability to recognize the eggs they have fertilized given the potential for stolen fertilizations by other males, unlike pupfish and fathead minnows, which have this ability (Sargent 1989).

FILIAL CANNIBALISM: REPRODUCTIVE MANAGEMENT?

Occurrence

Filial cannibalism is widely known in fish with male parental care, but it also occurs in species with other types of parental care. Biparental convict cichlids *Cichlasoma nigrofasciatum* are filial cannibals, at least under laboratory conditions (FitzGerald 1976). In this species, male, but not female convict cichlids consumed their own fry. Much individual variation in filial cannibalism occurs among male *Tilapia* (Schwanck 1986), although most males are capable of eating fry, some males were more cannibalistic than others. In many cichlids, both parents invest heavily in parental care over a protracted breeding season. During this period the parents experience an increased risk of predation and injury, and probably a lower growth rate. If the expected reproductive output from a given brood is low, early destruction of the brood by one or both parents might actually increase lifetime reproductive success (see Keenleyside 1979 for review).

Both heterocannibalism and filial cannibalism occur in mouth-brooding cichlids but the former type of cannibalism seems to be limited to the time of spawning and to the time after termination of mouthbrooding. Mrowka (1987) experimentally studied the occurrence of filial cannibalism in the iteroparous maternal mouth-brooding cichlid *Pseudocrenilabrus multicolor*. After spawning the female leaves the male's territory and carries the brood in her mouth for 10–11 days until the young have reabsorbed their yolk and are able to swim. The mothers consume no food during this time but may completely cannibalize small clutches of their own young. Many mothers have in their mouths a mixture of fertilized and unfertilized eggs. The unfertilized eggs are selectively swallowed from day 2 to day 5 after spawning. It is not known how females distinguish between fertilized and unfertilized eggs but a small portion of fertilized eggs (approximately 15 per cent) is also swallowed during mouth-brooding. Mrowka (1987) provides two explanations for swallowing fertilized eggs. First, some of the developing offspring may die because of developmental abnormalities or disease. Second, partial swallowing of the brood is used to offset food limitation. If food limitation is too great then the whole brood in the mouth may be eaten. Thus, to save the greater part of the brood a few offspring are sacrificed to ensure the survival of the rest. If, as a result of cannibalism, the size of the brood falls below approximately 20 per cent of the number of eggs spawned, the remaining eggs, fertilized or not, are swallowed. A third explanation, not considered by Mrowka (1987), is that partial swallowing of the brood may be accidental because eggs may be stuck together, although selection might not favour swallowing *any* eggs if the cost of partial swallowing is high.

Explanations of the adaptive significance of filial cannibalism assume that the parent sacrifices part of its present brood to undertake additional brood cycles. However, the pattern of forced starvation and brood cycling seen by custodial parents in sticklebacks and other fishes is not a pre-condition for the occurrence of filial cannibalism. Hoelzer (1988) and Peterson and Marchetti (1989) reported high levels of filial cannibalism (but not heterocannibalism) in the Cortez damselfish *Stegastes rectifraenum*. This species feeds primarily on algae and its spawning area is also its feeding territory, so it may not be food-limited. Interestingly, some males cannibalized their entire clutches. The benefits of complete clutch cannibalism may be long-term increases in growth and survival, but this has not been measured. Males preferentially ate smaller clutches and were more likely to cannibalize younger clutches (Peterson and Marchetti 1989; see also Ochi 1985). It is possible that filial cannibalism of complete clutches is common in other fish (see Table 2 in Peterson and Marchetti 1989).

Filial cannibalism as a gamete acquisition strategy

Rohwer (1978) explained cannibalism in fishes with male parental care in terms of inclusive fitness (Hamilton 1964). The key point is, that during the breeding season guardian males are confined to their territories and may be forced to fast because their movements are restricted. Any given male may have up to several thousand eggs in its nest as the males are able to collect several different clutches from one or more females. By effectively 'parasitizing' a part of the female's reproductive investment for food, a male may increase rather than decrease his lifetime reproductive success. Rohwer's study has stimulated considerable field and laboratory research on cannibalism in sticklebacks (e.g. Whoriskey and FitzGerald 1985*a*; Salfert and Moodie 1985; FitzGerald and van Havre 1987; Foster 1988; Foster *et al.* 1988; Ridgway and McPhail 1988; Vickery *et al.*1988; Hyatt and Ringler 1989*a*,*b*) and other species (e.g. Petersen and Marchetti 1989; Hoelzer 1988). While there is some evidence in support of Rohwer's predictions (see below), it is clear that his theory does not inadequately explain the existence of filial cannibalism. Contrary to Rohwer, its occurrence is not necessarily linked to the abundance of food supplies (e.g. Belles-Isles and FitzGerald 1991). Perhaps filial cannibalism is determined by *quality* (e.g. digestibility, nutritional content) rather than the quantity of available food. Conspecific eggs may contain some essential nutrient that is difficult to obtain in sufficient quantities by only foraging on other prey.

Nest-raiding and egg-stealing

In addition to explaining why certain species eat their own young, Rohwer (1978) also provides an explanation for nest-raiding and egg-stealing. Nesting male sticklebacks often raid the nests of conspecific males to steal eggs which are then placed in their own nests. This occurs in the both the laboratory (van den Assem 1967) and in natural populations (Whoriskey and FitzGerald 1985*b*), although its frequency in the latter is unknown. Rohwer (1978) argued that male sticklebacks steal eggs because females prefer to spawn in nests already containing eggs (e.g. Ridley and Rechten 1981; Jamieson and Colgan 1989; but see Belles-Isles *et al.* 1990). Consequently, males without clutches may fill their nests with stolen eggs. A female may minimize her losses to the filial cannibalistic male by spawning with males with eggs. The more eggs there are, the lower the chances of a particular egg being eaten.

Filial versus heterocannibalism

Rohwer (1978) emphasized that an individual should always prefer heterocannibalism to eating relatives, and offered specific predictions to distinguish between the two types of cannibalism. First, he argued that, if only heterocannibalism occurs, then guardian males with egg-guarding neighbours are more likely to have eggs in their stomachs than isolated guardians. Second, if filial cannibalism occurs, then eggs should be found in the stomachs of isolated guardians nearly as often as in the stomachs of guardian males with close neighbours. Salfert and Moodie (1985) tested Rohwer's prediction using parental brook sticklebacks. They used radioactive isotopes to label the eggs within the female prior to courtship and deposition of eggs in the nests of single males kept in aquaria. By monitoring changes in radioactivity levels of live males, they were able to test for the occurrence of filial cannibalism throughout the breeding cycle. Filial cannibalism was practised by all of the 41 males tested over one or two breeding cycles. There was also significant parental male heterocannibalism.

These results do not permit unequivocal rejection of Rohwer's hypothesis, since the solitary males were well-fed (see discussion in Hyatt and Ringler (1989*b*). Both Belles-Isles and FitzGerald (1991) and Hoelzer (1988) found that filial cannibalism of eggs was unrelated to food levels, contrary to Rohwer's prediction. Belles-Isles and FitzGerald (1991) argued that filial cannibalism may be more accurately distinguished from heterocannibalism by comparing territorial males tending eggs to those without eggs, rather than comparing territorial and non-territorial males. Parental males should be more cannibalistic than non-parental males if filial cannibalism is important since the former males can eat both their own and their neighbour's eggs, whereas the latter males can only eat their neighbour's eggs. These predictions were supported when tested with two species of sticklebacks (Belles-Isles and FitzGerald 1991).

Cannibalism in species with discrete brood cycles

Rohwer's (1978) theory of filial cannibalism combined with Trivers's (1974) theory of parent-offspring conflict, predicts that towards the end of the brood cycle, when only a few eggs or larvae remain in the nest, a male will benefit more by eating these eggs as an investment in future reproduction than if he guards them. Thus, if filial cannibalism takes place, it should result in males eating the last few eggs or larvae in the nest at the end of the cycle. Such behaviour has been reported in the laboratory on many occasions (e.g. van Iersel 1953; Salfert and Moodie 1985). However, this behaviour could be a laboratory artefact resulting from the intermittent feeding schedule typically applied to captive fish.

Seasonal changes in cannibalism

Rohwer (1978) predicted that filial cannibalism is less frequent later in the season because a male's chances of remating are lower. However, in the laboratory and field, the proportion of male threespine sticklebacks eating their own eggs, and the number of eggs eaten by these males were greater late in the season than earlier (Belles-Isles and FitzGerald 1991). Furthermore, parental males were more cannibalistic than non-parental males later in the season, indicating that filial cannibalism is relatively rare early in the breeding season but later in the season it becomes more common. In contrast, there was no seasonal change in filial cannibalism in parental male blackspotted sticklebacks.

Male versus female cannibalism

Rohwer (1978) argued that if filial cannibalism predominates over heterocannibalism, guardian males should eat more eggs than females because they have easier access to their own eggs. He did not consider that by using courtship tactics, females can approach egg masses closely without being attacked by the defending male. The evidence in support of Rohwer's prediction is equivocal; the studies of Hynes (1950), Semler (1971) and deMartini (1976; see also deMartini 1987) support his prediction, and more recent studies by Hyatt and Ringler (1989*b*) and Salfert and Moodie (1985) also showed that male threespine and male brook sticklebacks, respectively, were more cannibalistic than females. However Whoriskey and FitzGerald (1985*a*) working with different populations of threespine sticklebacks, showed that females were more cannibalistic than males, but that sympatric male blackspotted sticklebacks were more cannibalistic than the females. Based on behavioural observations in the field and a comparison of the developmental stage of eggs in the stomachs of males with eggs in their nests, Kynard (1979) concluded that most egg eating by threespine sticklebacks in Wapato Lake in Washington state (USA) was due to heterocannibalism rather than filial cannibalism. The differences in results among populations may be due to differences in food availability and quality, abundance of eggs, or both.

To conclude, we emphasize that Rohwer's (1978) explanations of male filial cannibalism, and subsequent studies that have attempted to test its predictions, have assumed that male protection and guarding of the young is energetically costly. This may be so, especially for parental fish that stop feeding, but both sticklebacks (Whoriskey and FitzGerald 1985*a*) and Cortez damselfish (Hoelzer 1988) feed during the parental phase and are filial cannibals. Starved male sticklebacks can easily rear one brood (FitzGerald *et al.*1989), but further studies are required to compare the *total* seasonal reproductive success of starved and non-starved parental males. Such studies are required to establish the benefits of filial cannibalism when feeding opportunities are restricted during the breeding season.

Heterocannibalism should be more prevalent than filial cannibalism in natural populations of fish, but this awaits a clear demonstration. Filial cannibalism should be most common when high quality food is limiting or the eggs of conspecific rivals are unavailable to steal. Female heterocannibalism in sticklebacks probably evolved because of the considerable nutritional advantages, but additional benefits such as improving the probability of mating with high quality males cannot be dismissed.

SIBLING CANNIBALISM

Occurrence

A well-known example of sibling cannibalism is *in utero* cannibalism in sharks; only one embryo survives in each oviduct because it eats all the other embryos and egg capusles (Wourms 1977). More recently, sibling cannibalism is reported in Midas cichlids *Cichlasoma citrinelli* by Valerio and Barlow (1986). In many fishes such as cichlids and sticklebacks, the fry remain together in shoals for considerable periods after hatching, which provides considerable opportunities for sibling cannibalism (e.g. Valerio and Barlow 1986; Hecht and Applebaum 1988). Larger fish are able to devour smaller ones because the larger ones swim faster (speed is generally a function of absolute size) and have a wide enough mouth gape to consume smaller conspecifics.

One factor facilitating sibling cannibalism is the uneven rates at which fish grow when reared together (e.g. Brown 1957), permitting some fish to be large enough to cannibalize smaller fish. However, morphological changes can reduce the capacity to cannibalize siblings. Valerio and Barlow (1986) speculated that as young Midas cichlids grow and gradually assume their adult morphology, they may lose their capacity to cannibalize siblings because their body shape becomes less well adapted for pursuit of fast swimming prey. Unfortunately, there is no information about sibling cannibalism among fry of this species in natural populations.

The effects of density, shelter, and food availability on the rate of sibling cannibalism has been investigated experimentally in the African sharptooth catfish *Clarias gariepinus* (Hecht and Appelbaum 1988). Cannibalism started 3.5 days after the onset of feeding. The cannibals had a mean total length of 8 mm at the start, and they continued their cannabilistic behaviour up to the size of 80 mm in length, 47 days after the onset of feeding. Two different types of cannibalism were evident, both of which were dependent upon the relationship between predator mouth width and prey head width. The first, which they termed Type I cannibalism, occurs in the length range from 8 to 45 mm, and is characterized by the prey being caught tail first and swallowed up to the head. Type II cannibalism is characterized by the prey being swallowed head first. The frequency of Type I cannibalism was higher than Type II. Sibling cannibalism was found to be positively density-dependent, was significantly suppressed by shelter, and negatively correlated with food availability.

Adaptive significance

Sibling cannibalism may be advantageous under conditions of low food availability.

Victims may be transitory 'food caches' (Alexander 1974) that store energy for their kin (see also Chapters by Crespi, Crump, Kukuk, Stanback and Koeing, and Waddell, this volume). Early food abundance followed by lower levels may also favour the evolution of the tendency to eat siblings. Ingestion of eggs or younger, smaller fry by older, larger siblings, allows them to survive critical early stages and accelerates their growth and development. In those species with parental care, the food-cache strategy also allows parents gradually to provision developing young. The production of expendable offspring to be eaten by siblings could be viewed as an energetic alternative to producing fewer eggs, each containing larger amounts of nutrients (Polis 1981). The best example of this in fish may be Wourm's (1977) study of *in utero* cannibalism in sharks.

ANTI-CANNIBALISM STRATEGIES

Offspring strategies

Few studies have directly addressed the question of what animals can do to avoid cannibalism (but see Chapters by Baur, Crump, Elgar, and Elwood this volume). As with predator-prey or host-parasite relationships, the prey or host may be expected to have evolved counter strategies to minimize losses to the predator. Some amphibians have eggs containing a substance that deters intraspecific predators (e.g. Durant *et al.* 1983), but there are few recorded examples of poisonous fish eggs. Fuhrman *et al.* (1969) critically reviewed the evidence for toxins in the roe of fishes. The most toxic and best known examples are pufferfishes (order Tetradontiformes). It is not known if cannibalism was the selective force responsible for the evolution of toxic gametes. However, it is noteworthy that so few of the approximately 20 000 fish species (Moyle and Cech 1982) have developed chemical defences for gametes in response to either intra- or interspecific predation. In fish it may be more important to produce many 'cheap' eggs than to invest in egg defences (Orians and Janzen 1974).

Fry and juveniles of many species form shoals, and shoaling may serve to reduce both intraspecific and interspecific predation (Pitcher 1986). Foster *et al.* (1988) argued that cannibalism was the driving force leading to the evolution of ontogenetic changes in habitat use in threespine sticklebacks. Small fry (<15 mm standard length) are concentrated in the vegetation whereas larger fry are pelagic. Because fry in all size classes feed primarily on zooplankton, even when in vegetation, Foster *et al.* (1988) argued that cannibalism of fry by adult conspecifics was responsible for the observed shift in habitat use with age (size). Fry are capable of rapidly assessing predation risk and of altering their behaviour accordingly.

Adult strategies

As previously mentioned, explanations of filial cannibalism assume that fish with male parental care parasitize the reproductive investment of females (Rohwer 1978). However little attention has been directed to how females might minimize their loss of eggs to males.

Female strategies

Females could reduce cannibalism on their eggs by spawning in nests containing eggs from previous females. There is some experimental evidence that this occurs in sticklebacks (Ridley and Rechten 1981; but see Belles-Isles *et al.* 1990) and the Cortez damsel fish (Peterson and Marchetti 1989). Perhaps cannibalistic females could also avoid eating their own eggs or avoid cannibalizing eggs in the nests where they deposited their own eggs. The ability of females to recognize and preferentially attack unrelated eggs (FitzGerald and van Havre 1987), may only occur in habitats where females are likely to re-encounter the nest in which they had previously spawned. It would be interesting to determine if females' discriminatory abilities vary with habitat types.

Male Strategies

As previously mentioned, in many populations of sticklebacks (e.g. Kynard 1979; Synder 1984; Whoriskey and FitzGerald 1985; Foster *et al.* 1988; Ridgway and McPhail1988), female and non-territorial male fish often form shoals to raid nests and consume the eggs. Raids are initiated by a single individual, but are usually successful only if other fish join the group. For males, the frequency of nest-raiding and egg-cannibalism by gravid females poses a serious risk. A guardian (parental) male defends his nest against groups of less than six or seven females by biting and chasing them off his territory. Males also use distraction displays to lure groups of females away from their nests (Whoriskey and FitzGerald 1985*a*; Ridgway and McPhail 1988; Foster 1988). The stickleback display is the only reported occurrence of this behaviour in fish, and the only report from any taxa of its use as an anti-cannibal strategy. Decoy displays are not always successful and males must sometimes court to obtain eggs. Many raids on nests may be triggered when males court. Males with eggs minimize this risk by courting isolated females and rarely risk courtship advances when females shoal (Belles-Isles *et al.* 1990). Males without eggs also rarely court groups of females. Perhaps males do not court a female in a group because if a group member spawns in its nest the others may immediately eat her eggs. Vickery *et al.* (1988) suggested that a male may refuse to court females which have previously raided his nest. To do this, males must recognize the females that raided his nest and forego the additional offspring he might produce by mating with these females. Unfortunately we do not know if males can recognize individual females. Males could control the number of eggs in their nests, but theoretically they should not refuse additional eggs because doing so could increase the number of reproductive, potentially raiding females remaining in the local population (Vickery *et al.* 1988).

RECAPITULATION AND PROSPECTS

In most of the studies reviewed, cannibalism is assumed to be adaptive. However, it is unlikely that all occurrences of cannibalism are adaptive (but see Glass *et al.* 1985). Cannibalism caused by social pathology (stress of overcrowding) could be especically prevalent in laboratory situations. For example, the sonoran topminnow *Poeciliopsis occidentalis occidentalis* is highly cannibalistic in aquaria but not in nature (Meffe 1984).

There is strong evidence that cannibalism has a genetic basis in fish, since different strains of one species have different heritable cannibalistic tendencies (Dominey and Blumer 1984). It would be interesting to discover if there are differences in the frequency and type of cannibalism among different populations. As previously mentioned, in some populations of threespine sticklebacks, males are more cannibalistic than females, whereas the opposite is true in other populations. Foster (ms. under review) has described differences in the occurrence and kind of male distraction display among geographically distinct populations of sticklebacks and showed that inter-population differences are associated with different patterns of cannibalism. Males in some populations exhibit the full repertoire of male courtship behaviour as described in some ethological textbooks (e.g. Slater 1985) and adult cannibalism of eggs guarded by males was never observed. In populations where this form of cannibalism is frequent, male courtship patterns were less conspicuous, and displays that divert cannibalistic conspecifics from offspring were frequently performed by guardian males. Diversionary displays were not performed by males at the approach of conspecifics in the non-cannibalistic populations.

The factors determing the type and frequency of cannibalism in closely-related species also requires investigation. Poeciliid fishes, which bear live young, include a spectrum from highly cannibalistic species to apparently non-cannibalistic species (Meffe 1984). For some species in the genus *Poeciliopsis*, a genetic component of cannibalism is apparent.

In summary, cannibalism is common and widespread in fishes and plays a major role in the reproductive biology of some species such as sticklebacks. However, why the patterns of cannibalism differ between closely-related species and why different populations of the same species have different patterns of cannibalism is still poorly understood. Answers to these questions will significantly advance our understanding of the evolutionary basis of cannibalism in fish.

ACKNOWLEDGEMENTS

The research on cannibalism in sticklebacks of both authors was supported by the National Sciences and Engineering Research Council of Canada (NSERC) and 'Fonds pour la Formation de Chercheurs et l'Aide la Recherche' (FCAR). We thank Jean-Claude Belles-Isles, Julian Dodson, France Dufresne, Susan Foster, Helga Guderley, Anne Magurran, Craig Sargent, Gary Meefe, and the editors for comments on various versions of the manuscript.

REFERENCES

Alexander, R.D. (1974). The evolution of social behaviour. *Annual Review of Ecology and Systematics*, **5**, 325–83.

Alheit, J. (1987). Egg cannibalism versus egg predation: their significance in anchovies. *South African Journal of Marine Science*, **5**, 467–70.

Assem, J. van den (1967). Territory in the three-spined sticklebacks *Gasterosteus aculeatus* L. An experimental study in intra-specific competition. *Behaviour Supplement*, **16**, 1–164.

Belles-Isles, J.-C., Cloutier, D., and FitzGerald, G.J. (1990). Female cannibalism and male courtship tactics in threespine sticklebacks. *Behavioural Ecology and Sociobiology*, **26**, 363–8.

Belles-Isles, J. C. (1991). De la valeur adaptative de l'ovocannibalisme chez l'épinoche à trois épines *Gasterosteus aculeatus*. Thèse de doctorat. Université Laval, Québec, Canada.

Belles-Isles, J.C. and FitzGerald, G.J. (1991). Filial cannibalism in sticklebacks: a reproductive management strategy? *Ethology Ecology and Evolution*, **3**, 49–62.

Brown, M.E. (1957). Experimental studies on growth. In *The physiology of fishes*, Vol 1, (ed. M.E. Brown), pp. 361–400. Academic Press, New York.

deMartini, E.E. (1976). The adaptive significance of territoriality and egg cannibalism in the painted greenling *Oxylebius pictus* Gill, a northeastern Pacific marine fish. Ph.D. thesis, University of Washington.

DeMartini, E.E. (1987). Paternal defense, cannibalism and polygamy: factors influencing the reproductive success of painted greenling (Pisces, Hexagrammidae). *Animal Behaviour*, **35**, 1145–58.

Dionne, M. (1985). Cannibalism, food availability, and reproduction in the mosquitofish (*Gambusia affinis*): a laboratory experiment. *American Naturalist*, **126**, 16–23.

Dominey, W. and L.S. Blumer (1984). Cannibalism of early life stages in fishes. In *Infanticide: comparative and evolutionary perspectives*, (ed. G. Hausfater and S.B. Hrdy), pp. 43–64. Aldine Publishing Company, New York.

Durant, J.P., Bouillon, M., and Parzefall, J. (1983). Réponse de *Proteus anguinius* L. amphibien cavernicole à des stimulations chimiques provenant de la ponte. *Mémoire de Biospéléologie*, **10**, 395–99.

FitzGerald, G.J. (1976). The effects of density of adult fish on reproduction and parental behavior in the convict cichlid *Cichlasoma nigrofasciatum* (Günther). Ph.D thesis, University of Western Ontario.

FitzGerald, G.J. (1983). The reproductive ecology and behaviour of three sympatric sticklebacks (Gasterosteidae) in a saltmarsh. *Biology of Behaviour*, **8**, 67–79.

FitzGerald, G.J. and Havre, N. van (1987). The adaptive significance of cannibalism in sticklebacks (Gasterosteidae: Pisces). *Behavioural Ecology and Sociobiology*, **20**, 125–8.

FitzGerald, G.J., Guderley, H., and Picard, P. (1989). Hidden reproductive costs in the threespined stickleback (*Gasterosteus aculeatus*). *Experimental Biology*, **48**, 295–300.

Foster, S.A. (1988). Diversionary displays of paternal sticklebacks: defenses against cannibalistic groups. *Behavioural Ecology and Sociobiology*, **22**, 335–40.

Foster, S.A., Garcia, V.B., and Town, M.Y. (1988). Cannibalism as the cause of an ontogenetic shift in habitat use by fry of the threespine stickleback. *Oecologia*, **74**, 577–85.

Fox, L. (1975). Cannibalism in natural populations. *Annual Review of Ecology and Systematics*, **6**, 87–106.

Fuhrman, F.A., Fuhram, G.A., Dull, D.L., and Mosher, H.S. (1969). Toxins from eggs of fishes and amphibia. *Journal of Agricultural Food Chemistry*, **17**, 417–24.

Glass, G.E., Holt, R.D., and Slade, N.A. (1985). Infanticide as an evolutionary stable strategy. *Animal Behaviour*, **33**, 384–91.

Gross, M.R. and Shine, R. (1981). Parental care and mode of fertilization in ectothermic vertebrates. *Evolution*, **35**, 775–93.

Hamilton, W. (1964). The genetical evolution of social behavior I, II. *Journal of Theoretical Biology*, **7**, 1–52.

Hecht, T. and Appelbaum, S. (1988). Observations on intraspecific aggression and coeval sibling cannibalism by larval and juvenile *Clarias gariepinus* (Clariidae: Pisces) under controlled conditions. *Journal of Zoology*, **214**, 21–44.

Hoelzer, G. (1988). Filial cannibalism in a non-brood cycling marine fish. *Environmental Biology of Fishes*, **21**, 309–13.

Hrdy, S.B. (1979). Infanticide among animals: a review, classification, and examination of the implications for the reproductive strategies of females. *Ethology and Sociobiology*, **1**, 13–40.

Hyatt, K.D. and Ringler, N. H. (1989*a*). The role of nest raiding and egg predation in regulating population density of threespine sticklebacks (*Gasterosteus aculeatus*) in a coastal British Columbia lake. *Canadian Journal of Fisheries and Aquatic Sciences*, **46**, 372–83.

Hyatt, K.D. and Ringler, N.H. (1989*b*). Egg cannibalism and the reproductive strategies of threespine sticklebacks (*Gasterosteus aculeatus*) in a coastal British Columbia lake. *Canadian Journal of Zoology*, **67**, 2036–46.

Hynes, H.B. N. (1950). The food of freshwater sticklebacks (*Gasterosteus aculeatus* and *Pygosteus pungitius*) with a review of methods used in the studies of the food of fishes. *Journal of Animal Ecology*, **19**, 36–58.

Iersel, J.J.A. Van (1953). An analysis of parental behaviour of the male three-spined stickleback (*Gasterosteus aculeatus* L.). *Behaviour Supplements*, **3**, 1–159.

Jamieson, I.G. and Colgan, P.W. (1989). Eggs in the nests of males and their effect on mate choice in three-spined stickleback. *Animal Behaviour*, **38**, 859–65.

Keenleyside, M.H.A. (1979). *Diversity and adaptation in fish behaviour*, Springer Verlag, New York.

Kynard, B. (1979). Breeding behaviour of a lacustrine population of threespine sticklebacks (*Gasterosteus aculeatus* L). *Behaviour*, **67**, 178–207.

Loadman, N.L., Moodie, G.E.E., and Mathias, J.A. (1986). Significance of cannibalism in larval walleye *Stizosteidion vitreum*. *Canadian Journal of Fisheries and Aquatic Sciences*, **43**, 613–18.

Loekle, D.M., Madison, D.M., and Christian, J.J. (1982). Time dependency and kin recognition of cannibalistic behaviour among poeciliid fishes. *Behavioural and Neural Biology*, **35**, 315–18.

Loiselle, P.V. (1983). Filial cannibalism and egg recognition by males of the primatively custodial teleost *Cyprinodon maculatus californiensis* Girard (Antherinomorpha: Cyprinodontidae). *Ethology and Sociobiology*, **4**, 1–9.

Marconato, A. and Bisazza, A. (1988). Mate choice, egg cannibalism and reproductive success in the river bullhead, *Cottus gobio* L. *Journal of Fish Biology*, **33**, 905–16.

Meffe, G. (1984). Density-dependent cannibalism in the endangered Sonoran topminnow (*Poeciliopsis occidentalis*). *Southwestern Naturalist*, **29**, 500–3.

Meffe, G. (1986). Cannibalism, food availability, and reproduction in mosquitofish: a critique. *American Naturalist*, **127**, 897–901.

Meffe, G. and Crump, M.L. (1987). Possible growth and reproductive benefits of cannibalism in the mosquitofish. *American Naturalist*, **129**, 203–12.

Moyle, P. B. and Cech, J.J., Jr. (1982). *An introduction to ichthyology*. Prentice-Hall Inc., Englewood Cliffs, NJ.

Mrowka, W. (1987). Filial cannibalism and reproductive success in the maternal mouthbrooding cichlid fish *Pseudocrenilabrus multicolor*. *Behavioural Ecology and Sociobiology*, **21**, 257–65.

Ochi, H. (1985). Termination of parental care due to small clutch size in the temperate damselfish, *Chromis notata*. *Environmental Biology of Fishes*, **12**, 155–60.

Orians, G. and Janzen, D.H. (1974). Why are embryos so tasty? *American Naturalist*, **108**, 581–92.

Paller, M.H. and Lewis, W.H. (1987). Effects of diet on growth depensation and cannibalism among intensively cultured larval striped bass. *Progressive Fish-Culturist*, **49**, 270–5.

Petersen, C.W. and Marchetti, K. (1989). Filial cannibalism in the Cortez Damselfish *Stegastes rectifraenum*. *Evolution*, **43**, 158–68.

Pitcher, T.J. (1986). Functions of shoaling behaviour in teleosts. In *The behaviour of teleost fishes*, (ed. T.J. Pitcher), pp. 294–337. Croom Helm Inc, London and Sydney.

Polis, G. (1981). The evolution and dynamics of intraspecific predation. *Annual Review of Ecology and Systematics*, **12**, 125–51.

Poulin, R. and FitzGerald, G.J. (1989). Early life histories of three sympatric sticklebacks in a salt-marsh. *Journal of Fish Biology*, **34**, 207–9.

Reebs, S.G., Whoriskey, F.G., and FitzGerald, G.J. (1984). Diel patterns of fanning activity and nocturnal behaviour of male three-spined sticklebacks *Gasterosteus aculeatus* L. (*f. trachurus*). *Canadian Journal of Zoology*, **62**, 329–34.

Ridgway, M. and McPhail, J.D. (1988). Raiding shoal size and a distraction display in male sticklebacks (Gasterosteus). *Canadian Journal of Zoology*, **66**, 201–5.

Ridley, M. and Rechten, C. (1981). Female sticklebacks prefer to spawn with males whose nests contain eggs. *Behaviour*, **76**, 152–61.

Rohwer, S. (1978). Parent cannibalism of offspring and egg raiding as a courtship strategy. *American Naturalist*, **112**, 429–40.

Salfert, I. G. and Moodie, G.E.E. (1985). Filial egg cannibalism in the brook stickleback, *Culaea inconstans* (Kirkland). *Behaviour*, **93**, 82–100.

Sargent, R.C. (1989). Allopaternal care in the fathead minnow, *Pimephales promeleas* stepfathers discriminate against their adopted eggs. *Behavioural Ecology and Sociobiology*, **25**, 379–85.

Schwanck, E. (1986). Filial cannibalism in *Tilipia mariae*. *Sonderdruck aus Journal of Applied Ichtyology*, **2**, 65–74.

Semler, D.E. (1971). Some aspects of adaptation in a polymorphism for breeding colours in the threespine stickleback (*Gasterosteus aculeatus*). *Journal of Zoology*, **165**, 291–302.

Slater, P.J.B. (1985). *An introduction to ethology*. Cambridge University Press.

Smith, R.S. and Whoriskey, F.G. (1988). Multiple clutches: female sticklebacks lose the ability to recognize their own eggs. *Animal Behaviour*, **36**, 1838–9.

Synder, R. J. (1984). Seasonal variation in the diet of the threespine stickleback, *Gasterosteus aculeatus* in Contra Costa County, California. *California Fish and Game*, **70**, 167–72.

Trivers, R. (1974). Parent-offspring conflict. *American Zoologist*, **14**, 249–64.

Valdes, E.S.,Shelton, P.A., Armstrong, M.J., and Field, J.G. (1987). Cannibalism in South African anchovy: egg mortality and egg consumption rates. *South African Journal of Marine Science*, **5**, 613–22.

Valerio, M. and Barlow, G.W. (1986). Ontogeny of young Midas cichlids: a study of feeding, filial cannibalism and agonism in relation to differences in size. *Biology of Behaviour*, **11**, 16–35.

Vickery, W., Whoriskey, F.G., and FitzGerald, G.J (1988). On the evolution of nest raiding and male defensive behaviour in sticklebacks (Pisces: Gasterosteidae). *Behavioural Ecology and Sociobiology*, **22**, 185–3.

Ward, G. and FitzGerald, G.J. (1983). Macrobenthic abundance and distribution in tidal pools of a Quebec salt marsh. *Canadian Journal of Zoology*, **61**, 1071–85.

Ward, G. and FitzGerald, G.J. (1988). Effects of sex ratio on male behaviour and reproductive success in a field population of threespine sticklebacks (*Gasterosteus aculeatus*) (Pisces: Gasterosteidae). *Journal of Zoology* (London), **215**, 597–610.

Whoriskey, F.G. and FitzGerald, G.J. (1985*a*). Sex, cannibalism and sticklebacks. *Behavioural Ecology and Sociobiology*, **18**, 15–18.

Whoriskey, F.G. and FitzGerald, G.J. (1985*b*). Nest sites of the threespine stickleback: can site characteristics alone protect the nest against egg predators and are nest sites a limiting resource? *Canadian Journal of Zoology*, **63**, 301–7.

Worgan, J.P. and FitzGerald, G.J. (1981). Diel activity and diet in three sympatric sticklebacks in tidal salt marsh pools. *Canadian Journal of Zoology*, **59**, 2375–9.

Wourms, J. (1977). Reproduction and development in chondrichthyan fishes. *American Zoologist*, **17**, 379–410.

12

Cannibalism in amphibians

Martha L. Crump

The complex life cycle of most salamanders and anurans (frogs and toads), involving an aquatic larval stage and a semi-terrestrial or terrestrial adult stage, creates a complicated web of potential intraspecific interactions that include cannibalism. Cannibalism is defined here as the killing and consumption of live conspecifics, and therefore will exclude from discussion the consumption of already dead conspecifics or parts of a conspecific (e.g., tails, limbs, and gills, Borland and Rugh 1943; Petranka 1989).

Cannibalism is both taxonomically and biogeographically widespread among amphibians, reported in at least 12 of the 21 families of anurans, seven of the nine families of salamanders, and one family of caecilians. Amphibians cannibalize both genetically related individuals (filial or sibling cannibalism, *sensu* Rohwer 1978) and genetically unrelated individuals (heterocannibalism, *sensu* Rohwer 1978). Generally individuals prey upon their own or younger age classes. Thus, adults may eat eggs, larvae, subadults (post-metamorphic individuals that are not yet sexually mature), or other adults. Subadults eat eggs, larvae, or other subadults, and larvae eat eggs or other larvae. Although cannibalism is widespread, in many cases the behaviour has not been well documented under natural field conditions, and the ecological and evolutionary significance remains unclear. For example, reports of cannibalism in many species of amphibians are based on captive animals. Stress or overcrowding may result in unnatural behaviour. For other species, evidence for cannibalism is based on one conspecific discovered in the stomach of another animal.

The present review of cannibalism in amphibians will be based primarily on species studied under field conditions and for which cannibalism appears to be a normal part of the animal's behavioural repertoire. Thus, I have omitted a large body of literature that describes annecdotal field or incidental laboratory observations. In this chapter I discuss types of cannibalism in amphibians, summarize the ecological circumstances under which cannibalism occurs or is expected to occur, discuss the fitness costs and benefits associated with eating conspecifics, and speculate on the evolution of cannibalism in amphibians. For previous reviews of cannibalism in amphibians see Crump (1983), Simon (1984), and Polis and Myers (1985).

TYPES OF CANNIBALISM WITHIN THE AMPHIBIA

Oophagy

Oophagy is a common form of cannibalism in amphibians (Simon 1984), practiced by all post-egg life history stages but most commonly by larvae (Table 12.1). Eggs provide an energy-rich resource, are high in calcium and phosphorus, and are especially vulnerable because they are immobile. Two distinct forms of conspecific oophagy are exhibited by amphibian larvae. One type is opportunistic feeding on eggs by tadpoles (anuran larvae) and salamander larvae, generally under conditions where food is limited (e.g. Muedeking and Heyer 1976; Crump 1983). This type of oophagy usually does not involve specializations by the cannibal for eating eggs (Crump 1983). The second type of oophagy involves consumption of unfertilized eggs (trophic or nutritive eggs) that are provided to the tadpoles by the mother (Weygoldt 1980; Zimmermann and Zimmermann 1981; Jungfer 1985; Wassersug *et al.* 1981; Lannoo *et al.* 1987). All tadpoles known or suspected to exhibit trophic feeding are arboreal and live in small, isolated bodies of water such as in the water-filled axils of plants or tree holes. These tadpoles generally have short digestive tracts associated with carnivory, and buccal modifications for oophagy (Noble 1929, 1931; Wassersug *et al.* 1981; Lannoo *et al.* 1987). Because this behaviour is sometimes referred to as 'cannibalism', it is mentioned here, but will not be considered further. Trophic feeding is not considered as cannibalism here, since 'victims' are inviable eggs rather than viable individuals.

Oophagy by subadult and adult conspecifics is less widespread than oophagy by larvae, presumably because few species of amphibians live as subadults and adults in the habitat where eggs develop. Filial cannibalism is found in some salamanders in which egg-brooding parents eat their own diseased or fungus-infected eggs (e.g. Tilley 1972; Forester 1979), thus increasing survival of the remaining eggs. Heterocannibalism occurs in both salamanders and anurans, where adults cannibalize others' viable eggs (e.g. Kaplan and Sherman 1980; Morin 1983; Townsend *et al.* 1984). In some species (e.g. *Taricha torosa*, Kaplan and Sherman 1980; Marshall *et al.* 1990) conspecifics eat eggs directly from a female's cloaca.

Cannibalism on larval stages

Larval–larval cannibalism is the most frequent type of cannibalism among amphibians (Table 12.1). Among salamanders, *Ambystoma tigrinum* larvae are particularly cannibalistic (Rose and Armentrout 1976; Lannoo and Bachmann 1984; Holomuzki 1986), as are *Hynobius nebulosus* (Kusano *et al.* 1985) and *Salamandra salamandra* (Warburg *et al.* 1979). Tadpole–tadpole cannibalism has been well documented under field conditions for numerous species (Table 12.1), many of which have large mouths, hypertrophied jaw musculature, and aggressive predatory feeding behaviour (e.g. *Hoplophryne rogersi*, *Ceratophrys*,

Table 12.1 Records of cannibalism within the Amphibia, based primarily on field observations

Taxa	Pred.	Prey	Source
Caudata			
Ambystomatidae			
Ambystoma annulatum	L	L	Hutcherson *et al.* 1989
Ambystoma tigrinum mavortium	L	E, L	Rose and Armentrout 1976
	A	E	Norris 1989
Ambystoma tigrinum nebulosum	L	E, L	Holomuzki 1986
Ambystoma tigrinum tigrinum	L	L	Lannoo and Bachmann 1984
Cryptobranchidae			
Andrias japonicus	A	E, L, S	Tago 1927
Cryptobranchus alleganiensis	A	E, S	Bishop 1927; Nickerson and Mays 1973
Dicamptodontidae			
Dicamptodon ensatus	L	L	Nussbaum and Clothier 1973
	A	A	Anderson 1960
Hynobiidae			
Hynobius nebulosus	L	L	Kusano *et al.* 1985
Plethodontidae			
Desmognathus fuscus fuscus	A (F)	E	Baldauf 1947; Wood and Clarke 1955
Desmognathus monticola	A	S	Colley *et al.* 1989
Desmognathus ochrophaeus	A (M, F)	E, L, S	Wood and Wood 1955; Tilley 1972
Plethodon cinereus	A	S	Heatwole and Test 1961
Plethodon glutinosus	A (F)	S	Powders 1973
Proteidae			
Necturus maculosus	A (F)	E	Bishop 1927
Salamandridae			
Notophthalmus viridescens	A	E, L	Morin 1983
Notophthalmus v. viridescens	A	L	Burton 1977
Salamandra salamandra	L	L	Degani *et al.* 1980; Warburg *et al.* 1979
Taricha torosa	A	E	Kaplan and Sherman 1980; Marshall *et al.* 1990
Anura			
Bufonidae			
Bufo arenarum	L	E	Crump, in prep.
Bufo boreas halophilus	A	S	Cunningham 1954
Bufo vulgaris	L	L	Nagai *et al.* 1971

Table 12.1 *(continued)*

Taxa	Pred.	Prey	Source
Dendrobatidae			
Dendrobates auratus	A (F)	E	Summers 1989
	L	L	K. Summers, pers. comm.
Discoglossidae			
Bombina bombina	A	S	Heusser 1971
Bombina orientalis	A (M)	E	Marshall *et al.* 1990
Hylidae			
Hyla pseudopuma	L	E, L	Crump 1983
Hyla rosenbergi	L	E, L	Kluge 1981
Leptodactylidae			
Ceratophrys cornuta	L	L	Noble 1931
Ceratophrys ornata	L	L	Cei 1980
Ceratophrys pierottii	A	A	Cei 1955, 1980
	S	S	Cei 1980
Eleutherodactylus coqui	A	E	Townsend *et al.* 1984
Lepidobatrachus asper	A	A	Reig and Cei 1963
Lepidobatrachus llanensis	L	L, E	Crump, in prep.
Leptodactylus knudseni	L	E	Hero and Galatti 1990
Leptodactylus pentadactylus	L	E	Muedeking and Heyer 1976
Pleurodema nebulosa	L	L	Cei 1979
Microhylidae			
Cophixalus parkeri	A	E	Simon 1983
Hoplophryne rogersi	L	E	Noble 1929
Myobatrachidae			
Lechriodus fletcheri	L	L	Martin 1967
Pelobatidae			
Scaphiopus bombifrons	L	L	Bragg 1946, 1964, 1965
Scaphiopus hammondi	L	L	Bragg 1965
Scaphiopus holbrooki holbrooki	L	L	Ball 1936
	L	E	Richmond 1947
Scaphiopus holbrooki hurteri	L	L	Bragg 1964, 1965
Scaphiopus multiplicatus	L	L	Pomeroy 1981
Pipidae			
Xenopus laevis	A	E, L	McCoid and Fritts 1980
Ranidae			
Pyxicephalus adspersus	A	L, S	Loveridge 1950
	S	S	Grobler 1972; Pienaar *et al.* 1976

Table 12.1 (*continued*)

Pyxicephalus adspersus	L	L	Grobler 1972
Rana esculenta	A	L, S	Dushin 1974
Rana ridibunda	A	L, S	Dushin 1974
Rana tigrina	L	L	McCann 1932
Rhacophoridae			
Chirixalus eiffingeri	L	E	Ueda 1986
Rhinophrynidae			
Rhinophrynus dorsalis	L	L	Starrett 1960

A: adult (F: female, M: male), E: eggs, L: larva, S: subadult.

Lepidobatrachus, species of the *Rana rugulosa* group, some hylids, and the cannibalistic morph of some *Scaphiopus*; Orton 1954; Ruibal and Thomas 1988). Although tadpoles generally prey on smaller, younger stages, in at least four species of *Scaphiopus* (Bragg 1964, 1965; Pomeroy 1981) and in *Osteopilus septentrionalis* (Crump 1986) younger conspecifics cannibalize transforming individuals. Some aquatic and semi-aquatic adult amphibians such as *Rana* (Dushin 1974), *Xenopus* (McCoid and Fritts 1980), *Notophthalmus* (Burton 1977; Morin 1983), and *Desmognathus* (Wood and Wood 1955) live in the same habitat as conspecific larvae and prey upon them (Table 12.1).

Cannibalism on subadult and adult stages

The least common form of cannibalism among amphibians is adults feeding on adults or subadults, and subadults feeding on subadults. Presumably such cannibalism results from indiscriminate predation by generalist foragers. Generally the prey individual is smaller than the cannibal. For many species the evidence for this type of cannibalism is based on only a few observations in nature (e.g. *Bufo boreas halophilus*, Cunningham 1954; *Dicamptodon ensatus*, Anderson 1960; *Plethodon glutinosus*, Powders 1973), even though in some cases the stomach contents of many individuals have been examined. Cannibalism in these species presumably is not an important source of mortality. In contrast, in a few species, subadults and adults frequently prey on conspecifics, especially when densities are high (e.g. *Chacophrys*, Cei 1955, 1980; *Pyxicephalus*, Grobler 1972; Pienaar *et al.* 1976; *Rana*, Dushin 1974). In these species although cannibalism may be a significant source of mortality, there is no evidence suggesting that conspecifics are preferred prey items; they are simply available and of appropriate size.

In utero cannibalism

Like trophic egg feeding, *in utero* cannibalism is a way in which a parent can provision its young with nutrients and energy. Also like trophic feeding, this behaviour does not fit the definition of cannibalism because the victims are not viable individuals. For example, in *Salamandra atra*, intra-uterine larvae feed on nutritive eggs. Only one or two eggs in each oviduct are fertilized; another 20–30 eggs degenerate and the yolk is ingested by the fetuses after they exhaust their own yolk reserves (references cited in Duellman and Trueb 1986).

Cannibalistic polyphenism

Polyphenism (discrete phenotypic differences in morphology, behaviour, and growth rates between individuals within a population) in relation to cannibalism has been well documented for *Ambystoma tigrinum* (tiger salamanders) and several species of *Scaphiopus* (spadefoot toads). Certain larvae grow larger, develop a distinctive morphology, and become cannibalistic, while others retain the 'typical' characteristics and usually do not prey on live conspecifics.

Three subspecies of the *Ambystoma tigrinum* complex (*A. t. mavortium, A. t. nebulosum,* and *A. t. tigrinum*) have both cannibalistic and 'typical' larval morphs (Collins *et al.* 1980; Lannoo and Bachmann 1984). Characteristics of the cannibalistic morph include a disproportionately large head, wide mouth, slender body form, slit-like eyes, and elongate vomerine teeth (Rose and Armentrout 1976). These exaggerated features presumably facilitate the capture and swallowing of large prey (Pierce *et al.* 1983). Cannibalistic morphs are macrophagous carnivores; they eat conspecifics when available, but are not obligate cannibals. These morphs are usually the largest individuals in the population, and their growth rates are generally faster than for non-cannibal morphs (Rose and Armentrout 1976; Lannoo and Bachmann 1984). In all polyphenic populations of *A. tigrinum* studied, the cannibalistic morph is relatively rare.

Factors responsible for induction of the cannibalistic morph are not entirely clear. Pierce *et al.* (1981) found differences in allozyme frequencies between cannibal and non-cannibal morphs of *A. t. mavortium* inhabiting the same pond and suggested that cannibal morphology might have some genetic basis, although allozyme differences were not consistent among different populations. Collins and Cheek (1983) demonstrated experimentally that polyphenism in *A. t. nebulosum* is in large part environmentally controlled. Their experiments suggest that high density of conspecifics, not food shortage, may induce the cannibal morphology in certain individuals. Lannoo and Bachmann (1984) also suggested, based on field observations, that for *A. t. tigrinum* high larval density, not food shortage, induces the expression of cannibalistic morph traits.

Lannoo *et al.* (1989) described a third morph of *A. t. tigrinum* with characteristics intermediate between typical and cannibal morphs. Under laboratory conditions, this morph is cannibalistic (although it lacks the specialized anatomy of the cannibal morph) and experiences subsequent accelerated growth and early metamorphosis even after eating only one conspecific. This intermediate morph is genetically indistinguishable from the typical morph, as determined by electrophoresis (Lannoo *et al.* 1989). These results are interesting, but it is necessary to determine (1) whether this observation is an artifact of captivity; (2) how widespread the occurrence of a third morph is within this subspecies and in the other two subspecies of *A. tigrinum* known to have cannibal morphs; and (3) environmental correlates associated with this behavioural plasticity.

Polyphenism is also found in tadpoles of four species of *Scaphiopus* (Orton 1954; Bragg 1956; Bragg and Bragg 1959; Bragg 1964; Pomeroy 1981). Cannibalistic morphs (referred to alternatively as carnivorous morphs) develop enlarged jaw muscles, larger and more serrate beaks, and generally larger body size than do omnivorous morphs. Tadpoles of the omnivorous morph are primarily scraping suspension feeders (which eat detritus and algae) whereas the cannibalistic or carnivorous morphs are highly predaceous on invertebrates and tadpoles. In some species of *Scaphiopus* (e.g. *S. multiplicatus*) a third form is found that is intermediate in morphology, foraging behaviour, and diet between cannibalistic and omnivorous morphs. Although Pomeroy (1981) referred to this form as a morph, more recent studies by Pfennig suggest that *S. multiplicatus* tadpoles remain somewhat totipotent throughout ontogeny with regard to morph determination; the intermediates appear to be animals in the process of transforming from the omnivore morph to the carnivore morph or vice versa (D. Pfennig, personal communication). Although there may be genetic variation in the propensity to become a cannibal in *Scaphiopus* (Hampton and Volpe 1963; Pomeroy 1981), morph determination in *S. multiplicatus* appears to be environmentally induced (Pomeroy 1981; Pfennig 1990*b*). In this species, ingestion of live fairy shrimp triggers transformation from the omnivorous morph to the carnivorous morph (Pfennig 1990*b*).

FACTORS CONTRIBUTING TO CANNIBALISM

Appropriate feeding morphology and behaviour

Feeding morphology must permit capture, handling, and ingestion of the prey in order for cannibalism to occur. The lack of appropriate morphology may explain the apparent absence of cannibalism in some species (see also Stanback and Koenig, Baur, and Hiraiwa-Hasegawa this volume). Tadpoles that are strictly microphagous suspension feeders cannot handle or ingest whole conspecific eggs or tadpoles, whereas species that have macrophagous feeding morphology can manipulate eggs and tadpoles. Predaceous larvae that feed on invertebrates can also catch younger conspecifics. Subadults and adults that normally feed on other amphibians can also eat conspecifics.

Availability of vulnerable prey

Four classes of amphibians are particularly vulnerable to predation. High fecundity and abundance, concomitant with small offspring size and vulnerability, typical of most amphibians may facilitate cannibalism of (1) the immobile egg stage and (2) the relatively immobile hatchling stage as has been suggested for fish (Dominey and Blumer 1984). In addition, (3) injured or otherwise weak or unhealthy larvae are readily consumed by conspecific larvae. (4) Metamorphosing anurans are extremely vulnerable to predation before they have completely resorbed their tails (Wassersug and Sperry 1977; Arnold and Wassersug 1978; Crump 1984). During metamorphosis, a tadpole passes through an awkward stage when it can neither swim as well as a younger conspecific nor hop as well as a fully metamorphosed froglet (Wassersug and Sperry 1977).

Low levels of food and high density of conspecifics

Cannibalism in many amphibians seems to be a response to low food levels. Examples of environments where food is often limited include restricted bodies of water (tree holes and in axils of leaves, Noble 1929, 1931; Wassersug *et al.* 1981), deserts (Blair 1976), salt flats (Cei 1980), and ephemeral rain pools (Degani *et al.* 1980; Crump 1983). Under such circumstances cannibalism could be an important means of sequestering otherwise limited food resources in an environment imposing a constant race against habitat evaporation (Blair 1976). Amphibians that have larvae with the potential to become cannibalistic (e.g. *Lechriodus fletcheri*, Martin 1967; *Scaphiopus*, Bragg 1965; ceratophrynine leptodactylids, Cei 1980) can exploit breeding habitats not available to other species due to lack of food.

Cannibalism in nature is also often associated with high densities of conspecifics. The high fecundity of many species of amphibians creates high densities of vulnerable age classes; evaporation of breeding sites leads to severe overcrowding of larvae and increases the frequency of contacts. The environmentally-induced developmental switch from omnivore morph to cannibal morph in *Ambystoma tigrinum* larvae seems to be cued by a high density of conspecifics (Collins and Cheek 1983). In drying habitats the proximate factor for increased cannibalism could be either food shortage or high density of conspecifics, or both (Bragg 1946).

CONSEQUENCES OF CANNIBALISM

Costs of cannibalism

Risk of injury or death

For some species the risk of injury or death to the cannibal by a conspecific defending itself may be a potential cost of cannibalism. For most amphibians, however, this risk is presumably negligible because they lack the means of aggressively defending themselves. In addition, because the cannibal is generally much larger than the victim, the risk of suffocation is minimal. Exceptions include adults of aggressive species such as *Pyxicephalus, Chacophrys, Ceratophrys,* and *Lepidobatrachus* that have odontoids (large fang-like teeth on the bottom jaw) and extremely large mouths enabling them to attempt to eat individuals of almost the same size. Cannibals have been known to suffocate themselves in at least *Pyxicephalus* (Grobler 1972) and *Lepidobatrachus* (R. Ruibal, personal communication; M. Crump, personal observation).

Developmental characteristics

Cannibalistic larvae may be disadvantaged by slower developmental rates and delays in the onset of metamorphosis. For example, Rose and Armentrout (1976) noted that the cannibalistic morph of *Ambystoma tigrinum* may be at a selective disadvantage relative to the typical morph under conditions of a drying habitat because cannibals are slow to metamorphose. Slower development of the cannibal morph is not a general pattern, however, as the opposite trend occurs under both laboratory and field conditions: larval *A. t. nebulosum* seem to metamorphose before the omnivore morph (D. Pfennig, personal communication). Similarly, Lannoo *et al.* (1989) found that cannibalistic *A. t. tigrinum* larvae metamorphosed faster than non-cannibalistic individuals. Spadefoot toads likewise exhibit extensive variability in developmental rates of cannibals; although cannibalism confers greater growth rate than does omnivory, cannibalism may result in either decreased or increased developmental rates (Bragg 1964, 1965; Pomeroy 1981). An additional potential cost of the cannibal morph in *S. multiplicatus* is a smaller lipid reserve relative to omnivores that transform at the same time (Pomeroy 1981). A newly metamorphosed individual with a smaller lipid reserve presumably would be at a disadvantage if the terrestrial habitat were food-limited.

Risk of eating a relative

The probability of eating a relative and thus incurring a cost to the cannibal because of lowered inclusive fitness, depends on many factors. For example, if adults are philopatric then larvae are likely to encounter close relatives. If subadults disperse to great distances from their natal pond after metamorphosis, then the likelihood of their offspring cohabiting with relatives is lower. The higher the density of oviposition sites available, the less often larvae will encounter relatives. Variation in breeding synchrony, clutch size, and spatial distribution of eggs per clutch deposited at a site likewise should affect encounter rate of relatives.

The cost in terms of inclusive fitness of eating a relative may be small or large, depending upon the predictability of the habitat and probability of survival. The cost would be small if resources in a drying pond were limited and the relative had a low probability of surviving; eating a relative may be the only means of surviving to metamorphosis. The cost would be greater in a situation where the relative has a high probability of surviving, such as in a long-lasting pond with abundant alternate prey. Cannibalism is generally more common among species that live in ephemeral sites than among species living in permanent bodies of water. The risk of eating a relative could be reduced if individuals recognized kin, as has been documented for certain tadpoles (Waldman 1988). Blaustein and O'Hara (1982) suggested that perhaps cannibalistic amphibian larvae can recognize kin and avoid eating them, as has been documented in certain fish (Loekle *et al.* 1982; Loiselle 1983; see also FitzGerald and Whoriskey, this volume), waterstriders (Nummelin 1989), and rodents (Elwood, this volume). However, Pfennig (1990*a*) reported that omnivorous morph *Scaphiopus multiplicatus* tadpoles cannibalized siblings and non-siblings at an equal rate in the laboratory, even when well fed (Pfennig, unpublished data; see also Baur, this volume, for gastropods). No data are available concerning association between kin recognition and prey preference for the cannibalistic morph in this species.

Benefits of cannibalism

At one end of a spectrum, in harsh environments where food is otherwise limited, survival may depend upon cannibalism if conspecifics are the only food available. At the other end of the continuum, conspecifics may be exploited opportunistically as a food source simply because they are readily available and easily caught and handled; the benefit of cannibalism may be only the additional energy and nutrients gained and consequently no different from any other act of predation. For species between the two ends of this continuum, cannibalism may confer advantages (e.g. nutrition, competitive ability, and increased mating opportunities) to the cannibal over and above simply obtaining a meal.

Nutrition

Numerous studies have suggested that cannibals experience faster growth or developmental rates, larger size, higher survivorship, or enhanced reproduction relative to non-cannibals (reviewed in Fox 1975; Polis 1981). However, to demonstrate that cannibalism provides a benefit beyond that of simply providing additional food, experiments must control for energy content and dry mass of food. If under such experimental conditions a diet of conspecifics results in significantly enhanced growth, development, or survival, then it might be concluded that conspecifics provide a better balance of nutrients than do heterospecific foods (Crump 1986; Meffe and Crump 1987; Crump 1990). For example, Nagai *et al.* (1971) demonstrated that *Bufo vulgaris* tadpoles fed conspecifics exhibited more efficient conversion of amino acids than did tadpoles fed a diet of artificial feed or fish meal. Experimental studies demonstrated that *Hyla pseudopuma* tadpoles fed conspecifics grew significantly larger than individuals fed heterospecific tadpoles of the same dry weight or energy content (Crump 1990). Additional studies are needed to determine specific nutritional benefits of cannibalism as compared with consumption of heterospecific prey. If such benefits are shown to be widespread, cannibalism in amphibians would be expected to be more common than currently recognized.

Ingestion early in development of only one conspecific triggers greatly accelerated growth in *A. t. tigrinum* larvae (Lannoo *et al.* 1989) and in *Lepidobatrachus laevis* and *L. llanensis* tadpoles (personal observation). The reason for this acceleration is not clear. There may be a window of time during which any large prey item eaten will accelerate growth, or perhaps only a conspecific confers acceleration due to maximum digestive efficiency.

Eickwort (1973) suggested that cannibalism should be favoured at that point in an animal's life cycle where the benefit is greatest, that is where both nutritional benefits and age-specific mortality are high. The larval stages of amphibians may represent such a point (Crump 1983). Age-specific mortality is highest at premetamorphic stages (reviewed in Wilbur 1980; Crump 1982). Also, amphibian larvae living in transient and unpredictable aquatic developmental sites are under intense selection to metamorphose as quickly as possible; because large size confers advantages (reviewed in Crump 1989), selection may favour transformation at as large a body size as possible. Conspecifics may better provide the nutrients necessary for growth and development than alternate foods (Crump 1990).

Competition

One consequence of cannibalism is the elimination of potential competitors, which results in more food or space for the cannibal and its offspring (Fox 1975; Polis 1981). This consequence may be especially important for many larval amphibians whose unpredictable developmental sites often become resource-limited. Presumably, a gain in additional resources is most crucial in volumetrically-restricted developmental sites such as tree holes and in ephemeral rain pools; in these situations, resources may become more limited as the larvae grow and develop, and as the developmental sites begin to dry.

Interference with survivorship of conspecific offspring is a form of intraspecific competition that may enhance the reproductive success of one individual at the expense of another. For example, when a male *Eleutherodactylus coqui* eats eggs fertilized by another male (Townsend *et al.* 1984), he eliminates the other male's current reproductive investment, thus increasing his own fitness. Captive dendrobatid frogs have been observed eating another's eggs (Wells 1978; Weygoldt 1980). Many species of salamanders also eat the eggs of conspecifics (Tilley 1972, Wood and Goodwin 1954; Morin 1983; Kaplan and Sherman 1980). In fact, cannibalistic oophagy is so common among certain amphibians that it is thought to have been a strong selective pressure in the evolution of egg-brooding behaviour by both males and females (Salthe and Mecham 1974).

Increased mating opportunities

By cannibalizing another's offspring, an individual may be able to obtain a mate sooner than it would otherwise (Trivers 1972; Hrdy 1979). Weygoldt (1980) suggested that some instances of cannibalistic oophagy by adult amphibians may be related to intense intrasexual competition for mating opportunities. In dendrobatid frogs, the sex that provides parental care is frequently unavailable for mating because of its extensive parental duties of egg brooding and tadpole transport (Wells 1981). Thus, if a male eats the eggs fertilized by another male (observed in *Dendrobates pumilio*, Weygoldt 1980), then the female that had been brooding that clutch could become available for mating more quickly. In *Dendrobates auratus* the male provides parental care. Thus, if a female eats another female's eggs, then the male would be released from his paternal duties and would be available to fertilize additional clutches sooner. However, in the only test of this hypothesis for amphibians the data suggest that female *D. auratus* compete (including eating another's eggs) to monopolize the parental care of particular males rather than to monopolize as many mates as possible (Summers 1989).

Population consequences

Fox (1975) and Polis (1981) suggested that cannibalism strongly affects population parameters in diverse groups of animals. Many authors have speculated that cannibalism in amphibians contributes to population regulation (e.g. Savage 1963; Bragg 1965; Heyer *et al.* 1975), but quantitative data are lacking. Although no long-term field studies have been performed in species known to be extremely cannibalistic in nature (e.g. *Scaphiopus, Chacophrys, Ceratophrys, Lepidobatrachus,* and *Pyxicephalus*) this behaviour almost certainly has the potential to significantly affect population density and demography. In contrast, other species known to be occasionally cannibalistic in nature may not exert enough predation pressure to affect population size. For example, although both adult and larval *Notophthalmus viridescens dorsalis* cannibalized larvae in laboratory experiments, Harris (1987) found no indication that cannibalism was a population regulation mechanism in small artificial ponds.

MECHANISMS THAT DETER CANNIBALISM

There are numerous mechanisms that may reduce the frequency of cannibalism. Thick, gelatinous egg capsules, unpalatability, and toxicity of some eggs and larvae are assumed to deter invertebrate and fish predators but may also reduce cannibalism (Crump 1983). A primary advantage of habitat separation of aquatic eggs and larvae from terrestrial adults is generally assumed to be the elimination of intraspecific competition, but such separation would also reduce the likelihood that indiscriminately foraging adults will eat conspecific young (Simon 1984). Amphibians have evolved numerous morphological, physiological, and behavioural adaptations for non-aquatic egg deposition, including viviparity and direct development, which completely eliminate the vulnerable free-swimming larval stage (Goin 1960). One consequence of these reproductive modes is reduced risk of cannibalism on eggs and/or larvae. Egg-brooding behaviour by a male or female parent can reduce intraspecific predation (e.g. Tilley 1972; Townsend *et al.* 1984). Kin recognition by tadpoles may reduce cannibalism, although tadpoles that remain in their natal site may simply avoid cannibalism rather than have the ability to discriminate relatives *per se* (D. Pfennig, personal communication). Alteration in spatial and temporal patterns of habitat use by smaller conspecifics in the presence of larger conspecific predators may reduce the level of cannibalism, although this was shown not to be the case in laboratory experiments with *A. t. nebulosum* (Holomuzki 1986).

The egg deposition patterns may also reduce the risk of cannibalism. Deposition of eggs singly and distributed widely over the substrate (as opposed to clusters of eggs patchily distributed) is advantageous for cannibalistic species because it reduces the rate of encounters with conspecifics (e.g. *Notophthalmus viridescens*, Wood and Goodwin 1954; *Ceratophrys* and *Lepidobatrachus*, M. Crump, personal observation). Deposition of eggs singly and egg-brooding behaviour seem to be alternative tactics that deter cannibalism for species that exhibit oophagy. Since size variation among individuals may increase the frequency of cannibalism (e.g. *Hynobius nebulosus*, Kusano *et al.* 1985), synchronous oviposition by members in a population may reduce the frequency of cannibalism. For example, *Lepidobatrachus llanensis* readily cannibalize smaller conspecifics if available, but if breeding is synchronous and therefore size variation of tadpoles is minimal, the tadpoles rarely feed on conspecifics (M. Crump, unpublished data). Finally, the choice of oviposition site by females to avoid conspecific oophagic tadpoles already present in the habitat would reduce cannibalism (e.g. *Hyla pseudopuma*, Crump 1991).

DISCUSSION

Cannibalism in amphibians is widespread taxonomically, common in some species and appears to be facultative in all species studied to date. (Trophic egg feeding by tadpoles and *in utero* oophagy in salamanders are obligate, but these behaviours should not be considered true cannibalism.) The most commonly eaten prey are eggs and larvae, and the most common cannibals are larvae. Cannibalism occurs most frequently in unpredictable, ephemeral environments. Amphibian larvae are under strong selection to grow as large as possible and develop as quickly as possible to escape their transient environment. They should eat the highest quality food possible (Crump 1983; Simon 1984). Overcrowding may lead to limited food resources and ultimately to cannibalism. There are no published data demonstrating that individuals prefer conspecifics to heterospecifics. In fact, the evidence suggests that most cannibalism results from indiscriminate predation. In most instances, cannibalism occurs simply because conspecific prey are an available source of food that is abundant or easy to catch and handle. Predatory behaviour may have evolved in larvae of certain species as a survival mechanism in unpredictable environments with limited food (e.g. Bragg 1965; Cei 1979; Warburg *et al.* 1979), but consumption of conspecifics is just a by-product of the predatory behaviour.

The assumption of indiscriminate predation does not imply that cannibalism is unimportant to amphibians. On the contrary, cannibalism can have significant implications for the ecology of amphibians on the individual, population, and community levels. Cannibalism can make the difference between death and survival for some larvae in overcrowded situations where food resources are limited.

Cannibalism by larvae can increase growth rates and decrease developmental time to metamorphosis. Oophagy of fungus-infected eggs can increase survival of remaining eggs within a clutch. Cannibalism can reduce competition for limited space and food resources, and increase fitness of the cannibal by reducing a competitor's reproductive success. Differential age-specific cannibalism can influence demographic parameters and ultimately affect community-wide interactions.

Thus, it is clear that consumption of conspecifics can impart benefits beyond simply providing a meal to the cannibal. But is cannibalism subject to natural selection? Theoretically, for cannibalism to evolve and be maintained in a population, the behaviour must be heritable and the fitness gains for the cannibal must outweigh the costs. Few supporting data are available for either of these conditions. There is no evidence for a purely heritable basis for cannibalistic polyphenism. Cannibalistic behaviour seems to be under partial genetic control (Hampton and Volpe 1963; Pierce *et al.* 1981; Pomeroy 1981), but environmentally induced (Pomeroy 1981; Collins and Cheek 1983; Pfennig 1990*b*). The costs and benefits associated with cannibalism vary among years because of differing probabilities of encountering relatives and because of the variable nature of the unpredictable environments utilized by amphibians. For example, if the pond is long-lived, providing sufficient time for metamorphosis, non-cannibalistic individuals may have an advantage because of their large lipid reserves relative to cannibals transforming at the same time (Pomeroy 1981). In a neighbouring short-lived pond, some cannibals may do well because of their faster growth and development rates while other cannibals with slower developmental rates are left stranded in the dry pond after most omnivores have transformed (Pomeroy 1981). Pfennig (1990*b*) found that morph differentiation in *Scaphiopus multiplicatus* tadpoles depends on the number of fairy shrimp ingested; above a critical number the tadpole develops into a carnivore (diet includes conspecifics) with faster development rates, and below this critical number the tadpole develops into an omnivore by default. Pfennig (1990*b*) argued that use of shrimp density as a proximate cue to induce morph differentiation enables a *Scaphiopus* tadpole to respond to its environment adaptively; density of shrimp is greatest in highly ephemeral ponds, the habitat where the faster-developing carnivore morph should be favoured. Currently there is no evidence that cannibalism has evolved in amphibians as a consequence of trade-offs between associated costs and benefits, but neither has any investigator attempted to measure quantitatively the costs and benefits of cannibalism for any species of amphibian.

Why does cannibalism occur in some species and in some populations of amphibians but not in others? The answer in part relates to morphological and behavioural constraints, but also involves environmental constraints. Species that possess or can facultatively develop appropriate jaw and digestive morphology and utilize predatory foraging tactics are most likely to become cannibals. Furthermore, these species can only be cannibalistic if they encounter younger, smaller, or otherwise vulnerable conspecifics. If the breeding pulses do not produce divergent age classes that overlap in time and if variability in growth and developmental rates is minimal, cannibalism will be rare. Can we predict which species will be cannibalistic? To some extent we can, with knowledge of the morphology and behaviour of the animals and knowledge of local environmental parameters. The best predictors of cannibalism are a combination of appropriate feeding morphology, predatory foraging behaviour, and an environment with a high density of asynchronously produced conspecifics, otherwise limited food, and unpredictable habitat duration.

Additional observations of cannibalism under natural field conditions will expand our knowledge of ecological correlates of the behaviour, but long-term field studies on species known to be strongly cannibalistic will be essential for insight into the evolution and maintenance of the behaviour. The most valuable species for further study are not the ones that exhibit cannibalism only rarely, but rather those that demonstrate polyphenism and environmentally-induced developmental switches (e.g. *Ambystoma tigrinum* Collins and Cheek 1983; *Scaphiopus multiplicatus* Pfennig 1990*b*). Future research with these and similar species might consider the questions in the following areas. First, what is the relative influence of heritability versus environmental induction of cannibalistic behaviour? What proximate environmental cues trigger polyphenism, and under what circumstances? Is variability in morph frequency among populations explained by genotypic or environmental differences? Second, what are the costs and benefits of cannibalism? Do these costs and benefits vary for different morphs in contrasting environments (e.g. highly ephemeral versus longer-lasting ponds)? How do growth, development, and survival covary with cannibalism? Why, in some species, do cannibals experience accelerated growth and development rates compared with non-cannibals? Do cannibals prefer conspecific to other prey items? Are there diseases or parasite transmission costs of cannibalism? Third, how is cannibalistic polyphenism maintained within populations? What are the selective factors favouring one morph over another in contrasting environments? What are the relative fitnesses of different morphs within a habitat? And finally, is there evidence that cannibalism has been a selective pressure in the evolution of kin recognition, parental care, habitat use, and oviposition behaviours? Can any variation between populations in these behaviours be explained by the presence or absence of cannibalism?

ACKNOWLEDGEMENTS

During manuscript preparation I was partially supported by a Faculty Development Award from the University of Florida. I thank R. F. Laurent and M. Halloy de Grosse of the Instituto de Herpetologia, Fundación Miguel Lillo, in Tucumán, Argentina, for providing work space, access to library facilities, and hospitality during the writing of this chapter. For critical reading and helpful comments on various drafts of the manuscript, I thank B. Baur, B. Crespi, M. Elgar, F. Hensley, P. Feinsinger, R. Jaeger, G. Meffe, D. Pfennig, B. Pierce, M. Simon, C. Toft, and R. Wassersug.

REFERENCES

Anderson, J. D. (1960). Cannibalism in *Dicamptodon ensatus. Herpetologica*, **16**, 260.

Arnold, S. J. and Wassersug, R. J. (1978). Differential predation on metamorphic anurans by garter snakes (*Thamnophis*): social behavior as a possible defense. *Ecology*, **59**, 1014–22.

Baldauf, R. (1947). *Desmognathus f. fuscus* eating eggs of its own species. *Copeia*, **1947**, 66.

Ball, S. C. (1936). The distribution and behavior of the spadefoot toad in Connecticut. *Transactions of the Connecticut Academy of Arts and Sciences*, **32**, 351–79.

Bishop, S. C. (1927). The amphibians and reptiles of Allegany State Park. *New York State Museum Handbook*, No. 3, 5–141.

Blair, W. F. (1976). Adaptation of anurans to equivalent desert scrub of North and South America. In *Evolution of desert biota,* (ed. D. W. Goodall), pp. 197–222. Texas Press, Austin.

Blaustein, A. R. and O'Hara, R. K. (1982). Kin recognition in *Rana cascadae* tadpoles: maternal and paternal effects. *Animal Behaviour*, **30**, 1151–7.

Borland, J. R. and Rugh, R. (1943). Evidences of cannibalism in the tadpoles of the frog *Rana pipiens. American Naturalist*, **77**, 282–5.

Bragg, A. N. (1946). Aggregation with cannibalism in tadpoles of *Scaphiopus bombifrons* with some general remarks on the probable evolutionary significance of such phenomena. *Herpetologica*, **3**, 89–97.

Bragg, A. N. (1956). Dimorphism and cannibalism in tadpoles of *Scaphiopus bombifrons (Amphibia, Salientia). Southwestern Naturalist,* **1**, 105–8.

Bragg, A. N. (1964). Further study of predation and cannibalism in spadefoot tadpoles. *Herpetologica*, **20**, 17–24.

Bragg, A. N. (1965). *Gnomes of the night.* University of Pennsylvania Press, Philadelphia.

Bragg, A. N. and Bragg, W. N. (1959). Variations in the mouth parts in tadpoles of *Scaphiopus* (*Spea*) *bombifrons* Cope (Amphibia: Salientia). *Southwestern Naturalist*, **3**, 55–69.

Burton, T. M. (1977). Population estimates, feeding habits and nutrient and energy relationships of *Notophthalmus v. viridescens*, in Mirror Lake, New Hampshire. *Copeia*, **1977**, 139–43.

Cei, J. M. (1955). Chacoan batrachians in central Argentina. *Copeia*, **1955**, 291–3.

Cei, J. M. (1979). Albinismo e canibalismo en larvas de *Pleurodema nebulosa* de la region precordillerana de la provincia de San Juan, Argentina (Anura, Leptodactylidae). *Publicaciones ocasionales Instituto de Biologia animal Universidad nacional Cuyo*, **11**, 1–3.

Cei, J. M. (1980). *Amphibians of Argentina. Monitore Zoologico Italiano Monografia*, No. 2, 1–609.

Colley, S. A., Keen, W. H., and Reed, R. W. (1989). Effects of adult presence on behavior and microhabitat use of juveniles of a desmognathine salamander. *Copeia*, **1989**, 1–7.

Collins, J. P. and Cheek, J. E. (1983). Effect of food and density on development of typical and cannibalistic salamander larvae in *Ambystoma tigrinum nebulosum. American Zoologist,* **23**, 77–84.

Collins, J. P., Mitton, J. B., and Pierce, B. A. (1980). *Ambystoma tigrinum*: a multispecies conglomerate? *Copeia*, **1980**, 938–41.

Crump, M. L. (1982). Amphibian reproductive ecology on the community level. In *Herpeto-*

logical communities, (ed. N. J. Scott, Jr.), pp. 21–36. *U.S. Fish and Wildlife Service, Wildlife Research Report* No. 13.

Crump, M. L. (1983). Opportunistic cannibalism by amphibian larvae in temporary aquatic environments. *American Naturalist*, **121**, 281–9.

Crump, M. L. (1984). Ontogenetic changes in vulnerability to predation in tadpoles of *Hyla pseudopuma*. *Herpetologica*, **40**, 265–71.

Crump, M. L. (1986). Cannibalism by younger tadpoles: another hazard of metamorphosis. *Copeia*, **1986**, 1007–9.

Crump, M. L. (1989). Effect of habitat drying on developmental time and size at metamorphosis in *Hyla pseudopuma*. *Copeia*, **1989**, 794–7.

Crump, M. L. (1990). Possible enhancement of growth in tadpoles through cannibalism. *Copeia*, **1990**, 560–4.

Crump, M. L. (1991). Choice of oviposition site and egg load assessment by a treefrog. *Herpetologica*, **47**, 308–15.

Cunningham, J. (1954). A case of cannibalism in the toad *Bufo boreas halophilus*. *Herpetologica*, **10**, 166.

Degani, G., Goldenberg, S., and Warburg, M. R. (1980). Cannibalistic phenomena in *Salamandra salamandra* larvae in certain water bodies and under experimental conditions. *Hydrobiologia*, **75**, 123–8.

Dominey, W. J. and Blumer, L. S. (1984). Cannibalism of early life stages in fishes. In *Infanticide: comparative and evolutionary perspectives*, (ed. G. Hausfater and S. B. Hrdy), pp. 43–64. Aldine Publishing Company, Hawthorne.

Duellman, W. E. and Trueb, L. (1986). *Biology of amphibians*. McGraw-Hill Book Co., New York.

Dushin, A. I. (1974). Diet of two frog species in the fisheries of the Mordovian ASSR. *Ékologiya*, **6**, 88–90. (English translation pp. 583–5.)

Eickwort, K. R. (1973). Cannibalism and kin selection in *Labidomera clivicollis* (Coleoptera: Chrysomelidae). *American Naturalist*, **107**, 452–3.

Forester, L. C. (1979). The adaptiveness of parental care in *Desmognathus ochrophaeus* (Urodela: Plethodontidae). *Copeia*, 1979, 332–41.

Fox, L. R. (1975). Cannibalism in natural populations. *Annual Review of Ecology and Systematics*, **6**, 87–106.

Goin, C. J. (1960). Amphibians, pioneers of terrestrial breeding habits. *Report of the Smithsonian Institution*, 1959-1960, 427–45.

Grobler, J. H. (1972). Observations on the amphibian *Pyxicephalus adspersus* Tschudi in Rhodesia. *Arnoldia (Rhodesia) Miscellaneous Publications*, **6**, 1–4.

Hampton, S. H. and Volpe, E. P. (1963). Development and interpopulation variability of the mouthparts of *Scaphiopus holbrooki*. *American Midland Naturalist*, **70**, 319–28.

Harris, R. N. (1987). An experimental study of population regulation in the salamander, *Notophthalmus viridescens dorsalis* (Urodela: Salamandridae). *Oecologia* , **71**, 280–5.

Heatwole, H. and Test, F. H. (1961). Cannibalism in the salamander, *Plethodon cinereus*. *Herpetologica*, **17**, 143.

Hero, J. M. and Galatti, U. (1990). Characteristics distinguishing *Leptodactylus pentadactylus* and *L. knudseni* in the Central Amazon rainforest. *Journal of Herpetology*, **24**, 226–8.

Heusser, H. (1971). Laich Raubern und kannibalismus bei sympatrischen anurans Kaulquappen. *Experientia*, **27**, 474–5.

Heyer, W. R., McDiarmid, R. W., and Weigmann, D. L. (1975). Tadpoles, predation and pond habitats in the tropics. *Biotropica*, **7**, 100–11.

Holomuzki, J. R. (1986). Intraspecific predation and habitat use by tiger salamanders (*Ambystoma tigrinum nebulosum*). *Journal of Herpetology*, **20**, 439–41.

Hrdy, S. B. (1979). Infanticide among animals: A review, classification, and examination of the implications for the reproductive strategies of females. *Ethology and Sociobiology*, **1**, 13–40.

Hutcherson, J. E., Peterson, C. L., and Wilkinson, R. F. (1989). Reproductive and larval biology of *Ambystoma annulatum*. *Journal of Herpetology*, **23**, 181–3.

Jungfer, K. H. (1985). Beitrag zur Kenntnis von *Dendrobates speciosus* O. Schmidt, 1857 (Salientia: Dendrobatidae). *Salamandra*, **21**, 263–80.

Kaplan, R. H. and Sherman, P. W. (1980). Intraspecific oophagy in California newts. *Journal of Herpetology*, **14**, 183–5.

Kluge, A. G. (1981). The life history, social organization, and parental behavior of *Hyla rosenbergi* Boulenger, a nest-building gladiator frog. *Miscellaneous Publications of the Museum of Zoology University of Michigan* No.160, 1–170.

Kusano, T., Kusano, H., and Miyashita, K. (1985). Size-related cannibalism among larval *Hynobius nebulosus*. *Copeia*, **1985**, 472–6.

Lannoo, M. J. and Bachmann, M. D. (1984). Aspects of cannibalistic morphs in a population of *Ambystoma t. tigrinum* larvae. *American Midland Naturalist*, **112**, 103–9.

Lannoo, M. J., Lowcock, L., and Bogart, J. P. (1989). Sibling cannibalism in noncannibal morph *Ambystoma tigrinum* larvae and its correlation with high growth rates and early metaphorphosis. *Canadian Journal of Zoology*, **67**, 1911–14.

Lannoo, M. J., Townsend, D. S., and Wassersug, R. J. (1987). Larval life in the leaves: arboreal tadpole types, with special attention to the morphology, ecology, and behavior of the oophagous *Osteopilus brunneus* (Hylidae) larva. *Fieldiana Zoology New Series*, **38**, 1–31.

Loekle, D. M., Madison, D. M., and Christian, J. J. (1982). Time dependency and kin recognition of cannibalistic behavior among poeciliid fishes. *Behavioral and Neural Biology*, **35**, 315–18.

Loiselle, P. V. (1983). Filial cannibalism and egg recognition by males of the primitively custodial teleost *Cyprinodon macularius californiensis* Girard (Antherinomorpha: Cyprinodontidae). *Ethology and Sociobiology*, **4**, 1–9.

Loveridge, A. (1950). History and habits of the East African Bullfrog. *East Africa Natural History Society Journal for 1947-1948*, **19**, 253–5.

Marshall, C. J., Doyle, L. S., and Kaplan, R. H. (1990). Intraspecific and sex-specific oophagy in a salamander and a frog: reproductive convergence of *Taricha torosa* and *Bombina orientalis*. *Herpetologica*, **46**, 395–9.

Martin, A. A. (1967). The biology of tadpoles. *Australian Natural History*, **15**, 326–30.

McCann, C. (1932). Notes on Indian batrachians. *Journal of the Bombay Natural History Society*, **36**, 152–80.

McCoid, M. J. and Fritts, T. H. (1980). Notes on the diet of a feral population of *Xenopus laevis* (Pipidae) in California. *Southwestern Naturalist*, **25**, 272–5.

Meffe, G. K. and Crump, M. L. (1987). Possible growth and reproductive benefits of cannibalism in the mosquitofish. *American Naturalist*, **129**, 203–12.

Morin, P. J. (1983). Competitive and predatory interactions in natural and experimental populations of *Notophthalmus viridescens dorsalis* and *Ambystoma tigrinum*. *Copeia*, **1983**, 628–39.

Muedeking, M. H. and Heyer, W. R. (1976). Descriptions of eggs and reproductive patterns of *Leptodactylus pentadactylus* (Amphibia: Leptodactylidae). *Herpetologica*, **32**, 137–9.

Nagai, Y., Nagai, S., and Nishikawa, T. (1971). The nutritional efficiency of cannibalism and an artificial feed for the growth of tadpoles of Japanese Toad (*Bufo vulgaris* sp.). *Agricultural and Biological Chemistry*, **35**, 697–703.

Nickerson, M. A. and Mays, C. E. (1973). The hellbenders: North American 'giant salamanders'. *Milwaukee Public Museum Publications in Biology and Geology*, **1**, 1–106.

Noble, G. K. (1929). The adaptive modifications of the arboreal tadpoles of *Hoplophryne* and the torrent tadpoles of *Staurois*. *Bulletin of the American Museum of Natural History*, **58**, 291–334.

Noble, G. K. (1931). *The biology of the Amphibia*. McGraw-Hill, New York.

Norris, D. O. (1989). Seasonal changes in diet of paedogenetic tiger salamanders (*Ambystoma tigrinum mavortium*). *Journal of Herpetology*, **23**, 87–9.

Nummelin, M. (1989). Cannibalism in waterstriders (Heteroptera: Gerridae): is there kin recognition? *Oikos*, **56**, 87–90.

Nussbaum, R. A. and Clothier, G. W. (1973). Population structure, growth, and size of larval *Dicamptodon ensatus* (Eschscholtz). *Northwest Science*, **47**, 218–27.

Orton, G. L. (1954). Dimorphism in larval mouthparts in spadefoot toads of the *Scaphiopus hammondi* group. *Copeia*, **1954**, 97–100.

Petranka, J. W. (1989). Density-dependent growth and survival of larval *Ambystoma*: evidence from whole-pond manipulations. *Ecology*, **70**, 1752–67.

Pfennig, D. W. (1990*a*). 'Kin recognition' among spadefoot toad tadpoles: a side-effect of habitat selection? *Evolution*, **44**, 785–98.

Pfennig, D. W. (1990*b*). The adaptive significance of an environmentally-cued developmental switch in an anuran tadpole. *Oecologia*, **85**, 101–7.

Pienaar, U. de V., Passmore, N. I., and Carruthers, V. C. (1976). *The frogs of the Kruger National Park*. National Parks Board of South Africa.

Pierce, B. A., Mitton, J. B., and Rose, F. L. (1981). Allozyme variation among large, small and cannibal morphs of the tiger salamander inhabiting the Llano Estacado of west Texas. *Copeia*, **1981**, 590–5.

Pierce, B. A., Mitton, J. B., Jacobson, L., and Rose, F. L. (1983). Head shape and size in cannibal and noncannibal larvae of the tiger salamander from West Texas. *Copeia*, **1983**, 1006–12.

Polis, G. A. (1981). The evolution and dynamics of intraspecific predation. *Annual Review of Ecology and Systematics*, **12**, 225–51.

Polis, G. A. and Myers, C. A. (1985). A survey of intraspecific predation among reptiles and amphibians. *Journal of Herpetology*, **19**, 99–107.

Pomeroy, L. V. (1981). Developmental polymorphism in the tadpoles of the spadefoot toad *Scaphiopus multiplicatus*. PhD dissertation, University of California, Riverside.

Powders, V. N. (1973). Cannibalism by the slimy salamander, *Plethodon glutinosus* in eastern Tennessee. *Journal of Herpetology*, **7**, 139–40.

Reig, O. A. and Cei, J. M. (1963). Elucidación morfologico-estadistica de las entidades del género *Lepidobatrachus* Budgett (Anura, Ceratophrynidae), con consideraciones sobre la extensión del distrito chaqueño del dominio zoogeográfico subtropical. *Physis*, **24**, 181–204.

Richmond, N. D. (1947). Life history of *Scaphiopus holbrookii holbrookii* (Harlan). Part I. Larval development and behavior. *Ecology*, **28**, 53–67.

Rohwer, S. (1978). Parent cannibalism of offspring and egg raiding as a courtship strategy. *American Naturalist*, **112**, 429–40.

Rose, F. L. and Armentrout, D. (1976). Adaptive strategies of *Ambystoma tigrinum* inhabiting the Llano Estacado of West Texas. *Journal of Animal Ecology*, **45**, 713–29.

Ruibal, R. and Thomas, E. (1988). The obligate carnivorous larvae of the frog, *Lepidobatrachus laevis* (Leptodactylidae). *Copeia*, **1988**, 591–604.

Salthe, S. N. and Mecham, J. S. (1974). Reproductive and courtship patterns. In *Physiology of the Amphibia*, (ed. B. Lofts), pp. 309–521. Academic Press, New York.

Savage, R. M. (1963). A speculation on the pallid tadpoles of *Xenopus laevis*. *British Journal of Herpetology*, **3**, 74–6.

Simon, M. P. (1983). The ecology of parental care in a terrestrial breeding frog from New Guinea. *Behavioral Ecology and Sociobiology*, **14**, 61-7.

Simon, M. P. (1984). The influence of conspecifics on egg and larval mortality in amphibians. In *Infanticide: comparative and evolutionary perspectives*, (ed. G. Hausfater and S. B. Hrdy), pp. 65–86. Aldine Publishing Co., Hawthorne.

Starrett, P. (1960). Descriptions of tadpoles of middle American frogs. *Miscellaneous Publications of the Museum of Zoology Michigan*, **110**, 1–37.

Summers, K. (1989). Sexual selection and intra-female competition in the green poison-dart frog, *Dendrobates auratus*. *Animal Behaviour*, **37**, 797–805.

Tago, K. (1927). Notes on the habits and life history of *Megalobatrachus japonicus*. *X Congres Internationale de Zoologie Tenu a Budapest* 1927, 828–38.

Tilley, S. G. (1972). Aspects of parental care and embryonic development in *Desmognathus ochrophaeus*. *Copeia*, **1972**, 532–40.

Townsend, D. S., Stewart, M. M., and Pough, F. H. (1984). Male parental care and its adaptive significance in a neotropical frog. *Animal Behaviour*, **32**, 421–31.

Trivers, R. L. (1972). Parental investment and sexual selection. In *Sexual selection and the descent of man 1871-1971*, (ed. B. Campbell), pp. 136–79. Aldine, Chicago.

Ueda, H. (1986). Reproduction of *Chiraxalus effingeri* (Boettger). *Scientific Report of The Laboratory of Amphibian Biology, Hiroshima University*, **8**, 109–16.

Waldman, B. (1988). The ecology of kin recognition. *Annual Review of Ecology and Systematics*, **19**, 543–71.

Warburg, M. R., Degani, G, and Warburg, I. (1979). Growth and population structure of *Salamandra salamandra* (L.) larvae in different limnological conditions. *Hydrobiologia*, **64**, 147–55.

Wassersug, R. J. and Sperry, D. G. (1977). The relationship of locomotion to differential predation on *Pseudacris triseriata* (Anura: Hylidae). *Ecology*, **58**, 830-9.

Wassersug, R. J., Frogner, K. J., and Inger, R. F. (1981). Adaptations for life in tree holes by rhacophorid tadpoles from Thailand. *Journal of Herpetology*, **15**, 41–52.

Wells, K. D. (1978). Courtship and parental behavior in a Panamanian poison-arrow frog (*Dendrobates auratus*). *Herpetologica*, **34**, 148–55.

Wells, K. D. (1981). Parental behavior of male and female frogs. In *Natural selection and social behavior*, (eds. R. D. Alexander and D. W. Tinkle), pp. 184–97. Chiron Press, New York.

Weygoldt, P. (1980). Complex brood care and reproductive behavior in captive poison-arrow frogs, *Dendrobates pumilio* O. Schmidt. *Behavioral Ecology and Sociobiology*, **7**, 329–32.

Wilbur, H. M. (1980). Complex life cycles. *Annual Review of Ecology and Systematics*, **11**, 67–93.

Wood, J. T. and Clarke, R. F. (1955). The dusky salamander: oophagy in nesting sites. *Herpetologica*, **11**, 150–1.

Wood, J. T. and Goodwin, O. K. (1954). Observations on the abundance, food, and feeding behavior of the newt, *Notophthalmus viridescens viridescens* (Rafinesque), in Virginia. *Journal of the Elisha Mitchell Scientific Society*, **70**, 27–30.

Wood, J. T. and Wood, F. E. (1955). Notes on the nests and nesting of the Carolina mountain dusky salamander in Tennessee and Virginia. *Journal of the Tennessee Academy of Science*, **30**, 36–9.

Zimmermann, H. and Zimmermann, E. (1981). Sozialverhalten, Fortpflanzungsverhalten und Zucht de Farberfrosche, *Dendrobates histrionicus* und *D. lehmanni* sowie einiger anderer Dendrobatiden. *Zeitschrift des Kolner Zoo*, **24**, 83–99.

13

Cannibalism in birds

Mark T. Stanback and Walter D. Koenig

Cannibalism in birds has generally been relegated to the realm of the anecdote. Any fitness benefits of cannibalism were often thought to accrue at the species level via population regulation (e.g. Wynne-Edwards 1962). This limited view of the significance of intraspecific killing and cannibalism changed following the publication of Hamilton's (1964) theory of kin selection. Inclusive fitness might be increased not only by killing and eating conspecifics, but even, in particular circumstances, by being eaten by a closely related conspecific (e.g. Alexander 1974; O'Connor 1978).

Avian cannibalism is often closely linked to other behaviours, particularly infanticide (Mock 1984). Consequently, our discussion of cannibalism necessarily draws from the infanticide literature, especially the reviews by Hrdy (1979), Mock (1984), and Rohwer (1986).

Previous workers have identified five potential ways in which killing of conspecifics may enhance the inclusive fitness of an individual:

(1) providing nutritional benefits;

(2) increased availability of resources for the killer or the killer's kin;

(3) providing additional breeding opportunities;

(4) parental manipulation leading to increased residual reproductive value (e.g. foregoing higher current reproductive success in exchange for higher survivorship); and

(5) preventing accidental adoption (Hrdy 1979; Pierotti and Murphy 1987).

These potential benefits of intraspecific killing and cannibalism are not mutually exclusive. For example, by eating a conspecific chick, an adult may benefit by

(1) nourishing itself and its progeny;

(2) lessening the chance of accidentally adopting the chick;

(3) lowering the reproductive success of competitors; and

(4) lowering population density, thereby reducing intraspecific competition for space and resources.

For this reason it is difficult to distinguish between the following scenarios: (a) selection that favours cannibalism, for which infanticide is often a necessary prerequisite, and (b) selection that favours infanticide, for which cannibalism is a subsequent option. Sometimes the context of the attack may provide clues as to its probable primary significance; for example, cannibalism of chicks at a nearby colony is unlikely to reduce the probability of accidental adoption. However, considerable observational and experimental data are needed in order to reject or support any of the above benefits as being important to particular cannibalistic events.

Although interest in avian cannibalism has advanced in recent years, empirical data remains scarce, making hypothesis testing difficult in most cases (but see Watanuki 1988; Bortolotti *et al.* 1991). In this review we first summarize reported cases of avian cannibalism according to the life-history stage at which they occur and their apparent fitness benefits. We then discuss factors promoting and limiting the evolution of cannibalism in birds and speculate on why cannibalism is relatively uncommon in this taxon.

THE DISTRIBUTION OF AVIAN CANNIBALISM

We define cannibalism as the consumption of all or a significant portion of a conspecific egg or individual. While this definition is similar to that used by Bortolotti *et al.* (1991), it contrasts with the definition employed by McNicholl (1977) and Mock (1984), which excludes cases in which individuals eat an already-dead conspecific. Taken to the extreme, however, this latter definition reduces cannibalism to only those cases in which the victim is eaten alive. Although we recognize that scavenging may be purely fortuitous, as in the less yellow-headed vulture *Cathartes burrovianus* (J. Eitniear, personal communication), it is difficult to discriminate such cases from those in which conspecifics are specifically targeted for death and subsequent consumption.

Adult–adult cannibalism

Adult–adult cannibalism appears to be rare among wild birds and apparently represents either opportunistic nutritional exploitation or abnormally vicious aggression. Paullin (1987) made repeated observations of American coots *Fulica americana* cannibalizing conspecifics killed by severe weather and avian cholera. In our search of the literature, we found the following anecdotal reports of birds killing and eating adult conspecifics: common moorhen *Gallinula chloropus* (Cawston 1983), pied wagtail *Motacilla alba* (Joslin 1964), great grey owl *Strix nebulosa* (Fisher 1975), and red-tailed hawk *Buteo jamaicensis* (Clevenger and Roest 1974). Several additional reports exist of adult birds viciously attacking, but not necessarily eating, conspecific corpses (e.g. chaffinches *Fringilla coelebs* (Wilkins 1964), pied wagtails (Joslin 1966), and European blackbirds *Turdus merula* (Coleman 1985)).

Adult–adult cannibalism is considered to be a ubiquitous threat in commercial poultry operations; debeaking is the standard procedure used to discourage this 'abnormal' behaviour (e.g. Ferguson 1968). However, a great deal of the literature on this phenomena involves various behavioural precursors of cannibalism such as feather, toe, and vent pecking rather than cannibalism *per se*, and is consequently difficult to interpret.

Adult cannibalism of eggs or young

In contrast to adult–adult cannibalism, adults commonly eat conspecific eggs and young (see Tables 13.1 and 13.2). Such inter-generational cannibalism may involve an adult unrelated to the victim, a related but non-parental adult, a replacement mate, or a parent. In this review, we use the term 'heterocannibalism' to refer to cases in which the cannibal is not normally a close relative of the victim, where close relatives are defined as individuals for which the coefficient of relatedness is ≥ 0.25; 'filial cannibalism' to cases in which parents eat their own offspring; 'sexually selected cannibalism' to cases involving replacement mates; and 'within-group cannibalism' to cases involving individuals, related or not, living together in a social unit. These categories are not always distinct; 'within-group' cannibalism may, for example, involve either close relatives or not (hence heterocannibalism).

Heterocannibalism

Heterocannibalism has been reported in at least seven avian families (Table 13.1), and is most frequent among colonial seabirds. In particular, many gulls, especially the larger species of *Larus*, are known to eat both the eggs and young of conspecifics (see Table 13.1). Several other colonial species for which heterocannibalism has been reported are also listed in Table 13.1.

Among passerines, heterocannibalism has been documented in the Australian magpie *Gymnorhina tibicen*, the European blackbird, and several members of the family Corvidae (Table 13.1). In both the group-living Florida scrub jay *Aphelocoma coerulescens* (G.E. Woolfenden, unpublished data) and the group-living grey-breasted (Mexican) jay *A. ultramarina* (Trail *et al.* 1981), heterocannibalism involving non-group members apparently occurs rarely, while heterocannibalism may be common in the territorial carrion crow *Corvus corone* (Yom-Tov 1974) and the colonial jackdaw *C. monedula* (Antikainen 1981). Given their omnivorous diets and well-known habit of eating bird's eggs, it is likely that cannibalism among corvids is more extensive than these records indicate.

In gulls, heterocannibalism is often a combination of a foraging strategy and an

Table 13.1 Distribution of heterocannibalism in birds

Taxa	Common name	Source
Fregatidae		
Fregata magnificens	Magnificent frigatebird	Bent 1922
Pelecanidae		
Pelecanus occidentalis	Brown pelican	Murphy 1936; Palmer 1962
Stercorariidae		
Catharacta maccormicki	South polar skua	Young 1963; Spellerberg 1971; Procter 1975
Laridae		
Larus argentatus	Herring gull	Ward 1906; Deusing 1939; Paynter 1949; Paludan 1951; Tinbergen 1953; Parsons 1971; Brown 1967
Larus ridibundus	Black-headed gull	Kirkman 1937
Larus fuscus	Lesser black-backed gull	Davis and Dunn 1976
Larus glaucescens	Glaucous-winged gull	Hunt and McLoon 1975; Hunt and Hunt 1976
Larus schistisagus	Slaty-backed gull	Watanuki 1988
Larus dominicanus	Dominican gull	Fordham 1970
Xema sabini	Sabine's gull	Ashmole 1963
Sterna paradisaea	Arctic tern	Pettingill 1939
Sterna fuscata	Sooty tern	Sprunt 1948
Rynchops niger	Black skimmer	Safina and Burger 1983
Corvidae		
Aphelocoma coerulescens	Florida scrub jay	Woolfenden, unpubl. data
Aphelocoma ultramarina	Gray-breasted jay	Trail et al. 1981
Corvus corone	Carrion crow	Yom-Tov 1974
Corvus monedula	Jackdaw	Antikainen 1981
Cracticidae		
Gymnorhina tibicen	Australian magpie	Rowley 1973
Turdidae		
Turdus merula	European blackbird	Woodward 1962

Table 13.2 Distribution of filial cannibalism in birds

Taxa	Common name	Source
Ciconiidae		
Ciconia ciconia	White stork	Schüz 1957
Accipitridae		
Accipiter nisus	Sparrowhawk	Moss (in Newton 1978)
Aquila verreauxi	Black eagle	Gargett 1978
Falconidae		
Falco tinnunculus	European kestrel	Schüz 1957
Falco sparverius	American kestrel	Willoughby and Cade 1964; Porter and Wiemeyer 1970; Bortolotti *et al.* 1991
Falco mexicanus	Prairie falcon	Holthuijzen *et al.* 1987
Rallidae		
Gallinula chloropus	Common moorhen	Gibbons, unpubl. data
Stercorariidae		
Catharacta maccormick	South polar skua	Young 1963
Laridae		
Larus argentatus	Herring gull	Goethe 1956; Chardine and Morris 1983
Cuculidae		
Geococcyx californianus	Roadrunner	Ohmart 1973
Tytonidae		
Tyto alba	Barn owl	Siegenthaler 1956
Strigidae		
Strix aluco	Tawny owl	Siegenthaler 1956
Picidae		
Melanerpes formicivorus	Acorn woodpecker	Stanback, unpubl. data
Motacillidae		
Motacilla alba	White wagtail	Schüz 1957
Laniidae		
Lanius collurio	Red-backed shrike	Schüz 1957
Corvidae		
Aphelocoma ultramarina	Gray-breasted jay	Trail *et al.* 1981
Pica pica	Black-billed magpie	Schüz 1957
Turdidae		
Turdus merula	European blackbird	Walsh 1964

attempt to prevent accidental adoption. Cannibalism is known to be quite frequent in some species, constituting a major source of mortality and even potentially a factor regulating population size. This possibility is most evident for colonial nesters such as the lesser black-backed gull *Larus fuscus* and the herring gull *L. argentatus*. In lesser black-backed gulls the single most important source of egg mortality is egg cannibalism by neighbours that had recently lost their own young (Davis and Dunn 1976). Parsons (1971) reported that 23.3 per cent of eggs and chicks in a large herring gull colony (over 900 pairs) were claimed by cannibalism. Both Deusing (1939) and Brown (1967) found heterocannibalism to be responsible for much of the pre-fledging mortality in their gull colonies. Several gull studies have even demonstrated a link between food availability and heterocannibalism (Hunt and McLoon 1975, McNicholl unpublished data). However, many chicks were killed and not eaten in these studies, indicating that at least some chicks were not killed because of food stress.

Comparable data on the extent and potential significance of heterocannibalism in non-colonial species are rare. A notable exception is Charles' (1972; in Yom-Tov 1974) estimate that cannibalism of eggs and chicks by intruding carrion crows in northeastern Scotland caused up to 75 per cent of the total egg loss and nestling mortality in this species.

Sexually selected cannibalism

The potential for replacement breeders to both kill and eat eggs or offspring exists in a wide variety of species (see Rohwer 1986), but we are aware of only two species in which sexually selected cannibalism has been documented. One is of a replacement male barn swallow *Hirundo rustica* that killed and apparently partially ate the nestlings of the widowed female with whom he presumably then mated (Sherman 1952). The second consists of two cases, one natural and one experimental, in which replacement female acorn woodpeckers *Melanerpes formicivorus* apparently killed and cannibalized young nestlings, thereby gaining additional breeding opportunities (Stacey and Edwards 1983). Sexually selected infanticide without subsequent cannibalism has been documented in several other species of birds (see references in Rohwer 1986 and Emlen *et al.* 1989).

Within-group cannibalism

Cooperative breeding occurs when more than one pair of mated birds cooperate to raise young at a single nest. Of particular interest here are 'plural-breeding' species (Brown 1987) in which two or more pairs of birds breed within a single social unit and 'joint-nesting' or 'mate-sharing' species (Koenig and Mumme 1987) in which two or more birds of one or both sexes share mates of the opposite sex. In both of these cases there is considerable potential for competition among breeders within groups that may result in infanticide and cannibalism. In some cases, cannibalism is perpetrated by genetic relatives other than parents or siblings, whereas in others the cannibals are apparently unrelated to their victims. We have combined these examples on the basis of their social context, but recognize that they provide an indistinct bridge between heterocannibalism and filial cannibalism (e.g., Curry 1988).

The most extensively documented example of within-group cannibalism occurs in the acorn woodpecker. In California, this species lives in groups comprised of a breeding core of up to seven birds, including one to four (rarely more) genetically related males and one to three genetically related females (Koenig and Mumme 1987). Because all co-breeding females lay their eggs in a single cavity and the number of offspring that can be fledged by a group is limited, there is intense competition both among breeding males to inseminate females and among females to have their eggs incubated rather than those of their co-breeders. In groups containing joint-nesting females, the first female to lay eggs invariably has one or more eggs removed from the nest by her co-breeder(s). Eggs then may be eaten by any or all members of the group, including both the female that removed and the female that laid the egg (Mumme *et al.* 1983).

Egg destruction and subsequent cannibalism occurs regularly and leads to the destruction of about 35 per cent of all eggs laid in groups containing two or more co-breeding females (Mumme *et al.* 1983). Overall, only 20 per cent of groups contain more than a single breeding female, but this form of egg destruction is nonetheless a major source of mortality and accounts for the loss of an estimated 8.4 per cent of all eggs laid, or 26.1 per cent of eggs lost to all causes (Koenig and Mumme 1987).

In addition to the 48 eggs removed by breeding females documented by Mumme *et al.* (1983), three eggs were removed by breeder males. The circumstances surrounding egg destruction in these latter circumstances was unknown, but one possibility has subsequently been investigated by Koenig (1990) in experiments in which one of two or more co-breeding males was temporarily removed and thus denied access to females during their fertile period. Removed males that were known or were likely to have been dominant to their co-breeders destroyed the nest by removing the eggs upon their release (whether the eggs were eaten was not known). This suggests that co-breeding males, like females, may engage in egg destruction and cannibalism of non-descendant kin in natural situations in which they are denied opportunity of parentage.

Besides the viable eggs that are removed and subsequently eaten by group members, female acorn woodpeckers breeding jointly, commonly lay undersized, abnormal eggs as the first egg of a clutch. These so-called 'runt' eggs may be removed by the non-parental female, as described above, or, more frequently, are removed and subsequently consumed by the parental female herself. Interestingly, a bird removing her own runt egg frequently removes a co-breeder's normal egg at the same time (Mumme *et al.* 1983). Overall, at least 4 per cent of all eggs laid by acorn woodpeckers are runt eggs, a frequency nearly an order of magnitude higher than that known for any other bird species (Koenig 1980). Whatever the physiological causes or the functional utility, if any, of runt eggs, they appear to be the nearest functional equivalent to trophic eggs (eggs produced for consumption by kin) known for any bird.

Another reason for egg cannibalism by males within breeding groups has been suggested for moorhens by Gibbons (1986). In this species males are occasionally bigamous and both the mates of a male lay in the same nest. The male does much of the incubating and, therefore, it is in his interest to ensure that both females lay synchronously, as this reduces the length of incubation time. Destruction and eating of an egg by a bigamous male has been observed and small clutches (of one or two eggs) frequently disappear on these territories prior to the 'main' clutch being laid. The bigamous male may destroy clutches until both females lay synchronously. The moorhen has an omnivorous diet and occasionally takes the eggs of other species.

Within-group infanticide and egg cannibalism also occurs in the grey-breasted jay, a species that lives and breeds in social units containing multiple pairs (Trail *et al.* 1981; Brown 1987). Such behaviour appears advantageous to dominant breeders, as it (1) removes competitors for food resources within groups and (2) makes available the victimized breeders to help at the nest of the more dominant group members. As killers were often immigrants, Trail *et al.* (1981) suggest that low within-group relatedness may facilitate the occurrence of this behaviour.

Similarly intense reproductive competition leading to within-group egg destruction has also been reported in several other cooperative breeders including the Arabian babbler *Turdoides squamiceps* (Zahavi 1990), noisy miner *Manorina melanocephala* (Whitmore 1986), white-winged chough *Corcorax melanorhamphos* (Heinsohn 1988), groove-billed ani *Crotophaga sulcirostris* (Vehrencamp 1977), smooth-billed ani *C. ani* (Loflin 1982), and the guira cuckoo *Guira guira* (R. Macedo, in preperation). However, cannibalism has not been observed in these species. This is surprising given the desirability of eggs as a nutritional resource (Polis 1981) and their lack of chemical defences (Orians and Janzen 1974). The lack of egg eating is particularly perplexing in the anis, where joint-nesting and egg-tossing is common (Koford *et al.* 1990) and where the diet is known to include birds eggs at least occasionally (Skutch 1983). Perhaps the acorn woodpecker's unique dexterity with acorns facilitates egg-eating in a way not present in anis or other species in which egg destruction regularly occurs.

Filial cannibalism

The consumption of eggs or young by parent birds, dubbed 'kronism' by Schüz (1957), is not well understood, despite having been reported in 13 avian families (Table 13.2). This is partly because actual observations of filial cannibalism are rare enough to have been attributed to social pathology or some other abnormality such as captivity or human disturbance. Filial cannibalism has been reported from birds faced with extreme environmental situations in which food is critically short and parents opt to eat their doomed offspring (e.g. herring gulls (Ward 1906; Goethe 1956); tawny owls *Strix aluco* (Siegenthaler 1956), and barn owls *Tyto alba* (Siegenthaler 1956)).

Regular filial cannibalism has been reported only for the roadrunner *Geococcyx californianus* (Ohmart 1973) and the American kestrel *Falco sparverius* (Bortolotti *et al.*, 1991). In the former, adults eat weak, undersized chicks within their nest. In the American kestrel, deceased nestlings are eaten in up to 60 per cent of nests which contain them, usually before siblings are capable of dismembering the carcass. The degree to which adults use the dead chicks to feed the living is unknown. While kestrel filial cannibalism is more common in food stressed nests, it is not associated with nest failure (Bortolotti *et al.*, 1991). In the acorn woodpecker filial cannibalism is suggested by

(1) the often immediate disappearance of very late hatching (and therefore doomed) nestlings (Stanback 1991);

(2) observations of acorn woodpeckers eating nestlings of other species (Fajer *et al.* 1987); and

(3) an observation of cannibalism and caching of a hatchling by a breeding pair in captivity (Stanback, personal observation).

While filial cannibalism may not be common among birds, additional work, particularly with raptors, will probably increase the frequency with which it is reported.

Sibling cannibalism

While siblicide is well documented in many avian families (see Mock 1984), active cannibalism by nestmates is less widespread, being best known in raptors (Table 13.3). There are several reasons for this: raptors often have a high degree of hatching asynchrony, they are carnivorous, and juveniles are capable of siblicide and are largely free from gape limitation, by virtue of their habit of consuming prey piecemeal. In contrast, in thousands of hours of observations on various siblicidal ardeid species, in only one case (in the great blue heron *Ardea herodias)* was a senior chick able to swallow a junior chick successfully (D. Mock, unpublished data).

The extent of sibling cannibalism (and filial cannibalism) following brood reduction in raptors is disputed. Even when partially eaten remains are found, it is usually unclear whether parents, siblings, or both are responsible (Pickwell 1948; Roest 1957; White and Cade 1971; Reese 1972; McNicholl 1976). Ingram (1959), for example, contended that cannibalism is the rule rather than the exception in such situations. Following up on this possibility, Alexander (1974) even proposed that doomed young might be temporarily kept alive specifically as a food resource for their siblings (the 'icebox hypothesis').

Although indisputable records of sibling cannibalism have surfaced over the years (see Table 13.3), cannibalism is not a universal outcome of brood reduction, even in raptors. The disappearance of deceased runts is often due to the removal of the corpse by one of the parents (Meyburg 1974; Gargett 1978; Stinson 1979).

Table 13.3 Distribution of sibling cannibalism in birds

Taxa	Common name	Source
Ardeidae		
Ardea herodias	Great Blue Heron	Mock, unpubl. data
Laridae		
Rynchops niger	Black Skimmer	Safina and Burger 1983
Alcidae		
Cepphus columba	Pigeon Guillemot	Rasmussen 1988
Accipitridae		
Pernis apivorus	Honey buzzard	Krambrich 1954
Buteo lagopus	Rough-legged hawk	Barth 1952
Buteo buteo	Common buzzard	Gilbert Brook 1924; Mebs 1964
Buteo platypterus	Broad-winged hawk	Heintzelman 1966
Buteo swainsoni	Swainson's hawk	Pilz 1976; Pilz and Siebert 1978
Circus cyaneus	Hen harrier	Bannerman 1956; Scharf and Balfour 1971
Circus aeruginosus	Marsh harrier	Vincent 1936; Bengston 1967
Accipiter cooperii	Cooper's hawk	Snyder and Wiley 1976
Falconidae		
Falco tinnunculus	European kestrel	Bonin and Strenna 1986
Falco sparverius	American kestrel	Bortolotti *et al.* 1991
Tytonidae		
Tyto alba	Barn owl	Hawbecker 1945;Baudvin 1979
Strigidae		
Bubo virginianus	Great horned owl	Errington *et al.* 1940
Asio flammeus	Short-eared owl	Richmond 1958; Ingram 1962; Parmelee *et al.* 1967

However, as Bortolotti *et al.* (1991) note, the assertion that dead nestlings are not eaten comes primarily from observations of obligately siblicidal eagles. In these species siblicidal brood reduction occurs soon after hatching despite a proximate abundance of food. Filial cannibalism may be more common in facultatively siblicidal species, in which siblicidal aggression is stimulated by hunger. However, even among well-studied kestrels (Bortolotti *et al.*, 1991) the extent to which parents provided the cannibalistic meal was unclear.

FACTORS PROMOTING AND CONSTRAINING CANNIBALISM

Morphology and phylogeny

Mock (1984) pointed out two important morphological and phylogenetic features preventing cannibalism in many species of birds: (1) rapid, determinate growth and (2) the practice of swallowing food whole. In contrast to taxa in which large individuals are always capable of consuming juveniles, the period during which young birds are vulnerable to intraspecific predation is very brief. Exceptions to this 'gape limitation' are hawks, owls, and a few other predatory species such as shrikes and skuas that dismember prey with their bills.

Diet

Many birds restrict their diet to sources that entail neither morphological nor dietary preferences suited to cannibalism. For example, notwithstanding the cases mentioned earlier, it is unsurprising that swallows and sunbirds are infrequent cannibals. For most avian taxa, cannibalism may simply not be an option.

Even for many of those species for which cannibalism is at least a possibility, infanticide does not seem to be primarily a strategy for nutritional exploitation (Aguilera 1990), and in any case it generally occurs during the time of year when food resources are at their maximum. Consequently, the nutritional benefits of eating a victim of infanticide may often be low.

Social system

As a result of biparental care, offspring in monogamous species are generally better protected against infanticide and cannibalism from marauders, than are offspring in species with uniparental care (Mock 1984). The fact that most birds exhibit some form of biparental care (Lack 1968) probably decreases the potential for cannibalism in this taxon relative to taxa in which biparental care is exceptional. Mock (1984) also suggested that monogamy, by reducing the opportunity for multiple matings by males, might restrict sexually selected infanticide. However, sexually selected infanticide has subsequently been documented in several monogamous species (e.g. Crook and Shields 1985; Goldstein *et al.* 1986; Freed 1986; Robertson and Stutchbury 1988). A more recent review of replacement mates in birds by Rohwer (1986) suggests that other factors, including the probability of renesting both within and between seasons, dispersal following nesting failure, and the probability of within-season divorce, may be more important than the mating system *per se* in determining whether such birds adopt or attempt to kill nestlings of their new mates.

The colonial nesting habit may also override any protection against cannibalism that monogamy may provide. Cannibalism is relatively common among colonially nesting seabirds (see Table 13.1), where feeding forays result in frequent and extended absences by one or both parents. This is not to say that all colonially nesting seabirds are cannibals; burrow nesting and single egg clutches both appear to reduce the threat of accidental adoption that promotes cannibalism in some colonies.

Constraints on sibling cannibalism

As mentioned earlier, most reports of sibling cannibalism involve raptorial species (Table 13.3). Yet additional constraints may exist even among raptors. Chief among these is hatching asynchrony, which is often insufficient to yield older nestlings physically capable of either dismembering younger siblings or swallowing them whole. For example, in the obligately siblicidal black eagle *Aquila verreauxi,* the first-hatched chick begins attacking the second-hatched chick immediately following the hatching of the latter (Gargett 1978). Although the older chick is able to locomote sufficiently to kill the younger chick, it usually lacks the coordination necessary to dismember and devour it. In such cases, cannibalism following siblicide requires that the attending adult treat the dead younger chick as a prey item, an event that apparently occurs only rarely (Gargett 1978, but see Bortolotti *et al.,* 1991). Thus the timing of brood reduction can place limits on sibling cannibalism.

Oviparity is an additional constraint on sibling cannibalism because shells provide unhatched chicks with protection from recently hatched siblings. Indeed, in contrast to organisms in which females provide trophic eggs for their developing offspring (Gruber and Campagno 1981; see Baur, Crespi, Kukuk, this volume), no nestling birds are able to open and devour the contents of other eggs in the nest. Although intra-nest sibling cannibalism does have similarities to the intra-uterine sibling cannibalism observed in sand sharks *Odontaspis tauris* (Springer 1948), such behaviour by birds is also obviated by oviparity.

Finally, it appears that the benefits of siblicide *per se,* in terms of resource competition, are probably more important to the perpetrator than the nutritional benefits of any subsequent cannibalism (see references in Mock and Parker 1986). In all obligately siblicidal and many facultatively siblicidal species, hunger does not appear to drive siblicidal harrassment (Gargett 1978; Edwards and Collopy 1983; Mock *et al.* 1987). However, hunger-sensitive sibling aggression is well documented for other facultatively siblicidal species (Drummond and Chavelas 1989).

Constraints on filial cannibalism

By separating the mother from zygote at an early stage, oviparity influences the options for the secondary adjustment of parental investment. In some cases, oviparity may broaden the available options by allowing the mother physical access to zygotes (Mock 1984). Oviparity, however, eliminates the possibility of embryonic resorption, an important method of brood manipulation found in other taxa (see Haig, this volume). From a purely functional point of view, resorption is analogous to filial cannibalism: in both cases embryos are killed and consumed by the parent. Thus an entire venue for postzygotic brood manipulation is unavailable to birds.

As Dickemann (1984, pp. 428–9) points out, 'there is no clear line between conserving energy by birth spacing and by consumption of young: Both can be seen as attempts by the parent to maximize reproduction, but involving differing degrees of wastage.' Yet waste is critical: with embryonic resorption excluded, birds are left only with the more energetically wasteful methods of desertion, brood reduction, and infanticide for manipulating brood composition. Only filial cannibalism can be used for recouping nutritional losses associated with previous parental investment.

It is also probable that strong selective pressure inhibiting fatal aggression against an individual's own offspring (e.g. Yom-Tov 1976; Graves and Whiten 1980, but see Horsfall 1984; Aguilera 1990; Urrutia and Drummond 1990) serves in some cases to limit the extent of filial cannibalism. Particularly striking in this regard is the absence of filial cannibalism of deserted young (but see Curry 1988), even though desertion of nestlings in the face of poor environmental conditions is common in birds. Actual parental starvation is rarely the cause of nest desertion. Rather, desertion results when the probability of nesting success is so low that the residual reproductive benefits of abandonment compensate for the loss in current reproduction (e.g. Boucher 1977; Tait 1980; Sargent and Gross 1985; Fujioka 1986). The nutritional benefit of the single cannibalistic meal that is offered when desertion is imminent may be small compared with available alternative foods. However, Bortolotti *et al.* (1991) found in their study population of American kestrels that food availability does influence the probability that an adult will eat its brood reduced chick.

Other birds also exhibit the behavioural flexibility to successfully display parental and cannibalistic behaviour simultaneously. As mentioned above, roadrunner parents selectively devour their weak chicks (Ohmart 1973). Female acorn woodpeckers are able to switch within days from cannibalism to incubation of eggs within their nest. This is comparable to the striking and similarly precipitous behavioural change exhibited by related female prairie dogs that switch from infanticide and cannibalism of each other's pups to communal suckling (Hoogland *et al.* 1989, see also Kukuk, Crespi, this volume).

Additional considerations

Scavengers: are conspecifics avoided?

Although cannibalism via scavenging is seldom reported in birds, this is probably due to a lack of sufficient observations. Interestingly, some scavenging species may specifically avoid eating conspecifics. For example, in gull colonies, dead adults and large chicks will lie for weeks without being touched, despite the fact that gulls regularly cannibalize young chicks and scavenge from carcasses of mammals and fishes (R. Pierotti, unpublished data). A similar phenomenon may occur in both black-billed and yellow-billed magpies *Pica pica* and *P. nuttalli,* which perform a ritualized form of mobbing ('funerals') when groups of birds come upon a dead conspecific (Miller and Brigham 1988; C. H. Trost, unpublished data; M. D. Reynolds and W. D. Koenig, unpublished data). Participants do not cannibalize the corpse, despite scavenging by these species in other contexts. Why there is this apparent reluctance to scavenge conspecifics in these cases is unknown; avoidance of infection by parasites or pathogens might be one plausible explanation.

Adoption by replacement 'mates'

Rohwer (1986) discussed why replacement mates of some species are infanticidal, whereas in others they adopt and behave parentally toward nestlings of their new mate. Species in the latter category include Cooper's hawks *Accipiter cooperii,* goshawks *A. gentilis,* and peregrine falcons *Falco peregrinus,* that are morphologically capable of cannibalizing the offspring of their new mate. Rohwer (1986) suggested that the larger size of females in these species may serve to constrain aggression by replacement males, and that infanticide and cannibalism by new mates may actually inhibit renesting within the season.

Deception and adoption in gulls

Wandering chicks in gull and skua colonies are often adopted by potential, or even actual, infanticidal cannibals (Graves and Whiten 1980; Pierotti 1980). Such adoption occurs despite the potentially high cost for the adopting pair: not only must foster parents forage for an extra (unrelated) chick, but such chicks often cause the demise of legitimate nest occupants (Mock 1984; Pierotti and Murphy 1987). Not surprisingly, potential adoptees are often killed and eaten with little or no provocation. Despite these often dire consequences, junior gull chicks frequently attempt to be adopted, presumably because they have an even lower probability of surviving to fledging in their natal nest.

The mechanisms by which potential cannibals adopt unrelated chicks constitutes an interesting example of deception (Mock 1984). Chicks use the submissive crouching that acts as a releaser for normal parental behaviour. In this manner, wandering chicks may gain access to the nest of a potential foster parent. Once in the vicinity of the nest, chicks take advantage of the nest recognition (rather than egg or chick recognition) that typifies adult behaviour in the first week after hatching (Conover *et al.* 1980; Graves and Whiten 1980). Interestingly, chicks can recognize their parents several days before parental recognition of chicks is fully developed, suggesting a cross-generational evolutionary conflict that chicks may be winning (Pierotti 1980).

Non-consumptive intraspecific egg destruction

Though intraspecific egg destruction without consumption is best known for cooperative breeders (see above), several non-cooperative breeders also engage in similar activities. In house wrens *Troglodytes aedon* (Belles-Isles and Picman 1986) and marsh wrens *Telmatodytes palustris* (Picman 1977), nest destruction appears to be a manifestation of interference competition for food. In European starlings *Sturnus vulgaris,* egg destruction occurs in several contexts. First, intruding females sometimes kill nestlings or remove eggs, presumably to encourage nest-site abandonment (Romagnano *et al.* 1986). Second, intraspecific brood parasites may remove eggs of the resident female and then lay their own eggs. Third, resident females may remove eggs within their nest laid by other females (Stouffer *et al.* 1988). However, birds fail to eat the eggs or nestlings that they remove, which are generally found on the ground near the box (P. C. Stouffer, personal communication).

Cannibalistic polyphenisms

Cannibalistic specialists have long been recognized in gulls and skuas (e.g. Deusing 1939; Paludan 1951; Parsons 1971). Parsons (1971), for example, found that half of the extensive cannibalism he documented in a colony of 903 herring gull nests was performed by only four breeding males. This sort of specialization may be characteristic of a more widespread phenomena of individual specialization on particular food types (e.g. the Cocos finch *Pinaroloxias inornata,* Werner and Sherry 1987). Similar specialization has been documented in western gulls, *L. occidentalis*, where individuals frequently specialize on particular fish species, garbage, pigeons, or conspecific chicks (Pierotti and Annett 1987).

Little work has been done to explore the fitness consequences of cannibalism in species exhibiting such polyphenisms (see also Crump, Waddell, this volume). No increase in reproductive performance by slaty-backed gull *L. schistisagus* pairs, including cannibalistic individuals, was found by Watanuki (1988). However, the possible benefits gained by cannibalistic individuals by preventing adoption and reducing the reproductive success of others in the colony were not considered.

It is also not known whether cannibalistic polyphenisms are heritable. Dietary and other environmental factors may influence the extent of cannibalism in many cases, as suggested by work on captive populations of American kestrels (Porter and Wiemeyer 1970) and in commercial poultry (e.g. Ferguson 1968). There is also evidence from several other studies that cannibalism, or at least behaviours associated with it, has a genetic basis in chickens *Gallus gallus* (e.g. Cuthbertson 1980; Berrio *et al.* 1987), suggesting the possibility of a genetic basis to cannibalistic polyphenisms in some wild populations. Additional work in this area, particularly on gulls, would be valuable.

SUMMARY

Most kinds of cannibalism, including adult–adult, sexually selected, and sibling cannibalism, are rare in birds, relative to other vertebrates. Filial cannibalism has been reported in a variety of families but never appears to be common, with the exception of the roadrunner and possibly several raptors. Only heterocannibalism, the eating of eggs or young by unrelated adults, is both widespread and sometimes quite common, especially among colonially nesting seabirds.

Constraints on cannibalistic behaviour appear to be acting at several levels in birds. Oviparity not only obviates resorption of embryos, but also has numerous implications for the types of interactions available to siblings. Gape limitation and dietary habits also contribute to the rarity of cannibalism, even in cases where infanticide and cannibalism would have obvious fitness advantages. The mating/social system, dispersion of nests, brood size, and nest type also undoubtedly influence the probability of cannibalism in a species.

There is clearly a need for additional studies of cannibalism in birds. At a proximate level, the motivational basis of this phenomenon is almost entirely unknown. Similarly, little is known about the genetic basis and fitness consequences of cannibalism, an area that is particularly amenable to study in species, such as gulls, in which individuals specializing in cannibalism can be identified.

ACKNOWLEDGEMENTS

We thank Gary Bortolotti, David Gibbons, Willaim Iko, Martin McNicholl, Doug Mock, Ray Pierotti, Mark Reynolds, Philip Stouffer, Chuck Trost, Karen Wiebe, and Glen Woolfenden for sharing unpublished data. Help in translations was generously provided by Doug Bell. Bernie Crespi, Mark Elgar, Gerry FitzGerald, David Gibbons, Doug Mock, and Ray Pierotti provided numerous helpful comments on the manuscript. Our work on acorn woodpeckers has been supported by the National Science Foundation, most recently through a graduate fellowship to MTS and grant DEB87-04992 to WDK. MTS was also supported by a Betty S. Davis Fellowship provided through the generosity of Fanny Hastings Arnold.

REFERENCES

Aguilera, E. (1990). Parental infanticide by white spoonbills *Platalea leucorodia*. *Ibis*, **132**, 124–5.

Alexander, R. D. (1974). The evolution of social behavior. *Annual Review of Ecology and Systematics*, **5**, 325–83.

Antikainen, E. (1981). The breeding success of the jackdaw *Corvus monedula* in nesting cells. *Ornis Fennica*, **58**, 72–7.

Ashmole, N. P. (1963). The biology of the wideawake or sooty tern *Sterna fuscata* on Ascension Island. *Ibis*, **103**b, 296–364.

Barth, E. K. (1952). Breeding biology of the rough-legged buzzard. *Finlands Jaktvardsstiftelse Papers on Game Research*, **8**, 126.

Bannerman, D. A. (1956). *Birds of the British Isles*. Oliver and Boyd, Edinburgh.

Baudvin, H. (1979). Taille des pontes et reussite des nichées chez la chouette effraie, *Tyto alba* en Bourgogne. *Alauda*, **47**, 13–16.

Belles-Isles, J. C. and Picman, J. (1986). House wren nest destroying behavior. *Condor*, **88**, 190–3.

Bengston, S. A. (1967). Observations on the reproductive success in 26 nests of the marsh harrier *Circus aeruginosus* in Skane Province Sweden. *Oological Record*, **41**, 23–8.

Bent, A.C. (1922). Life histories of North American petrels and pelicans and their allies. *U.S. National Museum Bulletin* No. 121.

Berrio, I., Fraga, L., and Perez, P. (1987). Influence of the naked neck gene (Na) on the incidence of pecking and cannibalism in light poultry. *Cuban Journal of Agricultural Science*, **21**, 283–7. (In Spanish.)

Bonin, B. and Strenna, L. (1986). The biology of the kestrel *Falco tinnunculus* in Auxois, France. *Alauda*, **54**, 241–62.

Bortolotti, G. R., Wiebe, K. L., and Iko, W. M. (1991). Cannibalism of nestling American kestrels by their parents and siblings. *Canadian Journal of Zoology*, **69**, 1447–53.

Boucher, D. H. (1977). On wasting parental investment. *American Naturalist*, **111**, 786–8.

Brown, J. L. (1987). *Helping and communal breeding in birds*. Princeton University Press.

Brown, R. G. B. (1967). Breeding success and population growth in a colony of herring and lesser black-backed gulls *Larus argentatus* and *L. fuscus*. *Ibis*, **109**, 502–15.

Cawston, J. M. (1983). One moorhen killing and eating another. *British Birds*, **76**, 451.

Chardine, J. W. and Morris, R. D. (1983). Herring gull males eat their own eggs. *Wilson Bulletin*, **95**, 477–8.

Charles, J. (1972). Territorial behaviour and the limitation of population size in the crow, *Corvus corone* and *Corvus cornix*. Ph.D. thesis, Aberdeen University.

Clevenger, G. A. and Roest, A. I. (1974). Cannibalism in red-tailed hawk. *Auk*, **91**, 639.

Coleman, J. R. (1985). Cannibalism by blackbird. *British Birds*, **78**, 511–12.

Conover, M. R., Klopfer, F. D., and Miller, D. E. (1980). Stimulus features of chicks and other factors evoking parental protective behaviour in ring-billed gulls. *Animal Behaviour*, **28**, 29–41.

Crook, J. R. and Shields, W. M. (1985). Sexually selected infanticide by adult male barn swallows. *Animal Behaviour*, **33**, 754–61.

Curry, R. L. (1988). Group structure, within-group conflict and reproductive tactics in cooperatively breeding Galapagos mockingbirds, *Nesomimus parvulus*. *Animal Behaviour*, **36**, 1708–28.

Cuthbertson, G. J. (1980). Genetic variation in feather-pecking behaviour. *British Poultry Science*, **21**, 447–50.

Davis, J. W. F. and Dunn, E. K. (1976). Intraspecific predation and colonial breeding in lesser black-backed gulls *Larus fuscus*. *Ibis*, **118**, 65–77.

Deusing, M. (1939). The herring gulls of Hat Island, Wisconsin. *Wilson Bulletin*, **51**, 170–5.

Dickemann, M. (1984). Concepts and classification in the study of human infanticide: sectional introduction and cautionary notes. In *Infanticide: comparative and evolutionary perspectives*, (ed. G. Hausfater and S.B. Hrdy), pp. 427–38. Aldine, New York.

Drummond, H. and Chavelas, C. G. (1989). Food shortage influences sibling agression in the blue-footed booby. *Animal Behaviour*, **37**, 806–19.

Edwards, T. C. and Collopy, M. W. (1983). Obligate and facultative brood reduction: an examination of factors that influence fratricide. *Auk*, **100**, 630–5.

Emlen, S. T., Demong, N. J., and Emlen, D. J. (1989). Experimental induction of infanticide in female wattled jacanas. *Auk*, **106**, 1–7.

Errington, P. L., Hamerstrom, F., and Hamerstrom, F. N., Jr. (1940). The great horned owl and its prey in North-central United States. *Iowa Agricultural Experimental Station Research Bulletin*, **227**, 759–860.

Fajer, E. D., Schmidt, K. J., and Eschler, J. G. (1987). Acorn woodpecker predation on cliff swallow nests. *Condor*, **89**, 177–8.

Ferguson, W. (1968). Abnormal behavior in domestic birds. In *Abnormal behavior in animals*, (ed. M. W. Fox), pp. 188–207. Saunders, Philadelphia.

Fisher, R. M. (1975). Possible intra-specific killing by a great gray owl. *Canadian Field-Naturalist*, **89**, 71–2.

Fordham, R. A. (1970). Mortality and population change of Dominican gulls in Wellington, New Zealand. *Journal of Animal Ecology*, **39**, 13–27.

Freed, L. A. (1986). Territory takeover and sexually selected infanticide in tropical house wrens. *Behavioral Ecology and Sociobiology*, **19**, 197–206.

Fujioka, M. (1986). Infanticide by a male parent and by a new female mate in colonial egrets. *Auk*, **103**, 619–21.

Gargett, V. (1978). Sibling aggression in the black eagle in the Matopos, Rhodesia. *Ostrich*, **49**, 57–63.

Gibbons, D.W. (1986). Brood parasitism and cooperative nesting in the moorhen, *Gallinula chloropus*. *Behavioral Ecology and Sociobiology*, **19**, 221–32.

Gilbert, H. A. and Brook, A. (1924). *Secrets of bird life*. Arrowsmith, London.

Goethe, F. (1956). *Die Silbermöve*. Ziemsen, Wittenberg/Lutherstadt.

Goldstein, H., Eisikovitz, D., and Yom-Tov, Y. (1986). Infanticide in the Palestine sunbird. *Condor*, **88**, 528–9.

Graves, J. A. and Whiten, A. (1980). Adoption of strange chicks by herring gulls, *Larus argentatus* L. *Zeitschrift für Tierpsychologie*, **54**, 267–78.

Gruber, S. H. and Campagno, L. J. V. (1981). Taxonomic status and biology of the bigeye thresher, *Alopias superciliosus*. *Fisheries Bulletin*, **79**, 617–40.

Hamilton, W. D. (1964). The genetical evolution of social behaviour. I, II. *Journal of Theoretical Biology*, **7**, 1–52.

Hawbecker, A. C. (1945). Food habits of the barn owl. *Condor*, **47**, 161–6.

Heinsohn, R. G. (1988). Inter-group ovicide and nest destruction in cooperatively breeding white-winged choughs. *Animal Behaviour*, **36**, 1856–8.

Heintzelman, D.S. (1966). Cannibalism at a broad-winged hawk nest. *Auk*, **83**, 307.

Holthuijzen, A. M. A., Duley, P. A., Hager, J. C., Smith, S. A., and Wood, K. N. (1987). Piracy, insectivory, and cannibalism of prairie falcons *(Falco mexicanus)* nesting in southwestern Idaho. *Journal of Raptor Research*, **21**, 32–3.

Hoogland, J. L., Tamarin, R. H., and Levy, C. K. (1989). Communal nursing in prairie dogs. *Behavioral Ecology and Sociobiology*, **24**, 91–5.

Horsfall, J. A. (1984). Brood reduction and brood division in coots. *Animal Behaviour*, **32**, 216–25.

Hrdy, S. B. (1979). Infanticide among animals: a review, classification, and examination of the implications for the reproductive strategies of females. *Ethology and Sociobiology*, **1**, 13–40.

Hunt, G. L., Jr. and Hunt, M. W. (1976). Gull chick survival: the significance of growth rates, time of breeding and territory size. *Ecology*, **57**, 62–75.

Hunt, G. L., Jr. and McLoon, S. C. (1975). Activity patterns of gull chicks in relation to feeding by parents: their potential significance for density-dependent mortality. *Auk*, **92**, 523–7.

Ingram, C. (1959). The importance of juvenile cannibalism in the breeding biology of certain birds of prey. *Auk*, **76**, 218–26.

Ingram, C. (1962). Cannibalism by nestling short-eared owls. *Auk*, **79**, 715.

Joslin, D. (1964). Male pied wagtail killing another and persistently attacking the corpse. *British Birds*, **57**, 329–30.

Joslin, D. (1966). Pied wagtail persistently attacking the corpse of another. *British Birds*, **59**, 156–7.

Kirkman, F. B. (1937). *Bird behaviour*. Nelson, London.

Koenig, W. D. (1980). The incidence of runt eggs in woodpeckers. *Wilson Bulletin*, **92**, 169–76.

Koenig, W. D. (1990). Opportunity of parentage and nest destruction in polygynandrous acorn woodpeckers, *Melanerpes formicivorus*. *Behavioral Ecology*, **1**, 55–61.

Koenig, W. D. and Mumme, R. L. (1987). *Population ecology of the cooperatively breeding acorn woodpecker*. Princeton University Press.

Koford, R. R., Bowen, B. S., and Vehrencamp, S. L. (1990). Groove-billed anis: joint nesting in a tropical cuckoo. In *Cooperative breeding in birds: long-term studies of ecology and behavior*, (ed. P. B. Stacey and W. D. Koenig), pp. 333–55. Cambridge University Press.

Krambrich, A. (1954). Kannibalismus beim Wespenbussard. *Vogelwelt*, **75**, 26.

Lack, D. (1968). *Ecological adaptations for breeding in birds*. Methuen, London.

Loflin, R. K. (1982). Egg-burying behavior and intraspecific egg parasitism in smooth-billed anis *(Crotophaga ani)*. American Ornithologists' Union, Chicago. (Abstract).

Macedo, R. Reproductive patterns and social organization of the communal Guira Cuckoo *(Guira guira)*. (In preparation.)

McNicholl, M. K. (1976). Recent reports of cannibalism and scavenging in raptors. *Raptor Research Foundation Ethology Information Exchange*, **4**, 6.

McNicholl, M. K. (1977). Usage of the terms 'cannibalism' and 'scavenging' in ecological literature. *Canadian Field-Naturalist*, **91**, 416.

Mebs, T. (1964). Zur Biologie und Populations dynamik des Maussebussards *(Buteo buteo)*. *Journal of Ornithology*, **105**, 247-306.

Meyburg, B.-U. (1974). Sibling aggression and mortality among nestling eagles. *Ibis*, **116**, 224–8.

Miller, W.R. and Brigham, R.M. (1988). 'Ceremonial' gathering of black-billed magpies *(Pica pica)* after the sudden death of a conspecific. *Murrelet*, **69**, 78–9.

Mock, D. W. (1984). Infanticide, siblicide, and avian nestling mortality. In *Infanticide: comparative and evolutionary perspectives*, (ed. G. Hausfater and S. B. Hrdy), pp. 3–30. Aldine, New York.

Mock, D. W. and Parker, G. A. (1986). Advantages and disadvantages of egret and heron brood reduction. *Evolution*, **40**, 459–70.

Mock, D. W., Lamey, T. C., and Ploger, B. J. (1987). Proximate and ultimate roles of food amount in regulating egret sibling aggression. *Ecology*, **68**, 1760–72.

Mumme, R. L., Koenig, W. D., and Pitelka, F. A. (1983). Reproductive competition in the communal acorn woodpecker: sisters destroy each other's eggs. *Nature*, **306**, 583–4.

Murphy, R. C. (1936). *Oceanic birds of South America*. American Museum of Natural History, New York.

Newton, I. (1978). Feeding and development of sparrowhawk nestlings. *Journal of Zoology, London*, **184**, 465–87.

O'Connor, R. J. (1978). Brood reduction in birds: selection for fratricide, infanticide, and suicide? *Animal Behaviour*, **26**, 79–96.

Ohmart, R. D. (1973). Observations on the breeding adaptations of the roadrunner. *Condor*, **75**, 140–9.

Orians, G. H. and D. H. Janzen. (1974). Why are embryos so tasty? *American Naturalist*, **108**, 581–92.

Palmer, R.S. (1962). *Handbook of North American birds*, Vol. 1. Yale University Press, New Haven.

Paludan, K. (1951). Contributions to the breeding biology of *Larus argentatus* and *Larus fuscus*. *Videnskabelige Meddelelser fra Dansk Naturhistorisk Forening*, **114**, 1–128.

Parmelee, D. F., Stephens, H. A., and Schmidt, R. H. (1967). The birds of south-eastern Victoria island and adjacent small islands. *National Museum of Canada Bulletin* 22.

Parsons, J. (1971). Cannibalism in herring gulls. *British Birds*, **64**, 528–37.

Paullin, D. G. (1987). Cannibalism in American coots induced by severe spring weather and avian cholera. *Condor*, **89**, 442–3.

Paynter, R. A. (1949). Clutch size and egg and chick mortality of Kent Island herring gulls. *Ecology*, **30**, 146–66.

Pettingill, O. S., Jr. (1939). History of one hundred nests of arctic tern. *Auk*, **56**, 420–8.

Pickwell, G. (1948). Barn owl growth and behaviorisms. *Auk*, **65**, 359–73.

Picman, J. (1977). Intraspecific nest destruction in the long-billed marsh wren, *Telmatodytes palustris palustris*. *Canadian Journal of Zoology*, **55**, 1997–2003.

Pierotti, R. (1980). Spite and altruism in gulls. *American Naturalist*, **115**, 290–300.

Pierotti, R. and Annett, C. (1987). Reproductive consequences of dietary specialization and switching in an ecological generalist. In *Foraging behavior*, (ed. A.C. Kamil, J.R. Krebs, and H.R. Pulliam), pp. 417–41. Plenum, New York.

Pierotti, R. and Murphy, E.C. (1987). Intergenerational conflicts in gulls. *Animal Behaviour*, **35**, 435–444.

Pilz, W. R. (1976). Possible cannibalism in Swainson's hawk. *Auk*, **93**, 838.

Pilz, W. R. and Siebert, L. K. (1978). Fratricide and cannibalism in Swainson's hawk. *Auk*, **95**, 584–5.

Polis, G. A. (1981). The evolution and dynamics of intraspecific predation. *Annual Review of Ecology and Systematics*, **12**, 225–51.

Porter, R. D. and Wiemeyer, S. N. (1970). Propagation of captive American kestrels. *Journal of Wildlife Management*, **34**, 594–604.

Procter, D. L. C. (1975). The problem of chick loss in the south polar skua *Catharacta maccormicki*. *Ibis*, **117**, 452–9.

Rasmussen, P.C. (1988). Apparent sibling cannibalism by a nestling pigeon guillemot. *Wilson Bulletin*, **100**, 136.

Reese, J.G. (1972). A Chesapeake barn owl population. *Auk*, **89**, 106–14.

Richmond, K. (1958). *Wild venture: a bird-watcher in Scotland*. Geoffrey Bles, London.

Robertson, F. J. and Stutchbury, B. J. (1988). Experimental evidence for sexually selected infanticide in tree swallows. *Animal Behaviour*, **36**, 749–53.

Roest, A.I. (1957). Notes on the American sparrow hawk. *Auk*, **74**, 1–19.

Rohwer, S. (1986). Selection for adoption versus infanticide by replacement 'mates' in birds. *Current Ornithology*, **3**, 353–95.

Romagnano, L., Lombardo, M. P., Stouffer, P. C., and Power, H. W. (1986). Suspected infanticide in the starling. *Condor*, **88**, 530–1.

Rowley, I. (1973). The comparative ecology of Australian corvids. II. Social organization and behaviour. *CSIRO Wildlife Research*, **18**, 25–65.

Safina, C. and Burger, J. (1983). Effect of human disturbance on reproductive success in the black skimmer. *Condor*, **85**, 164–71.

Sargent, R. C. and Gross, M. R. (1985). Parental investment decision rules and the Concorde fallacy. *Behavioral Ecology and Sociobiology*, **17**, 43–5.

Scharf, W. C. and Balfour, E. (1971). Growth and development of nestling hen harriers. *Ibis*, **113**, 323–9.

Schüz, E. (1957). Das verschlingen eigener junger ('Kronismus') bei Vogeln und seine Bedeutung. *Vogelwarte*, **19**, 1–15.

Sherman, A. R. (1952). Birds near at hand. In *Birds of an Iowa dooryard*, (ed. F. J. Pierce), pp. 62–88. Christopher, Boston.

Siegenthaler, S. (1956). Beobachtungen am Waldkauz *(Strix aluco). Ornithologische Beobachter*, **53**, 10–13.

Skutch, A. F. (1983). *Birds of tropical America*. University of Texas Press, Austin.

Snyder, N. F. R. and Wiley, J. W. (1976). Sexual size dimorphism in hawks and owls of North America. *Ornithological Monographs*, **20**, 1–96.

Spellerberg, I. F. (1971). Breeding behaviour of the McCormick skua *Catharacta maccormicki* in Antarctica. *Ardea*, **59**, 189–230.

Springer, S. (1948). Oviphagous embryos of the sand shark, *Carcharias taurus*. *Copeia*, **1948**, 153–7.

Sprunt, A., Jr. (1948). The tern colonies of the Dry Tortugas Keys. *Auk*, **65**, 1–19.

Stacey, P. B. and Edwards, T. C., Jr. (1983). Possible cases of infanticide by immigrant females in a group-breeding bird. *Auk*, **100**, 731–3.

Stanback, M.T. (1991). Causes and consequences of nestling size variation in the cooperatively breeding acorn woodpecker. Ph.D. thesis, University of California at Berkeley.

Stinson, C. H. (1979). On the selective advantage of fratricide in raptors. *Evolution*, **33**, 1219–25.

Stouffer, P. C., Romagnano, L. C., Lombardo, M. P., Hoffenberg, A. S., and Power, H. W. (1988). A case of communal nesting in the European starling. *Condor*, **90**, 241–5.

Tait, D. E. N. (1980). Abandonment as a reproductive tactic–the example of grizzly bears. *American Naturalist*, **115**, 800–8.

Tinbergen, N. (1953). *The herring gull's world*. Collins, London.

Trail, P. W., Strahl, S. D., and Brown, J. L. (1981). Infanticide in relation to individual and flock histories in a communally breeding bird, the Mexican jay *(Aphelocoma ultramarina). American Naturalist*, **118**, 72–82.

Urrutia, L. P. and Drummond, H. (1990). Brood reduction and parental infanticide in Heermann's gull. *Auk*, **107**, 772–94.

Vehrencamp, S. L. (1977). Relative fecundity and parental effort in communally nesting anis *Crotophaga sulcirostris*. *Science*, **197**, 403–5.

Vincent, J. (1936). *The romance of nature*. Country Life, London.

Walsh, T. P. (1964). Blackbird eating own eggs. *British Birds*, **57**, 436.

Ward, H. L. (1906). Why do herring gulls kill their young? *Science*, **24**, 593–4.

Watanuki, Y. (1988). Intraspecific predation and chick survival: comparison among colonies of slaty-backed gulls. *Oikos*, **53**, 194–202.

Werner, T.K. and Sherry, T.W. (1987). Behavioral feeding specialization in *Pinaroloxias inornata*, the 'Darwin's Finch' of Cocos Island, Costa Rica. *Proceedings of the National Academy of Science, USA*, **84**, 5506–10.

Whitmore, M. J. (1986). Infanticide of nestling noisy miners, communally breeding honey-eaters. *Animal Behaviour*, **34**, 933–5.

White, C. M. and Cade, T. J. (1971). Cliff-nesting raptors and ravens along the Colville River in arctic Alaska. *Living Bird*, **3**, 75–96.

Wilkins, M. A. (1964). Male chaffinch attacking the body of another. *British Birds*, **57**, 471.

Willoughby, E. J. and Cade, T. J. (1964). Breeding behavior of the American kestrel (sparrow hawk). *Living Bird*, **3**, 75–96.

Woodward, I. (1962). Egg-eating by hen blackbird. *British Birds*, **55**, 591.

Wynne-Edwards, V.C. (1962). *Animal dispersion in relation to social behaviour*. Oliver & Boyd, Edinburgh.

Yom-Tov, Y. (1974). The effect of food and predation on breeding density and success, clutch size and laying date of the crow (*Corvus corone* L.). *Journal of Animal Ecology*, **43**, 479–98.

Yom-Tov, Y. (1976). Recognition of eggs and young by the carrion crow *(Corvus corone)*. *Behaviour*, **59**, 247–51.

Young, E. C. (1963). The breeding of the south polar skua *Catharacta maccormicki*. *Ibis*, **105**, 203–33.

Zahavi, A. (1990). Arabian babblers: the quest for social status in a cooperative breeder. In *Cooperative breeding in birds: long-term studies of ecology and behavior*, (ed. P. B. Stacey and W. D. Koenig), pp. 103–30. Cambridge University Press.

14

Pup-cannibalism in rodents: causes and consequences

Robert Elwood

Rodents are usually thought of as primarily vegetarian, but studies of the natural diet increasingly demonstrate the omnivorous nature of many species. For example, ground squirrels *Spermophilus beldingi,* hunt both insects and small mammals and feed on carrion (Sherman 1981); grasshopper mice *Onychomys* spp., feed on insects and small vertebrates; common hamsters *Cricetus cricetus,* hunt insects, lizards, frogs, mice, young birds and even snakes; doormice (family Gliridae) eat insects, spiders, and earthworms; and Wagner's gerbil *Gerbillus dasyurus,* feeds on snails, leaving piles of shells near burrow entrances (Macdonald 1984). Some rodents that specialize in aquatic habits use crustaceans and fish as their primary food source (e.g. fish-eating rats and mice, tribe Ichthyomyini) (Macdonald 1984). The use of animal matter in the diet thus appears to be widespread amongst rodents, even if it forms a minority constituent in most species (e.g. Prakash 1962; Hansson 1985).

If small animals are commonly utilized as food then cannibalism, especially of the small, relatively defenceless infants is also possible. Pup-cannibalism may provide a substantial gain in nutritional terms at little physical risk from the infant that is attacked and consumed. However, there might be costs of cannibalizing pups that are closely related to the cannibal, and in general, pup-cannibalism will be selected if the following inequality holds:

$$\frac{\text{Benefit to cannibal} + (\text{Benefit to surviving kin x } r)}{\text{Costs to cannibal} + (\text{Costs to kin x } r)} > 1$$

where r is the degree of relatedness. The cannibal will obtain benefit from the nutrition gained in terms of survival and future reproductive success. Benefits might also accrue to surviving kin in terms of their increased survivorship and future reproduction. Thus, a lactating female that cannibalizes a pup may be able to provide better-quality milk for her surviving infants or other kin when communal suckling takes place (e.g. prairie dogs, *Cynomys ludovicianus,* Hoogland *et al.* 1989). Clearly the level of benefit in terms of inclusive fitness (*sensu* Hamilton 1964) depends on the degree of relatedness (r) between the cannibal and the relatives that gain. The costs of cannibalism will include those of normal foraging, that is search and handling costs, and might also include the cost of intervention from conspecifics. There is also a potential cost in terms of the loss of shared genes, which depends on the degree of relatedness (*r*) between the cannibal and the victim. This cost, however, depends on the probability of survival of the victim had it not been eaten and, if this is low, the costs in terms of inclusive fitness of the cannibal will be low.

Animals that harm their own young will incur the greatest costs in terms of inclusive fitness, and thus there will have to be a substantial increase in the fitness of the cannibal and perhaps other close kin, which may usually be the survivors of the litter, to offset this cost (O'Connor 1978). Parental cannibalism is thus likely to be a special case. There are fewer contstraints on non-parental pup-cannibalism, when the cost to shared genes is likely to be considerably lower. Thus non-parental cannibalism may be common in cases where adults have access to unrelated infants, but some mechanism(s) may be required to ensure that it is directed selectively at non-kin, if it is to enhance the inclusive fitness of the perpetrator. Non-parental pup-cannibalism, however, is likely to reduce the fitness of the parents of the victims and mechanisms are expected that reduce the effectiveness of cannibalistic attacks on their young. Such mechanisms, are likely to influence social structure and population dynamics. These are the main themes discussed in the present review.

PARENTAL CANNIBALISM

Parental cannibalism occurs when a biological parent consumes some or all of the infants. For the most part, this section will be restricted to examples of maternal cannibalism rather than paternal cannibalism. This is because of the paucity of information on the latter, possibly due to either its rare occurrence or lack of study. There are accounts of males cannibalizing infants but it is unusual for fathers to do so (Labov *et al.* 1985 for review). Generally, the presence of fathers either enhances growth and development of young (e.g. gerbils *Meriones unguiculatus* Elwood and Broom 1978), survival of young (e.g. California mouse *Peromyscus californicus* Dudley 1974) or has no detrimental effect (McCarty and Southwick 1977; Hartung and Dewsbury 1979; Thomas and Birney 1979; Ostermeyer 1981).

In contrast, maternal cannibalism is common and well documented. It may be classified broadly as 'abnormal' or 'adaptive'. The former is attributed to some artificial circumstance of laboratory maintenance, whereas the latter, although also being observed in laboratory situations, is viewed as providing some benefit to the parent.

Abnormal maternal cannibalism

Females are particularly prone to cannibalize part or all of their first litter, especially if the female is young (Rowell 1961, but see Day and Galef 1977) and if she is disturbed at about the time of parturition (Ahroon and Fidura 1976; Ahroon 1979; DeSantis and Schmaltz 1984). The disturbance may be relatively minor as in handling of the new born pups by the experimenter (Rowell 1961; Lane-Petter 1968, but see Day and Galef 1977) or may involve a more drastic surgical manipulation of the female (Kimble *et al.* 1967; Gandelman *et al.* 1971; Malsbury *et al.* 1985) or may be due to the presence of other individuals. Goldman and Swanson (1975) reported that crowded hamsters *Mesocricetus auratus,* sometimes cannibalize young and that part of the cannibalism was due to the mothers (also see Diamond and Mast 1978). Crowding is clearly an artefact of laboratory conditions for hamsters, which are normally solitary. Poor survival of infants has also been noted in confined populations of mice *Mus musculus*, due partly to poor maternal care and maternal cannibalism (Southwick 1955). Pup-cannibalism may occur if the infants are subject to surgical manipulation causing them to behave differently from untreated infants (DeSantis and Schmaltz 1984). Finally, abnormal pup-cannibalism may occur when there is some congenital abnormality of the mother. Deaf mice, for example, are prone to cannibalize their young at birth. The mothers cannot hear the calls that appear to signal to the mother that the pup is being handled roughly and, after eating the placenta and cord, the mother may continue to eat the infant (Noirot 1972). Deaf females may also attack pups that are delivered and cleaned normally, apparently because the females cannot hear the calls that normally elicit maternal behaviour (Noirot 1972).

Some of the responses noted above may be adaptive. For example, severe disturbance of the mother and/or litter in natural conditions may reduce the chances of survival of the young thereby reducing the cost of parental cannibalism. However, most of the studies discussed above were not designed to test functional explanations and thus any conclusions about adaptation are tentative.

Adaptive maternal cannibalism

There are several situations in which it would benefit the parent to cannibalize some or all of the litter. If an infant died from natural causes, consumption of the body will recoup some of the energy investment (Reynolds 1981). Not all dead pups are consumed, however, and in gerbils *M. unguiculatus* (personal observation), *Peromyscus maniculatus* (Perrigo 1987) and hamsters *M. auratus* (Rowell 1961), some may remain in the nest for a day or more after death, although in hamsters, dead pups may be removed from the nest, placed on the food pile and eaten later. It might also benefit a parent to stop investing in young that are sickly and unlikely to survive (Reynolds 1981) and thus avoid further wasted investment. For this form of infanticide to be effective, however, mothers should have the ability to discriminate between sick and healthy pups and this appears to be the case (DeSantis and Schmaltz 1984). Females may therefore utilize pups as food, when there is a low probability of survival.

One interesting feature of maternal cannibalism in rodents concerns the optimal allocation of resources to young in an environment with unpredictable resource availability. There may be circumstances in which the death of one or more of the pups provides a net benefit to the surviving offspring and to the parent (O'Connor 1978). This will occur when available resources are (a) insufficient for the successful rearing of the entire litter or (b) insufficient to produce large, healthy offspring that are likely to reproduce (see also Stanback and Koenig, this volume). Two experimental approaches have been used to investigate optimal resource allocation in rodents. The first involves experimental manipulation of the litter size and the second involves manipulation of resource availabilty.

Mice allocate resources to their young in a selective manner (Konig and Markl 1987; Konig *et al.* 1988). Production of lightweight young reduces the subsequent reproductive success of those young (Konig and Markl 1987) and females fail to wean small pups (Konig *et al.* 1988). In experiments in which litter size was increased above the number naturally produced, infanticide and cannibalism increased markedly (Gandelman and Simon 1978). Mothers with experimentally-adjusted litters of 16 pups showed significantly more cannibalism than those with 12 pups, resulting in an average of nine pups weaned regardless of the intial experimental litter size. Most deaths occurred three to seven days post partum, most likely due to maternal infanticide rather than the pups dying of starvation (Gandelman and Simon 1978). Similar findings were obtained in a substantial study of litter size, parental investment, and reproductive success in mice (Fuchs 1981) that indicated a positive correlation between initial litter size and cannibalism. Maternal cannibalism was particularly common in low weight mothers. For those rearing 12 or 14 pups, the average weight of infanticidal mothers was significantly lower than that of non-infanticidal mothers (Fuchs 1981). As milk production is associated with body weight, it seems that mothers with low milk production reduce their litters to a number that may be provided with sufficient milk. Fuchs (1981) also showed that most killing occurred within three to four days post partum, suggesting that mothers were killing and usually cannibalizing pups early to minimize wasted maternal investment. Furthermore, a female apparently selects her victims according to their weight: pups of lower than average weight in each litter tend to be killed (Gandelman and Simon 1978) suggesting that females attempt to rear only high quality offspring at the expense of those with a lower chance of future reproduction.

Pup-cannibalism is particularly common in the hamster (Day and Galef 1977). In this species, cannibalism of part of a litter is common with an average loss of 2.7 pups. Experimental reduction of the litter size at parturition by two pups virtually eliminates cannibalism, whereas increasing litter size by four pups increases cannibalism to an average of 6.4 pups (Day and Galef 1977). These loses occur in the first three days post partum, and increasing the litter size after this time has little effect on subsequent cannibalism. This time-based shift in pup-cannibalism does not appear to be due to changes in the characteristics of the pups (Day and Galef 1977). Hamsters typically produce more pups than they can rear, and there is little relationship between the initial natural litter size and subsequent cannibalism (Day and Galef 1977). Why such a method of regulating reproductive effort should be employed is not clear since embryo resorption, or smaller initial litters, would appear to be more economical.

Day and Galef (1977) conclude that post partum cannibalism is also a mechanism by which females adjust their litter size in response to current changes in environmental conditions. Variation in resource availability was thus predicted to influence pup-cannibalism. Recent studies on hamsters are congruent with this prediction; litter size was less when resources were reduced (Labov *et al.* 1986) although how pups were lost was not determined. Infanticide and cannibalism should not always be inferred; for example, losses of pups in wood rats *Neotoma floridana,* in response to food restriction, appeared to be due to specific neglect of certain pups rather than to direct killing and cannibalism (McClure 1981). Moderate food restriction failed to induce pup-cannibalism in laboratory rats (Smart and Preece 1973) or deer mice *Peromyscus maniculatus* (Perrigo 1987).

Pup-cannibalism may be used to balance energy demands and reproductive success. Female mice *M. musculus,* subjected to either 24 or 48 hours of food deprivation from day 2 of lactation showed a greater tendency to kill and eat their young than those deprived on day 12; those subject to 48 hours of deprivation showed the greater degree of cannibalism than those deprived for just 24 hours (Bronson and Marsteller 1985). Furthermore, food restriction had greater effects on cannibalism by females with large litters than those with small litters (Marsteller and Lynch 1983). Thus, cannibalism appeared to provide an immediate food source as well as reducing the energetic demands of the litter. There seems to be a trade-off between the degree of deprivation faced by the mother and the chances that the young might survive a period of low milk supply. Pups that have already received substantial investment (day 12) may be valued more highly by the mother because the pups require less future investment prior to weaning and are less likely to be adversely affected by short-term decline in milk supply (Bronson and Marsteller 1985). Older pups, with their greater thermoregulatory ability, also might be better able to withstand the absence of their mother whilst she forages when food is scarce. Thus, older pups are less likely to be cannibalized even though short-term energy demands on the mother might be greater (Bronson and Marsteller 1985).

The weaning strategies of *P. maniculatus* and *M. musculus* were markedly different (Perrigo 1987). *Peromyscus* females only killed infants under conditions of extreme food deprivation, usually killing the entire litter. Females appeared to rear as many young as possible, even though the lactation period had to be extended in food deprived conditions. *Mus* mothers, on the other hand, increasingly used pup-cannibalism as a way of balancing energy demand when food was less accessible. Most cannibalism occurred in the first four days post partum; the few pups that died after 12 days showed no bite marks and were not cannibalized (Perrigo 1987). Thus with moderate deprivation, litters were reduced but with increased deprivation the entire litter might be cannibalized. Pup-cannibalism thus provided a short-term boost in recycled calories for the female and hence more food for any surviving pups. In this way *Mus* appeared to be maintaining pup quality at the expense of pup numbers.

Finally, females might improve their fitness by manipulating the sex ratio of the litter by selective cannibalism. Where the cost of rearing good quality males and females differs, and the variance in reproductive success is greater in the more expensive sex, it may pay a mother to produce the more expensive sex under good conditions and the less expensive sex under poor condutions (Trivers and Willard 1973). There is clear evidence that sex ratios change prior to parturition when pregnant females are energetically stressed during pregnancy in hamsters (Labov *et al.* 1986; Huck *et al.* 1988) and mice (Krackow and Hoek 1989) but it is possible that this is due to lower viability of male embryos rather than selective reabsorption by the mother. There is evidence also that the sex ratio of litters may change during the lactation period. Food restricted wood rats, for example, show differential investment in the sexes with female offspring gaining more weight than males (McClure 1981). Food restriction during lactation results in female-biased litters in mice, which might be due to cannibalism of male pups (Marsteller and Lynch 1983). In all these cases, there is the possibility that males are simply less viable in adverse conditions and to date there is no firm evidence that mothers kill pups to bias sex ratios in an adaptive manner.

NON-PARENTAL PUP-CANNIBALISM

There are at least four possible functions of non-parental infanticide:

(1) increased food acquisition;

(2) decreased resource competition by removal of unrelated infants;

(3) avoidance of misdirected parental care by killing unrelated young; and

(4) increased male reproductive success.

While cannibalism is the sole requirement for the first functional category, it is not for the other three, for which the primary gain is via the death of the infant(s) and cannibalism may bring additional gains. Pup-cannibalism, therefore, does not necessarily mean that food acquistion is the primary function. Nevertheless, when cannibalism is the primary function, the killer may also gain in one or more of the other ways. Thus, none of the above functions are mutually exclusive. Although only the first three may apply to females, all four may apply to males.

Food acquisition

Pup-cannibalism has been observed in natural populations of a variety of species. The clearest examples are studies of Belding's ground squirrels *Spermophilus beldingi* (Sherman 1981), black-tailed prairie dogs *Cynomys ludovicianus*

(Hoogland 1985) and white-footed mice *Peromyscus leucopus* (Wolff and Cicirello 1989). In Belding's ground squirrel both sexes commit infanticide but it is most common in non-resident, adult females and one-year-old males. Females are less likely than males to cannibalize their victims, suggesting that infanticide has different functions for the two sexes. That food acquisition is at least one important function of cannibalism in this species was inferred from observations of fights between males for the dead infants, which were immediately cannibalized by the winners, indicating that a carcass is a valuable resource (Sherman 1981). Observations of stalking other species and feeding on carcasses of animals killed on roads suggested that one-year-old males are more carnivorous than females. Prairie dogs of both sexes also kill infants but, in this case, females are typically cannibalistic (Hoogland 1985). Lactating females that killed and cannibalized other infants were more likely to wean a litter, they had larger litters containing heavier young at weaning, and were heavier themselves at weaning than were non-killers, suggesting that cannibalism may provide considerable benefits. A recent field experiment on white-footed mice involved placing infants in traps so that an adult entering the trap encountered a pup but could move to a second chamber from which the adult could not escape (Wolff and Cicirello 1989). Under these conditions 43 per cent of the pups that were killed were partially cannibalized and both sexes engaged in cannibalism. Cannibalism, however, was particularly likely by pregnant or lactating females that may have high energy demands (Wolff and Cicirello 1989).

Laboratory studies also suggest that acquisition of food is an important function of non-parental infanticide in other rodent species. Short-term deprivation of food increases the proportion of animals killing and cannibalizing infants by male gerbils (Elwood and Ostermeyer 1984*a*) and by both sexes in mice (Svare and Bartke 1978) and long-term deprivation (23 hours per day for 11 days) increases cannibalism by male rats (Paul and Kupferschmidt 1975). Furthermore, the manner in which gerbils and rats of both sexes kill infants suggests it may be primarily a means of food acquisition (Paul and Kupferschmidt 1975; Elwood 1977, 1980,1981; Peters and Kristal 1983; Mennella and Moltz 1988). Both species are described as simply eating pups rather than attacking them. These observations contrast with the manner of killing by *M. musculus*. Males attack pups with rapid bites to the head and thorax (vom Saal 1985; Elwood 1986) and cannibalism occurs in 50 per cent or less of cases (Labov 1980; McCarthy and vom Saal 1985). In mice, the function of non-parental cannibalism appears only to be partly for food acquisition. Curiously, male gerbils that encounter infants of other species in their nests with their own young, kill these alien young with rapid bites to the head and thorax (Elwood and Ostermeyer 1986) suggesting that in these circumstances the primary aim is to kill rather than to eat. There seem to be two functionally and motivationally separate forms of pup killing in gerbils resulting in different behaviours.

Resource competition

Infanticide by adult female Belding's ground squirrels *S. beldingi,* appears to function as a mechanism of resource competition (Sherman 1981). Immigrant females often kill more than one infant, but rarely cannibalize their victims, and then usually settle near to the victim's burrow. The loss of a litter to predators commonly causes a female to abandon the burrow and move to another area. Thus, by killing an entire litter an immigrant female might gain access to a territory. Similar explanations apply to infanticide by musk rats *Ondatra zibethica* (Errington 1963), female mice *M. musculus* (McCarthy and vom Saal 1985), female white-footed mice *P. maniculatus* (Wolff and Cicirello 1989) and female prairie dogs *C. ludovicianus* (Hoogland 1985).

Avoidance of misdirected parental care

Avoidance of misdirected parental care is frequently cited as a function of infanticide (e.g. Sherman 1981; Elwood and Ostermeyer 1984*b*) and it may be appropriate in two situations. First, infanticide may occur when infants are mobile and attempt to obtain parental care from non-relatives, as in elephant seals and colonial sea birds (reviewed by Sherman 1981). In rodents, this might explain infanticide in precocial species such as spiny mice *Acomys cahirinus*, or in cases of killing juveniles of altricial species (Paul and Kupferschmidt 1975). Second, avoidance of misdirected care as the basis for infanticide might occur when nests are parasitized, as in many species of birds that detect alien eggs and reject them (Rothstein 1982). Even conspecific eggs may be rejected if they appear in the nest at an inappropriate time (Stouffer *et al.* 1987). No cases of parasitism in rodents by dumping young are known. However, alien young are killed if sufficiently genetically different from own young (Elwood and Ostermeyer 1986; Paul 1986) or if they appear at an inappropriate time (e.g. Elwood 1977, 1981). Hoogland *et al.* (1989) reported that some female prairie dogs *C. ludovicianus,* engage in communal suckling but others do not. Those that fail to suckle communally often kill and cannibalize the young, perhaps in order to avoid giving milk to infants that are less closely related to the killer than are her own young. Whether or not infants are killed might depend on the ease of discriminating offspring from other young and the cost of mistakes (Hoogland *et al.* 1989).

Advancement of male reproduction

A male may advance his reproduction if he is able to induce oestrus in the mother of the infants he kills, mate with her, and sire young sooner than if the infants had not been killed. This appears to be the reason for many instances of infanticide by male rodents. Advancement of male reproduction might also occur due to the shorter gestation period of non-lactating females (Mallory and Brooks 1978; vom Saal and Howard 1982; Wolff and Cicirello, 1989).

MECHANISMS FOR SELECTION OF VICTIMS

Victims of pup-cannibalism are clearly not selected at random. Even in cases of parental cannibalism there is evidence that mothers cannibalize injured (DeSantis and Schmaltz 1984) and small pups (Gandelman and Simon 1978) in preference to others, but the mechanisms by which this selection occurs are unknown. In contrast, there has been extensive work investigating mechanisms by which both males and females avoid harming close kin. The sexes, however, differ in the mechanisms that mediate this ability.

How females avoid harming their own young

Virgin and early pregnant female hamsters (Richards 1966), gerbils (Elwood 1977) and wild house mice (McCarthy *et al.* 1986) normally cannibalize infants but cannibalism is inhibited prior to parturition, and instead conspecific infants elicit maternal behaviour. This inhibition of cannibalism appears to be mediated by the dramatic hormonal changes that occur during pregnancy resulting in a 'maternal state', when the female responds to all young in a maternal manner (Rosenblatt and Siegel 1983).

In general, the maternal state may enable the female to avoid harming all infants until her own young are weaned when she may resume pup-cannibalism (Elwood 1977). Prairie dogs *C. ludovicianus*, however, may kill the young of other females before their own young are weaned (Hoogland 1985), as do gerbils (Elwood 1981) and white-footed mice *P. maniculatus* (Wolff and Cicirello 1989). In the case of prairie dogs the discrimination between own young and others in the coterie has been suggested to be mediated by their location (Hoogland *et al.* 1989). Females may cannibalize young infants that have yet to leave their natal burrow. However, if their own young have left the burrow, they do not harm unrelated, mobile young which may be confused with their own. The killing of infants found outside the home burrow by white-footed mice may also be mediated by location-based kin-recognition (Wolff and Cicirello 1989).

Laboratory studies on gerbils, however, demonstrate a shift in maternal responses towards unrelated pups during the lactation period (Elwood 1981). When their own young are less than ten days old, female gerbils will readily foster unrelated day-old pups, but as their own young become older, females are more likely to cannibalize day-old pups. This may be due to the ease of discrimination between their own young that have fur and newborn that do not; the time that newborn pups may be accepted is extended if the female's own young are replaced with younger foster pups (Elwood 1981). If the female's litter is removed at birth, then there is a rapid return to cannibalism indicating that the presence of young pups inhibits cannibalism and maintains maternal responsiveness (Elwood 1981). Similar findings have been reported for hamsters (Siegel and Greenwald 1978).

One mechanism that appears to play little or no part in mediating selective infanticide by females is familiarity with the young. Theoretically, the female is ideally situated to interact with her infants during parturition, to learn the features of her offspring, and to kill pups that do not match those features. There is no evidence, however, that this occurs (Konig 1989). Females can discriminate between own and unrelated young, as judged by differences in their inspection of infants, but this does not lead to selective cannibalism (Ostermeyer and Elwood 1983).

How males avoid harming their own young

The mechanisms by which males avoid harming their own young are quite different from those of females. Females come into a parental state as a result of hormonal changes associated with gestation. Clearly the different physiology of males precludes this mechanism. A male might be present at his mate's parturition, gain information about his pups at that stage, and then respond in a paternal fashion only to his own young. Indeed, evidence indicates that male mice, housed with their litters, investigate alien pups more than their own pups, but do not harm the former (Ostermeyer and Elwood 1983). Naive male gerbils are less likely to harm pups of a different species than are males that have experience of their own young (Elwood and Ostermeyer 1986). Experience enables gerbils to discriminate between pups of different species but this ability does not seem to extend to discrimination between conspecific pups (Elwood and Ostermeyer 1986).

Males may be absent during parturition and thus fail to gain information about their offspring at that time. Some mechanism, other than familiarity with the pups, must be used for these males to be selective in pup-cannibalism. Males of a variety of species are selective when they first encounter pups (reviewed by Labov *et al.* 1985) and several ideas about the mediating mechanisms have been proposed (reviewed by Elwood 1985). These are of two main types: first, those that rely on the male recognizing some feature correlating with relatedness (or the probability of relatedness); and second, those that propose some physiological change specific to males.

There are three hypotheses that involve some aspect of recognition. These are recognition of

(1) some feature of the pup;
(2) a previous sexual partner; or
(3) a particular location.

Numerous studies demonstrate differential responsiveness towards unfamiliar kin and non-kin in rodents (reviewed by Hepper 1986). It is possible, therefore, that males could discriminate between their own and alien young at the first encounter, either by some recognition gene(s) or by phenotype matching (Hepper 1986). Matching might be to self, to familiar kin, or to a previous sexual partner. In this final case the mother might deposit her odours on the pup. There is one study that suggests that recognition of pups occurs in mice, with only unrelated pups being killed (Paul 1986), but numerous other studies have failed to repeat this result (e.g. Huck *et al.* 1982; Brooks and Schwarzkopf 1983; Elwood 1985; vom Saal 1985; Gubernick and Alberts 1989; Kennedy 1989; Parmigiani 1989; Perrigo and vom Saal 1989).

Secondly, males might avoid killing infants that are found in association with previous sexual partners. This mechanism relies on recognition of the female and allows a male to avoid killing his own infants. It would not enable him to discriminate between his own pups or those sired by another male, if the female had mated more than once. Male mice introduced into the cage of either their previous mate or a strange female were more likely to kill pups in the cage of a strange female, even if it contained foster pups actually fathered by the experimental male (Huck *et al.* 1982). Other studies, have failed to replicate these results (vom Saal and Howard 1982; Brooks and Schwarztkopf 1983; Parmigiani 1989; Kennedy 1989), but McCarthy and vom Saal (1986*a*) found an effect in one experiment but not in two others.

Finally, males might avoid killing pups found in locations where related young might also be expected (Beilharz 1975), but evidence from several studies has failed to support this idea (Brooks and Schwarzkopf 1983; Kennedy 1989; Parmigiani 1989).

Parmigiani (1989) recently investigated all three ideas by allowing male mice to cohabit with their mate for a period after copulation and then encounter own or unrelated pups in a variety of situations. No males killed pups in these experiments. Kennedy (1989) compared infanticide rates in two strains of male mice that had copulated but not subsequently cohabited with the female. There were low levels of infanticide of own pups and unrelated pups of both the same and different strains but no differences between the treatments. Males were further tested in (a) their home cage or the cage of a strange male, (b) the cage of a previous mate or of a strange female and (c) where those females were either absent or present, and (d) where the female was restrained or unrestrained (Kennedy 1989). None of these factors influenced the probability of infanticide, indicating that selectivity of infanticide and pup-cannibalism by males is not mediated by recognition of genotype, associated females, or location.

In contrast to mechanisms involving recognition there is considerable evidence demonstrating the existence of some physiological change (Elwood 1985) or 'state mediated recognition' (Elwood 1991) that enables males to be selective in their infanticide and pup-cannibalism. This was first demonstrated in the Mongolian gerbil, the males of which stop pup-cannibalism after they have mated and are housed with their mate during pregnancy (Elwood 1977). Similar findings have since been reported for mice (Labov 1980; Soroker and Terkel 1988), rats (Brown 1986), Siberian hamsters *Phodopus sungorus* (Gibber *et al.* 1984) and spiny mice *Acomys cahirinus* (Makin and Porter 1984). In these species infanticide by males is inhibited at about the time that their own young are likely to be born. Three hypotheses concern the mechanism that mediates this shift from infanticidal to non-infanticidal states: (a) copulation initiates a paternal state, (b) pheromones from the female are involved, and (c) female aggression towards the male initiates a non-infanticidal state.

Copulation has an inhibitory effect on infanticide in some strains of mice and rats (Elwood 1985; vom Saal 1985; Mennela and Moltz 1988; Soroker and Terkel 1988; Perrigo and vom Saal 1989) but not in other strains (Jakubowski and Terkel 1985*a*,*b*; Brown 1986; McCarthy and vom Saal 1986*b*; Kennedy and Elwood 1988), or in gerbils (Elwood 1980). Copulation in mice generally causes a reduction of infanticidal tendencies approximately three weeks later (i.e. when the pups should be born), but not all males are inhibited at this time. It is also less effective in inhibiting infanticide if the male has previous infanticidal experience (Elwood and Ostermeyer 1984*c*; Elwood 1986; McCarthy and vom Saal 1986*a*).

The other two hypotheses suggest that post-copulatory cohabitation effects inhibition of infanticide. Male gerbils, inhibited from pup-cannibalism, return to cannibalism if the pregnant female is removed from the cage (Elwood 1980). Furthermore, post-copulatory cohabitation virtually eliminates infanticide in mice that are not inhibited by copulation *per se* (Elwood and Ostermeyer 1984*c*; Elwood 1985, 1986; Kennedy and Elwood 1988; Soroker and Terkel 1988). Cohabitation with the pregnant female is also effective in rats (Brown 1986). This inhibition was assumed to be due to a pheromone from the pregnant female (Elwood 1980) but experimental evidence is not supportive for gerbils (Elwood and Ostermeyer 1984*b*) or mice (Elwood 1985; Kennedy 1989). In rats, however, soiled bedding from pregnant females partially inhibits infanticide in sexually-naive males (Menella and Moltz 1988) and maternal excretia maintains paternal responsiveness in the California mouse (Gubernick and Alberts 1989). Thus, in some species, pheromones may be produced by females that inhibit the cannibalistic tendencies of their mates.

Finally, post-copulatory interactions between the sexes may result in subordination of the male and this may inhibit his infanticidal tendencies. In gerbils, the female has to remain with the male to maintain his paternal responsiveness (Elwood 1980) and males that are subordinate to other males are also inhibited in pup-cannibalism (Elwood and Ostermeyer 1984*a*). In mice, subordinate males are less likely to be infanticidal than are dominant males (Huck *et al.* 1982; vom Saal and Howard 1982). Furthermore, previously infanticidal males that are subsequently defeated in fights with other males may become non-infanticidal whereas previously non-infanticidal males that win fights may switch to being infanticidal (Elwood 1986). Females become aggressive towards their mates shortly after mating and appear to be dominant to the males 24 hours later (Elwood 1985). Male mice only have to be housed with their mates for 24 hours after copulation for infanticide to be inhibited (Elwood 1985) and those with relatively large females are more likely to be inhibited in infanticide (Elwood 1986). These data suggest that social subordination inhibits infanticide in male mice and perhaps in gerbils.

Males of different species may rely on different mechanisms to inhibit pup-cannibalism and, within a single species, more than one mechanism may operate to ensure that a male does not kill his own offspring. Males show a temporal shift in infanticidal tendencies that only allows pup-cannibalism at times when their own young are unlikely to be encountered (Elwood 1977).

DEVELOPMENTAL ASPECTS OF PUP-CANNIBALISM

Pup-cannibalism is not committed by the two sexes with equal probability; females are more likely to show this activity in gerbils (Elwood 1977; 1980) and hamsters (Marques and Velenstein 1976) whereas males are more likely in laboratory mice (e.g. Jakubowski and Terkel 1982), rats (e.g. Jakubowski and Terkel 1985*a*), and Siberian hamsters (Gibber *et al.* 1984). There are differences between strains in mice (Svare *et al.* 1984) and rats (Jackubowski and Terkel 1985*b*), and even differences in the same strain when bred in different establishments (e.g. Brown 1986) or tested at different ages (Gandelman 1973; Elwood 1980).

In male mice (Gandelman 1973; McCarthy and vom Saal 1986*b*), Siberian hamsters (Giber *et al.* 1984), and gerbils (Elwood 1980) older males are more likely to cannibalize pups especially after the onset of puberty. Testosterone is important in the development of this activity in mice (reviewed by Svare *et al.* 1984) and rats (Rosenberg and Sherman 1975). Castration in adulthood reduces infanticide, and replacement therapy with testosterone reverses this effect (Gandelman and vom Saal 1975). Furthermore, testosterone mediates the temporary increase in infanticide shortly after copulation in mice but is not involved in the inhibition of infanticide two to three weeks after mating (Perrigo and vom Saal 1989).

Androgens also influence infanticide during the perinatal period (Svare *et al.* 1984), however, they appear to play an inhibitory role at this time. Castrated newborn males and females, given testosterone proprionate in adulthood, show an increase in infanticide compared with control animals (Samuels *et al.* 1981). Natural variation in exposure to foetal hormones may also influence infanticidal tendencies in adulthood. Male mice, that as embryos had females on either side (OM males), tend to have lower levels of testosterone during late gestation and show higher levels of infanticide in adulthood than do those that had males on either side (2M males) (vom Saal 1983; Perrigo and vom Saal 1989). Stressing the pregnant female removes this difference, and offspring of such females show the low level of infanticide typical of 2M males (vom Saal 1983). Difference in foetal hormones or responsiveness to such hormones may explain some of the variation in infanticide seen in different strains of mice (Svare and Broida 1982; McCarthy and vom Saal 1986*b*).

The post-weaning social environment may also influence pup-killing. Isolation of adults for 12–15 days enhances pup-cannibalism in gerbils (Elwood and Ostermeyer 1984*a*). In mice, however, isolation for prolonged periods either has no effect (Brooks and Schwarzkopf 1983) or inhibits infanticide (Ghiraldi and Svare 1986; McCarthy and vom Saal 1986*b*). Post-weaning exposure to males of the same age results in higher levels of attacks on pups than does exposure to females, possibly due to different types of play in single sex and mixed sex groups (Gibber and Terkel 1985). Post-weaning exposure to newborn infants inhibits infanticide in adulthood in mice (Jakubowski and Terkel 1982) but not Siberian hamsters (Gibber and Terkel 1985), or gerbils (Elwood 1980). Juvenile gerbils that remain with the parents and the subsequent litter are inhibited in pup-cannibalism at an age when isolated juveniles become cannibalistic. This suppression of infanticide is probably due to the subordination of the juveniles by the parents, and ensures that younger siblings are not harmed in the extended family systems that tend to form in this species (Ostermeyer and Elwood 1984).

PARENTAL DEFENCE AGAINST PUP-CANNIBALISM

Newborn pups represent a considerable investment to the mother and possibly the father. The loss of pups to infanticidal attacks by other adults may therefore be costly. This may have led to the evolution of a number of behavioural mechanisms that reduce infanticide, including territorial behaviour, maternal aggression, pregnancy blocking by females, and inducing parental states in non-parents.

Territoriality appears to be common in rodents and may serve a number of functions. One important function is protection of the young (Brooks 1984). In Belding's ground squirrel, for example, trespassing adults never entered a mother's burrow when the mother was in the territory (Sherman 1981), and if they attempted to enter they were attacked and chased by the burrow owner. Trespassing adults entered the burrow and killed and cannibalized infants only when the mother was absent. Similar findings were reported for prairie dogs (Hoogland 1985). In another species of ground squirrel *Spermophilus parryii*, territorial males appear to play an active role in the defence of the young (McLean 1983). Male defence of territories against intrusion by other males was most prevalent when infants were present. Infanticide was less likely if the territorial male had been resident during the mating period but if he was experimentally removed from the territory, incoming males killed the young. It seems that female ground sqirrels could not exclude new males from the burrow and resident males spent much of their time acting as lookouts against such intruders.

During pregnancy, maternal aggression, a specific type of aggression seen during the reproductive period, may serve to protect against pregnancy blocking (see below) or future infanticide. It also occurs during the lactation period and, at this time, attacks by females against intruders may protect the infants from infanticide (Mallory and Brooks 1978; Wolff 1985). The post-partum aggressive state is maintained by sucking stimulation from the young and, if the young are removed, maternal aggression declines (Svare *et al.* 1980).

Maternal aggression has been studied mainly in laboratory mice, which show a distinct 'target specificity' in attack. Males are more likely to be attacked than females (Goyens and Noirot 1977) and virgin females are more often attacked than pregnant or lactating females (Rosenson and Asheroff 1975; Goyens and Noirot 1977). Indeed, males that are rearing infants (paternal males) are less likely to be attacked than are virgin males that are possibly infanticidal (Parmigiani *et al.* 1988). Recent work has also demonstrated that females attack virgin infanticidal males more than virgin non-infanticidal males (Elwood *et al.* 1990*b*). Thus, aggressive mothers appear to discriminate between intruders and respond according to the risk they present to the infants. In laboratory experiments, however, maternal aggression does not always keep the intruder from the pups but in natural burrow systems maternal aggression might be more successful.

Newly pregnant mice may fail to implant the blastocysts if exposed to strange males (Bruce 1959, 1960), a phenomenon commonly called the 'Bruce effect' or 'pregnancy blocking'. The advantage to the male is clear, since he reduces the fitness of other males whilst gaining the opportunity to mate with a female. Females should also gain, since selection would otherwise act to reduce this effect of males (Bronson and Coquelin 1980). Various hypotheses have been put forward concerning female gain (Schwagmeyer 1979). One suggests that females lose a small amount of investment by pregnancy block in order to avoid the potentially greater loss of having the litter destroyed by an infanticidal male (Labov 1981*a*; Huck 1982,1984). This hypothesis implies that females should show pregnancy block only when the risk to the future litter is high as not all adults pose the same risk (see above).

Dominant male mice are both more likely to be infanticidal and more likely to cause pregnancy block than are subordinates (Huck 1982, but see Labov 1981*b*). Female hamsters are both more likely to be cannibalistic and to cause pregnancy block than are males (Huck *et al.* 1983; Huck 1984). Furthermore, female mice are more likely to show pregnancy block to males that have been experimentally identified as infanticidal, than they are to non-infanticidal males (Elwood and Kennedy 1990). Those females that persisted with the pregnancy in the presence of a previously infanticidal male were more likely to have their litter killed than were those that produced a litter in the presence of a previously non-infanticidal male.

The final mechanism that might be employed to reduce infanticide is that a female may induce a paternal state in a male that is not the father of her litter (Elwood 1991). If males use specific mechanisms that allow them to avoid harming their own young (see above) they could be used by females to induce a non-infanticidal state or even obtain paternal care for her young. For example, copulation is important in the normal inhibition of infanticide in males (vom Saal 1985) and there is evidence that females copulate with particular males simply to induce a non-infanticidal state. Female water voles *Arvicola terrestris,* shift nest location during pregnancy and some move into the home range of a new male (Jeppsson 1986). Females that move into the home range of a new male were found to have mating plugs, after mating with the new male, whereas those that remained with the original male did not. Pregnancy did not terminate when a new male was encountered. Rather, females came into pseudoestrus and mated apparently to induce a non-infanticidal state.

A second important factor in the onset of paternal care is subordination of the male (Elwood 1985). Maternal aggression (see above) may act partly to induce non-infanticidal states in new males. Males that are attacked and defeated by females may subsequently be parental (Mallory and Brooks 1978; Jackubowski and Terkel 1985*a*). Similar mechanisms that inhibit infanticide and induce paternal care occur in fish and birds (Elwood 1991 for review).

DEFENCE BY INFANTS TO REDUCE INFANTICIDE

Newborn rodents tend to be altricial, with limited sensory and communicatory ability (Elwood and McCauley 1983), and they might be considered to have little influence on their fate when encountered by a potential cannibal. However, infant rodents produce fewer ultrasonic vocalizations in the presence of odours from males than from females, perhaps in order to hide from potentially infanticidal males (Conely and Bell 1978; Lyons and Banks 1982; Ostermeyer and Elwood 1983). Experiments to determine whether infant mice can discriminate between infanticidal and non-infanticidal males by their odour provide contradictory results (Elwood *et al.* 1990*a*). Significantly more calls were produced when pups were exposed to odours from infanticidal males than from noninfanticidal males. Furthermore, infants became more agitated in the presence of odours from infanticidal males and were more likely to move away from those odours. Increased calls by infants may alert the parents, and moving away might reduce risk, perhaps by hiding among the nest material.

POPULATION DYNAMICS AND SOCIAL STRUCTURE

The prevalence of pup-cannibalism, or of other forms of infanticide, in natural populations is poorly understood (see Brooks 1984 for a review). It has been observed in large, diurnal species of rodents; infanticide accounts for an estimated eight per cent of infant mortality in Belding's ground squirrel (Sherman 1981) and some 47 per cent of prairie dog litters are wholly or partly killed by conspecifics (Hoogland 1985). It occurs in natural populations of small, nocturnal species (Woolf and Cicirello 1989) and in a wide variety of species tested in captivity (see above). Furthermore, the existence of mechanisms specifically inhibiting or disinhibiting infanticide in an adaptive manner suggest that it is a common trait. Little or no data, however, are available to indicate the proportion of infants killed by conspecifics in other species. Data from field studies involving trapping have estimated infant mortality (reviewed by Brooks 1984) and these commonly report figures of bewteen 30 and 80 per cent infant mortality, but ascribing causes to these deaths is difficult.

The studies cited in the present review permit some speculation concerning the effects of restricted food supply on cannibalism, and hence recruitment to the population. As a response to declining food supplies parents are likely to attempt to increase foraging effort but if insufficient food is obtained, they may resort to pup-cannibalism (Perrigo 1987). The alternative is for females not to cannibalize but to continue efforts to rear the pups with little or no chance of producing viable young. Thus, parental cannibalism will tend to have a stabilizing effect on the population because cannibalism of some of the pups will enable the resources to be recycled to any remaining pups (Bronson and Marsteller 1985) or, if all the pups are cannibalized, the female may be in a better physiological condition for future attempts at reproduction.

Laboratory studies indicate that declining food resources are likely to enhance non-parental cannibalism (e.g. Elwood and Ostermeyer 1984*a*) and attempts to gain access to nests containing pups may increase. Access to those nests is likely to be made easier by the increased foraging efforts of the parents. Because the costs, in terms of inclusive fitness, of cannibalism to non-parents are much lower than to parents it is likely that an increase in non-parental cannibalism will occur earlier during a period of food shortage than will parental cannibalism. Should lactating females cannibalize pups of other females, then their own pups may have an increased chance of survival. Non-parental cannibalism will generally result in a decline in recruitment to the adult population beyond that driven by parental cannibalism alone.

Several studies have monitored the growth and decline of populations within enclosures (e.g. Brown 1953; Calhoun 1962; Lloyd and Christian 1967) and attempts have been made to discover the causes of infant mortality (reviewed by Brooks 1984). In some studies, population growth slowed due to infant mortality (e.g. Calhoun 1962; Lidicker 1976) and much of this has been attributed to infanticide and cannibalism. There are great differences between these studies, some of which were attributed to the degree of inbreeding within the populations (Brooks 1984). Groups that originated from several pairs of unrelated animals typically showed increased aggression and infanticide at lower densities than occurred with groups starting with just one pair or small numbers of related individuals. This is congruent with the suggestion that individuals will cannibalize only when there is no detriment to inclusive fitness. One way of reducing the chances of infants being killed by conspecifics would be to remain within kinship groups, suggesting that infanticide might influence social structure. Indeed, kin may even assist in the protection of infants. Studies on prairie dogs, however, show that relatives still cannibalize related pups if the gain from doing so is high (Hoogland *et al.* 1989).

Infanticide might also influence social structure by increasing the utility of territorial behaviour, and the increased maternal aggression observed in laboratory studies might simply be an indication of increased territoriality at the time of reproduction. In addition to acting as a counter measure to infanticide, therefore, territorality might have direct consequences for population growth and regulation (Brooks 1984; Stenseth 1985). Clearly there is a gap in our understanding of the links between infanticide, cannibalism, social structure, and population dynamics.

ACKNOWLEDGEMENTS

Thanks are due to Bernie Crespi, Mark Elgar, Lawrence Frank, and Mariko Hiraiwa-Hasegawa for their considerable efforts to improve this chapter. Some of the work reported here was supported by the Science and Engineering Research Council of the United Kingdom of Great Britain and Northern Ireland.

REFERENCES

Ahroon, J. A. (1979). Does the male parent (*Meriones unguiculatus*) contibute to pup mortality? A reply. *Animal Behaviour,* **27**, 311–12.

Ahroon, J. K. and Fidura (1976). The influence of the male on maternal behaviour in the mongolian gerbil (*Meriones unguiculatus*). *Animal Behaviour*, **24**, 372–5.

Beilharz, R. G. (1975). The aggressive response of male mice (*Mus musculus* L.) to a variety of stimulus animals. *Zeitschrift für Tierpsychologie*, **39**, 141–9.

Bronson, F. H. and Coquelin, A. (1980). The modulation of reproduction by priming pheromones in house mice: speculations on adaptive function. In *Chemical signals, vertebrates and aquatic invertebrates,* (ed. D. Muller-Schwarze and R. M. Silverstein), pp. 243–65. Plenum Press, New York.

Bronson, F. H. and Marsteller, F. A. (1985). Effect of short-term food deprivation on reproduction in female mice. *Biology of Reproduction*, **33**, 660–7.

Brooks, R. J. (1984). Causes and consequences of infanticide in populations of rodents. In *Infanticide in animals and man: comparative and evolutionary perspectives,* (ed. G. Hausfater and S. B. Hrdy), pp. 331–48. Aldine Press, New York.

Brooks, R. J. and Schwarzkopf, L. (1983). Factors affecting incidence of infanticide and discrimination of related neonates in male *Mus musculus*. *Behavioral and Neural Biology*, **37**, 149–61.

Brown, R.E. (1986). Social and hormonal factors influencing infanticide and its suppression in adult Long-Evans rats (*Rattus norvegicus*). *Journal of Comparative Psychology*, **100**, 155–61.

Brown, R. Z. (1953). Social behaviour, reproduction, and population changes in the house mouse. *Ecological Monographs*, **23**, 217–40.

Bruce, H. M. (1959). An exteroceptive block to pregnancy in the mouse. *Nature*, **184**, 105.

Bruce, H. M. (1960). A block to pregnancy in the mouse caused by proximity of strange males. *Journal of Reproduction and Fertility*, **1**, 96–103.

Calhoun, J. B. (1962). Population density and social pathology. *Scientific American*, **206**, 139–48.

Conely, L. and Bell, R. W. (1978). Neonatal ultrasounds elicited by odor cues. *Developmental Psychobiology*, **11**, 193–7.

Day, C. S. and Galef, B. G. (1977). Pup cannibalism: one aspect of maternal behavior in Golden hamsters. *Journal of Comparative and Physiological Psychology*, **91**, 1179–89.

DeSantis, D. T. and Schmaltz, L. W. (1984). The mother-litter relationship in develomental rat studies: cannibalism vs caring. *Developmental Psychobiology*, **17**, 255–62.

Diamond, M. and Mast, M. (1978). Crowding, reproduction, and maternal behavior in the golden hamster. *Behavioral Biology*, **23**, 477–86.

Dudley, D. (1974). Contributions of paternal care to the growth and development of the young in *Peromyscus californicus*. *Behavioral Biology*, **11**, 155–66.

Elwood, R. W. (1977). Changes in the responses of male and female gerbils (*Meriones unguiculatus*) towards test pups during the pregnancy of the female. *Animal Behaviour*, **25**, 46–51.

Elwood, R. W. (1980). The development, inhibition and disinhibition of pup-cannibalism in the Mongolian gerbil. *Animal Behaviour*, **28**, 1188–94.

Elwood, R. W. (1981). Postparturitional re-establishment of pup-cannibalism in female gerbils. *Developmental Psychobiology*, **14**, 209–12.

Elwood, R. W. (1985). The inhibition of infanticide and the onset of paternal care in male mice, *Mus musculus*. *Journal of Comparative Psychology*, **99**, 457–67.

Elwood, R. W. (1986). What makes male mice paternal? *Behavioral and Neural Biology*, **46**, 54–63.

Elwood, R. W. (1991). Parental states as mechanism for kinship recognition and deception about relatedness. In *Kin recognition*, (ed. P. Hepper), pp. 289–307. Cambridge University Press.

Elwood, R. W. and Broom, D. M. (1978). The influence of litter size and parental behaviour on the development of Mongolian gerbil pups. *Animal Behaviour*, **36**, 438–54.

Elwood, R. W. and Kennedy, H. F. (1990). the relationship between infanticide and pregnancy block in mice. *Behavioral and Neural Biology*, **32**, 293–305.

Elwood, R. W. and McCauley, P. J. (1983). Communication in rodents: infants to adults. In *Parental behaviour of rodents*, (ed. R. W. Elwood), pp. 126–50. John Wiley and Sons, Chichester.

Elwood, R. W. and Ostermeyer, M. C. (1984*a*). The effects of food deprivation, aggression and isolation on infanticide in the male Mongolian gerbil. *Aggressive Behavior*, **10**, 293–301.

Elwood, R. W. and Ostermeyer, M. C. (1984*b*). Infanticide by male and female Mongolian gerbils: ontogeny, causation and function. In *Infanticide in animals and man: comparative and evolutionary perspectives*, (ed. G. Hausfater and S. B. Hrdy), pp. 367–86. Aldine Press, New York.

Elwood, R. W. and Ostermeyer, M. C. (1984*c*). Does copulation inhibit infanticide in rodents? *Animal Behaviour*, **32**, 293–305.

Elwood, R. W. and Ostermeyer, M. C. (1986). Discrimination between conspecific and allospecific infants by male garbils and mice before and after experience of their own young. *Developmental Psychobiology*, **19**, 327–34.

Elwood, R. W., Kennedy, H. F., and Blakely, H. M. (1990*a*). Responses of infant mice to odors from infanticidal noninfanticidal and paternal male mice. *Developmental Psychobiology*, **23**, 309–17.

Elwood, R. W., Nesbitt, A. A., and Kennedy, H. F. (1990*b*). Maternal aggression and the risk of infanticide in mice. *Animal Behaviour*, **40**, 1080–6.

Errington, P. L. (1963). *Muskrat populations*. Iowa State University Press, Aems.

Fuchs, S. (1981). Optimality of parental investment: the influence of nursing on reproductive success of mother and female young house mice. *Behavioural Ecology and Sociobiology*, **10**, 39–51.

Gandelman, R. (1973). The development of cannibalism in male Rockland-Swiss mice and the influence of olfactory bulb removal. *Developmental Psychobiology*, **6**, 159–64.

Gandelman, R. and Simon, N. G. (1978). Spontaneous pup-killing in response to large litters. *Developmental Psychobiology*, **11**, 235–41.

Gandelman, R. and vom Saal, F. S. (1975). Pup killing in mice? The effects of gonadectomy and testosterone administration. *Physiology and Behavior*, **15**, 647–51.

Gandelman, R., Zarrow, M., Denenberg, V., and Meyers, M. (1971). Olfactory bulb removal eliminates maternal behavior in the mouse. *Science*, **171**, 210–11.

Ghiraldi, L. L. and Svare, B. B. (1986). Postpubertal isolation decreases infanticide and elevates parental care in C57BL/6J male mice. *Physiology and Behavior*, **36**, 59–62.

Gibber, J. R., Piontkewitz, Y., and Terkel, J. (1984). Responses of male and female Siberian hamsters towards pups. *Behavioral and Neural Biology*, **42**, 177–82.

Gibber, J. R. and Terkel, J. (1985). Effects of postweaning social experience on response of Siberian hamsters (*Phodopus sungorus sungorus*) towards young. *Journal of Comparative Psychology*, **99**, 491–3.

Goldman, L. and Swanson, H. (1975). Population control in confined colonies of golden hamsters (*Mesocricetus auratus* Waterhouse). *Zeitschrift für Tierpsychologie*, **37**, 225–36.

Goyens, J. and Noirot, E. (1977). Intruders of differing reproductive status alter aggression differentially in early and late pregnant mice. *Aggressive Behavior*, **3**, 119–25.

Gubernick, D. J. and Alberts, J. R. (1989). Postpartum maintenance of paternal behaviour of the biparental California mouse *Peromyscus californicus*. *Animal Behaviour*, **37**, 656–64.

Hamilton, W. D. (1964). The genetical evolution of social behaviour. II. *Journal of Theoretical Biology*, **7**, 17–52.

Hansson, L. (1985). The food of bank voles, woodmice and yellow-necked mice. In *The ecology of woodland rodents, bank voles and wood mice*, (ed. J. R. Flowerdew, J. Gurnell, and J. H. W. Gipps), pp. 141–68. Oxford Scientific Publications.

Hartung, T. G. and Dewsbury, D. A. (1979). Paternal behavior in six species of muroid rodents. *Behavioral and Neural Biology*, **26**, 466–78.

Hepper, P. G. (1986). Kin recognition: functions and mechanisms a review. *Biological Reviews*, **61**, 63–93.

Hoogland, J. L. (1985). Infanticide in prairie dogs: lactating females kill offspring of close kin. *Science*, **230**, 1037–40.

Hoogland, J. L., Tamarin, R. H., and Levy, C. K. (1989). Communal nursing in prairie dogs. *Behavioural Ecology and Sociobiology*, **24**, 91–5.

Huck, U. W. (1982). Pregnancy block in laboratory mice as a function of male social status. *Journal of Reproduction and Fertility*, **66**, 181–4.

Huck, U. W. (1984). Infanticide and the evolution of pregnancy block in rodents. In *Infanticide in animals and man: comparative and evolutionary perspectives*, (ed. G. Hausfater and S. B. Hrdy), pp. 349–65. Aldine Press, New York.

Huck, U. W., Soltis, R. L., and Coopersmith, C. B. (1982). Infanticide in male laboratory mice: effects of social status, prior sexual experience, and basis for discrimination between related and unrelated young. *Animal Behaviour*, **30**, 1158–65.

Huck, U. W., Bracken, A. C., and Lisk, R. D. (1983). Female-induced pregancy block in the golden hamster. *Behavioral and Neural Biology*, **38**, 190–3.

Huck, U. W., Pratt, N. C., Labov, J. B., and Lisk, R. D. (1988). Effects of age and parity on litter size and offspring sex ratio in golden hamsters (*Mesocricetus auratus*). *Journal of Reproduction and Fertility*, **83**, 209–14.

Jakubowski, M. and Terkel, J. (1982). Infanticide and caretaking in non-lactating *Mus musculus*: influence of genotype, family group and sex. *Animal Behaviour*, **30**, 1029–35.

Jakubowski, M. and Terkel, J. (1985*a*). Transition from pup killing to parental behavior in male and virgin female albino rats. *Physiology and Behavior*, **34**, 683–6.

Jakubowski, M. and Terkel, J. (1985*b*). Incidence of pup killing and parental behavior in virgin female and male rats (*Rattus norvegicus*): differences between Wistar and Sprague-Dawley stocks. *Journal of Comparative Psychology*, **99**, 93–97.

Jeppsson, B. (1986). Mating by pregnant water voles (*Arvicola terrestris*): a strategy to counter infanticide by males? *Behavioural Ecology and Sociobiology*, **19**, 293–6.

Kennedy, H. F. (1989). Father-infant recognition in mice. PhD thesis, The Queen's University of Belfast.

Kennedy, H. F. and Elwood, R. W. (1988). Strain differences in the inhibition of infanticide in male mice (*Mus musculus*). *Behavioral and Neural Biology*, **50**, 349–53.

Kimble, D. P., Rogers, L., and Hendrickson, C. W. (1967). Hippocampal lesions disrupt maternal, not sexual behavior in the albino rat. *Journal of Comparative and Physiological Psychology*, **63**, 401–7.

Konig, B. (1989). Behavioural ecology of kin recognition in house mice. *Ethology, Ecology and Evolution*, **1**, 99–110.

Konig, B. and Markl, H. (1987). Maternal care in house mice: 1. The weaning strategy as a means of parental manipulations of offspring quality. *Behavioural Ecology and Sociobiology*, **20**, 1–9.

Konig, B., Riester, J., and Markl, H. (1988). Maternal care in house mice (*Mus musculus*): 11. The energy cost of lactation as a function of litter size. *Journal of Zoology, London*, **216**, 195–210.

Krackow, S. and Hoeck, H. N. (1989). Sex ratio manipulation, maternal investment and behaviour during concurrent pregnancy and lactation in house mice. *Animal Behaviour*, **37**, 177–86.

Labov, J. B. (1980). Factors influencing infanticidal behaviour in wild house mice (*Mus musculus*). *Behavioural Ecology and Sociobiology*, **6**, 297–303.

Labov, J. B. (1981*a*). Pregnancy blocking in rodents: adaptive advantages for females. *American Naturalist*, **118**, 361–71.

Labov, J. B. (1981*b*). Male social status, physiology and ability to block pregnancies in female house mice (*Mus musculus*). *Behavioural Ecology and Sociobiology*, **8**, 287–91.

Labov, J. B., Huck, U. W., Elwood, R. W., and Brooks, R. J. (1985). Current problems in the study of infanticidal behaviour of rodents. *Quarterly Review of Biology*, **60**, 1–20.

Labov, J. B., Huck, U. W., Vaswani, P., and Lisk, R. D. (1986). Sex ratio manipulation and decreased growth of male offspring of undernourished golden hamsters (*Mesocricetus auratus*). *Behavioural Ecology and Sociobiology*, **18**, 241–9.

Lane-Petter, W. (1968). Cannibalism in rats and mice. *Proceedings of the Royal Society of Medicine*, **61**, 1295–6.

Lidicker, W. Z. (1976). Social behaviour and density regulation in house mice living in large enclosures. *Journal of Animal Ecology*, **45**, 677–97.

Lloyd, J. A. and Christian, J. J. (1967). Relationships of activity and aggression to density in two confined populations of house mice (*Mus musculus*). *Journal of Mammalogy*, **48**, 262–9.

Lyons, D. M. and Banks, E. M. (1982). Ultrasounds in neonatal rats: novel, predator and conspecific odor cues. *Developmental Psychobiology*, **15**, 455–60.

Macdonald, D. (ed) (1984). *The encyclopaedia of mammals*. Vol 2. Equinox, Oxford.

Makin, J. W. and Porter, R. H. (1984). Paternal behavior in the spiney mouse (*Acomys cahirinus*). *Behavioral and Neural Biology*, **41**, 135–51.

Mallory, F. F. and Brooks, R. J. (1978). Infanticide and other reproductive strategies in the collard lemming, *Dicrostonyx groenlandicus*. *Nature*, **273**, 144–6.

Malsbury, C. W., Miceli, M. O., and Scouten, C. W. (1985). Neural basis of reproductive behavior. In *The hamster: reproduction and behavior,* (ed. H. I. Siegel), pp. 229–59. Plenum Press, New York.

Marques, D. M. and Valenstein, E. S. (1976). Another hamster paradox: more males carry pups and fewer kill and cannibalize young than do females. *Journal of Comparative and Physiological Psychology*, **90**, 653–7.

Marsteller, F. A. and Lynch, C. B. (1983). Reproductive consequences of food restriction at low temperature in lines of mice divergently selected for thermoregulatory nesting. *Behavior Genetics*, **13**, 397–410.

McCarthy, M. M. and vom Saal, F. (1985). The influence of reproductive state on infanticide by wild female mice. *Physiology and Behavior*, **35**, 843–849

McCarthy, M. M. and vom Saal, F. (1986*a*). Inhibition of infanticide after mating by wild male house mice. *Physiology and Behavior*, **36**, 203–9.

McCarthy, M. M. and vom Saal, F. (1986*b*). Infanticide by virgin CF-1 and wild male house mice (*Mus musculus*): effects of age, prolonged isolation, and testing procedure. *Developmental Psychobiology*, **19**, 279–90.

McCarthy, M. M., Bare J. E., and vom Saal, F. S. (1986). Infanticide and parental behaviour in wild female house: effects of ovariectomy, adrenalectomy and administration of oxytocin and prostaglandin F_2a. *Physiology and Behaviour*, **36**, 17–23

McCarty, R. and Southwick, C. H. (1977). Patterns of parental care in two cricetid rodents, *Onychomys torridus* and *Peromyscus leucopus*. *Animal Behaviour*, **25**, 945–8.

McClure, P. A. (1981). Sex-biased litter reduction in food-restricted wood rats (*Neotoma floridana*). *Science*, **211**, 1058–60.

McLean, I. G. (1983). Paternal behaviour and killing of young in arctic ground squirrels. *Animal Behaviour*, **31**, 32–44.

Mennella, J. A. and Moltz, H. (1988). Infanticide in rats: male strategy and female counter strategy. *Physiology and Behavior*, **42**, 19–28.

Noirot, E. (1972). Ultrasounds and maternal behaviour in small rodents. *Developmental Psychobiology*, **5**, 371–87.

O'Connor, R. J. (1978). Brood reduction in birds: selection for fratricide, infanticide or suicide? *Animal Behaviour*, **26**, 79–96.

Ostermeyer, M. C. (1981). The parental behaviour of the house mouse, *Mus musculus* L. PhD Thesis. The Queen's University of Belfast.

Ostermeyer, M. C. and Elwood, R. W. (1983). Pup recognition in *Mus musculus*: parental discrimination between own and alien young. *Developmental Psychobiology*, **16**, 75–82.

Ostermeyer, M. C. and Elwood, R. W. (1984). Helpers (?) at the nest in the mongolian gerbil, *Meriones unguiculatus*. *Behaviour*, **91**, 61–77.

Parmigiani, S. (1989). Inhibition of infanticide in male house mice (*Mus domesticus*): is kin recognition involved? *Ethology, Ecology and Evolution*, **1**, 95–8.

Parmigiani, S., Sgoifo, A., and Mainardi, D. (1988). Parental aggression displayed by female mice in relation to the sex, reproductive status and infanticidal potential of conspecific intruders. *Monitore Zoologia Italia*, **22**, 193–201.

Paul, L. (1986). Infanticide and maternal aggression: synchrony of male and female reproductive strategies in mice. *Aggressive Behavior*, **12**, 1–11.

Paul, L. and Kupferschmidt, J. (1975). Killing of conspecifics and mouse young by male rats. *Journal of Comparative and Physiological Psychology*, **88**, 755–63.

Perrigo, G. (1987). Breeding and feeding strategies in deer mice and house mice when females are challenged to work for their food. *Animal Behaviour*, **35**, 1298–316.

Perrigo, G. and vom Saal, F. S. (1989). Mating-induced regulation of infanticide in male mice: fetal programming of a unique stimulus-response. In *Ethoexperimental analysis of behavior*, (ed. R. J. Blanchard, P. F. Brain, and S. Parmigiani), pp. 320–36. Kluwer, Dordrecht.

Peters, L. C. and Kristal, M. B. (1983). Suppression of infanticide in mother rats. *Journal of Comparative Psychology*, **97**, 167–77.

Prakash, I. (1962). Ecology of the gerbils of the Rajasthan desert. *Mammalia*, **26**, 311–31.

Reynolds, R. D. (1981). Preventing maternal cannibalism in rats. *Science*, **213**, 1146.

Richards, M. P. M. (1966). Maternal behaviour in the golden hamster: responsiveness to young in virgin, pregnant, and lactating females. *Animal Behaviour*, **14**, 310–13.

Rosenberg, K. M. and Sherman, G. F. (1975). The role of testosterone in the organization, maintenance and activation of pup killing in the male rat. *Hormones and Behavior*, **6**, 173–9.

Rosenblatt, J. S. and Siegel, H. I. (1983). Physiological and behavioural changes during pregnancy and parturition underlying the onset of maternal behaviour in rodents. In *Parental behaviour of rodents*, (ed. R. W. Elwood), pp. 23–66. John Wiley and Sons, Chichester.

Rosenson, L. M. and Asheroff, A. F. (1975). Maternal aggression in CD-1 mice: influence of the hormonal condition of the intruder. *Behavioral Biology*, **15**, 219–24.

Rothstein, S. I. (1982). Mechanisms of avian egg recognition: which egg parameters elicit responses by rejector species? *Behavioural Ecology and Sociobiology*, **11**, 229–39.

Rowell, T. E. (1961). The family group in golden hamsters: its formation and break up. *Behaviour*, **17**, 81–94.

Samuels, O., Jason, G., Mann, M., and Svare, B. (1981). Pup killing behavior in mice: suppression by early androgen exposure. *Physiology and Behavior*, **26**, 473–7.

Schwagmeyer, P. L. (1979). The Bruce effect: an evaluation of male/female advantages. *American Naturalist*, **114**, 932–38.

Sherman, P. W. (1981). Reproductive competition and infanticide in Belding's ground squirrels and other animals. In *Natural selection and social behavior*, (ed. R. D. Alexander and D. W. Tinkle), pp. 311–31. Chiron Press, New York.

Siegel, H. I. and Greenwald, G. S. (1978). Effects of mother-litter separation on later maternal responsiveness in the hamster. *Physiology and Behavior*, **21**, 147–9.

Smart, J. L. and Preece, J. (1973). Maternal behaviour of undernourished mother rats. *Animal Behaviour*, **21**, 613–19.

Soroker, V. and Terkel, J. (1988). Changes in incidence of infanticidal and parental responses during the reproductive cycle of male and female wild mice *Mus musculus*. *Animal Behaviour*, **36**, 1275–81.

Southwick, C. H. (1955). Regulatory mechanisms of the house mouse populations: social behaviour affecting litter survival. *Ecology*, **36**, 627–34.

Stenseth, N. C. (1985). Models of bank vole and woodmouse populations. In *The ecology of woodland rodents, bank voles and woodmice*, (ed. J. R. Flowerdew, J. Gurnell, and J. H. W. Gipps), pp. 339–76. Oxford Scientific Publications.

Stouffer, P. C., Kennedy, E. D., and Power, H. W. (1987). Recognition and removal of intraspecific parasite eggs by starlings. *Animal Behaviour*, **35**, 1583–4.

Svare, B. and Bartke, A. (1978). Food deprivation induces conspecific pup-killing in mice. *Aggressive Behavior*, **4**, 253–61.

Svare, B. and Broida, J. (1982). Genotypic influences on infanticide in mice: environmental, situational and experimental determinants. *Physiology and Behaviour*, **28**, 171–5.

Svare, B., Mann, H., and Sammuels, O. (1980). Mice: suckling stimulation but not lactation important for maternal aggression. *Behavioral and Neural Biology*, **29**, 453–62.

Svare, B., Kinsley, C. H., Mann, H., and Broida, J. (1984). Infanticide: accounting for genetic variation in mice. *Physiology and Behavior*, **33**, 137–52.

Thomas, J. A. and Birney, E. C. (1979). Parental care and mating system of the prairie vole *Microtus ochrogaster*. *Behavioural Ecology and Sociobiology*, **5**, 171–86.

Trivers, R. L. and Willard, D. E. (1973). Natural selection of parental ability to vary the sex ratio. *Science*, **179**, 90–2.

vom Saal, F. S. (1983). Variation in infanticide and parental behavior in male mice due to prior intrauterine proximity to female fetuses: elimination by prenatal stress. *Physiology and Behavior*, **30**, 675–81.

vom Saal, F. S. (1985). Time-contingent change in infanticide and parental behavior induced by ejaculation in male mice. *Physiology and Behavior*, **34**, 7–15.

vom Saal, F. S. and Howard, L. S. (1982). The regulation of infanticide and parental behavior: implications for reproductive success in male mice. *Science*, **215**, 1270–2.

Wolff, J. O. (1985). Maternal aggression as a deterrent to infanticide in *Peromyscus leucopus* and *P. maniculatus*. *Animal Behaviour*, **33**, 117–23.

Wolff, J. O. and Cicirello, D. M. (1989). Field evidence for sexual selection and resource competition in white-footed mice. *Animal Behaviour*, **38**, 637–42.

15

Cannibalism among non-human primates

Mariko Hiraiwa-Hasegawa

Reports of intraspecific killing and cannibalism among non-human primates, both from wild populations and captive groups (Angst and Thommen 1977; Goodall 1986; Hausfater and Hrdy 1984), were originally thought to be abnormal behaviour resulting from factors such as overcrowding or changes in the social environment (e.g. Curtin and Dolhinow 1978; Boggess 1979). More recently, primatologists have sought adaptive explanations of intraspecific killing, especially infanticide (Hrdy 1979; Hausfater and Hrdy 1984).

Infanticide among non-human primates usually occurs in populations where male–male competition for mates is extremely high; populations that form one-male social units, such as the hanuman langur *Presbytis entellus*, or multi-male social units with copulations highly monopolized by the highest-ranking animal, such as the red howlers *Alouatta seniculus*. The killer is the male that is supposed to have had no opportunity to sire the victim infants, and the females resume their sexual receptivity shortly after the infanticidal attack and mate with the killer. In these cases, infanticide is usually interpreted as a male reproductive strategy that curtails the female's investment in the offspring of other males and thereby increases the chance of siring offspring of his own (Hiraiwa-Hasegawa 1987, 1988).

However, cannibalism of victims of intraspecific killing is rare among primates, with the exception of chimpanzees *Pan troglodytes*. The adaptive significance of cannibalism among non-human primates has received relatively little attention, perhaps because it is not a widely observed behaviour. This chapter investigates the costs and benefits of cannibalism among non-human primates, and examines in detail the idea that the nutritive value of conspecifics is the major cause of cannibalism following infanticide in chimpanzees.

DISTRIBUTION OF CANNIBALISM

Cannibalism has been recorded in natural populations of redtail monkeys *Cercopithecus ascanius*, chacma baboons *Papio cynocepharus ursinus*, olive baboons *P. c. anubis*, Japanese macaques *Macaca fuscata*, gorillas *Gorilla gorilla beringei*, and chimpanzees. With the exception of the olive baboon *P. c. anubis*, in which a mother ate the body of her own infant, all occurred following infanticide committed by individuals other than the mother. Although infanticide has been observed in 14 species of non-human primates in the wild, cannibalism following infanticide has been recorded in only five species. There are no records of adult primates being killed and eaten by conspecifics.

Most records of cannibalism in non-human primates in the wild are anecdotal observations. In the Kibale Forest, Uganda, Struhsaker (1977) observed an adult male redtail monkey take over a group of females. The new male subsequently killed and then fed on the posterior quarters and hindlegs of an infant that was born around the time of the take-over. One month later another infant was born in the group. The same male also killed this infant and ate its brains and body, leaving the skin and hair.

An adult, probably solitary, male chacma baboon *P. c. ursinus* from Kruger National Park, South Africa, seized an infant from its mother and cannibalized it (National Parks Board 1959, cited in Saayman 1971). However, cannibalism does not always accompany intraspecific killing in chacma baboons. Vincent (1969) reported two cases of an adult male killing a juvenile male, but cannibalism did not occur on either occassion. Out of the five cases of infant death following attacks or maltreatment recorded in Moremi Wildlife Reserve, Botswana, from 1977 to 1980, on two occasions (Collins *et al.* 1984) the victim infant was partially eaten by an adult male. In Gombe National Park, Tanzania, a female olive baboon was observed to eat part of her own infant, which had recently died (Goodall 1986). Infanticide by immigrant males has been observed several times in this population of olive baboons at Gombe, but cannibalism did not take place on any other occasion (Collins *et al.* 1984).

In Hakone National Park, a solitary male Japanese macaque killed an infant belonging to a troop he had been approaching, and then consumed part of the infant's body (Nihonzaru 1974). Finally, a female mountain gorilla and her offspring were suspected of killing and eating an infant of another female who belonged to the same group: an infant disappeared and fragments of gorilla bones and skins were found in the faeces of those individuals (Fossey 1984).

These observations represent all the records of cannibalism for non-human primates in the wild, other than for chimpanzees. As extensive observations have been made of these species of monkeys and apes, it is likely that the frequency of cannibalism in these species is extremely low.

Cannibalism has also been observed for a variety of species under captive conditions, including three species of bushbabies *Galago senegalensis* (Doyle *et al.* 1967; Izard and Simons 1986; Sauer 1967), *G. classicaudatus* (Izard and Simons 1986; Rosenson 1972; Sauer 1967) and *G. garnetti* (Izard and Simons 1986); the mouse lemur *Cheirogaleus medius* (Foerg 1982); the common marmoset *Callithrix jacchus* (Rothe 1975); and the squirrel monkey *Saimiri sciureus* (Manocha and Long 1977). Cannibalism of neonates seems to be a frequent phenomenon among captive prosimians and marmosets.

Cannibalism of neonate galagos by group members has been reported from several laboratories where galagos are kept in social groups (*G. crassicaudatus:* Izard and Simons 1986, Rosenson 1972, Sauer 1967; *G. senegalensis:* Doyle *et al.* 1967, Izard and Simons 1986, Sauer 1967; *G. garnetti:* Izard and Simons 1986). In one mixed colony of galagos, infant mortality that could be attributed to cannibalism as a percentage of total infant death ranged from 8.3 per cent in *G. crassicaudatus* to 50 per cent in *G. senegalensis* (Izard and Simons 1986). For this reason, many authors recommend isolation of pregnant females from other members of the group. However, the identity of the cannibal was not reported in these studies, being mentioned merely as 'other members of the group'. In one case of cannibalism in *G. senegalensis*, a newborn was eaten by its adult brother, which was reported to be highly stressed as a result of intense aggressive father–son interactions (Zimmermann 1989).

In one colony of mouse lemurs *Cheirogaleus medius* newborn (and probably stillborn) infants born into a family group were frequently cannibalized (Foerg 1982). However, the circumstances surrounding the cannibalistic attack, and the identity of the cannibals was not mentioned.

In the wild, it is reported that pregnant female galagos isolate themselves prior to parturition and for a period of time after the infants are born (Sauer and Sauer 1963). Rosenson (1972) suggested that the threat of cannibalism from other group members might be an explanation for maternal isolation in galagos. However, the rate of cannibalism towards newborn infants in three species of captive galagos, when the mothers were isolated from the group prior to parturition, was not significantly different from that of the mothers maintained together with the group (Izard and Simons 1986). This also means that the mother galagos are partially responsible for cannibalism. However, the phenomenon has not been investigated in detail under natural conditions.

Cannibalism of neonates by group members has also been reported in common marmosets *Callithrix jacchus* (Rothe 1975). Rothe (1975) reported that 'non-viable newborns were more likely to be cannibalized, especially when they appeared abnormal'. During placentophagia, in which most group members consumed the placenta, infants were frequently killed and within a few minutes were partially eaten (Rothe 1975). Mothers did not interfere with this behaviour unless the infant vocalized, and it seems that an infant is regarded as a piece of food by the family members if it fails to respond properly. This phenomenon may be widespread among callitrichids in captivity, although it is not usually described in the literature (W. C. McGrew personal communication), and has not been reported in wild populations.

Experiments with squirrel monkeys *Saimiri sciureus* (Manocha and Long 1977), in which mothers were fed a low protein diet (only eight per cent of the calories derived from protein) during their late pregnancy experienced a high rate of abortion and devoured their aborted fetus. Mothers fed a high protein diet (25 per cent of the calories derived from protein) did not eat the aborted fetus, although abortion itself was rare among them (Manocha and Long 1977).

In contrast to the species described above, cannibalism is frequently observed among chimpanzees in wild populations, and the victim of infanticide is almost always consumed by conspecifics. Cannibalism in chimpanzees is discussed below in more detail.

In summary, all the victims of cannibalism recorded among non-human primates were infants, and all the incidents occurred following infanticide. However, it is important to note that cannibalism does not always accompany infanticide. Thus, an understanding of cannibalism requires an evaluation of the circumstances under which infanticide victims are exploited as food.

CANNIBALISM AND PRIMATE CARNIVORY

Since cannibalism represents exploitation of conspecifics as food, it is likely to be of benefit only in those species that frequently exploit vertebrate meat as a source of protein. That not all species which commit infanticide also commit cannibalism suggests that cannibalism is related to dietary patterns.

Table 15.1 lists non-human primate species' records of infanticide, cannibalism, and dietary habits, according to whether they exploit vertebrates and invertebrates as their primary or secondary source of food. Among the 11 species that have a record of cannibalism, eight species exploit vertebrates or invertebrates as an important source of food. In contrast, vertebrate or invertebrate foods are not reported to be part of the diet for the eight species that commit infanticide but do not cannibalize their victims. These species, mostly colobines and howlers, are specialized leaf-eaters, and derive their protein almost entirely from leaves (Eisenberg 1978).

Nevertheless, there are 20 species that feed on a considerable amount of vertebrates or invertebrates but do not commit cannibalism (Table 15.1). Apparently, omnivory alone does not always lead to cannibalism. Consequently, exploitation of the flesh of victims as a source of protein is a nutritionally advantageous option for infanticidal species that exploit vertebrates or invertebrates as an important source of protein. While selection may have favoured infanticide for a variety of reasons in several species of non-human primates, few selection pressures have operated directly to favour subsequent cannibalism.

In this respect, cannibalism might be most frequently observed in species that frequently commit infanticide, and for whom vertebrate and invertebrate food represents an important part of their diet. A quantitative comparative analysis of the data in Table 15.1 may be misleading, since more accurate information on the species' dietary habits and an adequate phylogentic organization of the taxa are required. Nevertheless, the patterns described above are suggestive of an association between dietary habits and the occurrence of cannibalism following infancticide. The fact that chimpanzees, probably the most carnivorous non-human primate, most frequently cannibalizes victims of infanticide seems to be consistent with this prediction.

Table 15.1 Diet, infanticide and cannibalism among non-human primates

Species	Diet[1]		Infanticide[2]	Cannibalism[2]
	Vertebrate	Invertebrate		
Cheirogaleus major	–	+	–	–
C. medius	–	+	(y)	(y)
Microcebus coquereli	–	+	–	–
M. murinus	–	+	–	–
Arctocebus calabarensis	–	+	–	–
Loris tardigradus	*	+	–	–
Nicticebus concang	*	+	–	–
Galago alleni	*	+	–	–
G. crassicaudatus	*	+	(y)	(y)
G. demidovii	–	+	–	–
G. elegantulus	–	+	–	–
G. senegalensis	–	+	(y)	(y)
G. garnetti	–	+	(y)	(y)
Tarsius spectrum	*	+	–	–
Callithrix spp.	*	+	(y)	(y)
Cebuella pygmaea	*	+	–	–
Leontopithecus rosalia	*	+	–	–
Saguinus fascicollis	–	+	–	–
S. geofroyi	–	+	–	–
S. oedipus	*	+	–	–
S. tamarin	*	+	–	–
Cebus apella	–	+	–	–
C. capucinus	*	+	–	–
Saimiri oerstedii	–	+	–	–
S. sciureus	–	+	–	y**
Alouatta seniculus	–	–	y	–
A. palliata	–	–	y	–
Cercopithecus albigena	*	+	–	–
C. ascanius	–	+	y	y
C. mitis	*	*	y	–
C. campbelli	–	*	y	–

Table 15.1 (*continued*)

C. pogonias	*	+	–	–
Miopithecus talapoin	*	+	–	–
Macaca fuscata	*	*	y	y
Macaca mulatta	*	*	y	–
Papio cynocephalus	*	*	y	y
Colobus badius	–	–	y	–
C. guereza	–	–	y	–
Presbytis entellus	–	*	y	–
P. cristatus	–	–	y	–
Gorilla gorilla	–	*	y	y
Pan troglodytes	+	*	y	y

(From Harding 1981 and Smuts *et al.* 1987.)
[1] Diet records: –, not eaten; *, recorded but its relative importance not described; +, secondary source of food; ++, main source of food.
[2] Cannibalism and infanticide records: y, in natural populations; (y), in captivity; y**, under experimental conditions.

Although the importance of cannibalism in wild populations of prosimians and marmosets is still unknown, they may exploit dead or non-viable newborns as a source of protein. Interestingly, experiments with squirrel monkeys *Saimiri sciureus* (Manocha and Long 1977), that take invertebrates as a secondary source of food, show that they will readily eat an aborted fetus when they are deficient in protein. Baboons apparently prey on a wider variety of vertebrates than most primate species, but the frequency of predation on mammals is low and the contribution of meat to their diet is apparently insignificant (Rhine *et al.* 1986). Higher hunting rates have been observed during 1973 in a baboon population in Gilgil, Kenya (Strum 1981), but whether this is a long-term, established phenomenon is uncertain. The frequency of infanticide is generally very low in baboons, and the proportion of infanticides that leads to cannibalism is very low, which probably explains the negligible occurrence of cannibalism.

PRIMATE SOCIAL ORGANIZATIONS AND CANNIBALISM

The previous section argued that cannibalism is an option following infanticide, but not a direct cause of infanticide. The social organization of primates may be a possible reason for the infrequency of cannibalism among non-human primates. Most primates live in groups with highly developed social relationships among individuals. Individuals within groups have close bonds with relatives and, in the case of non-relatives, cooperative relationships based on reciprocal altruism are important for their reproductive success (Walters and Seyfarth 1987). Therefore, if an individual attempts both to kill and consume a conspecific, it might expect strong resistance from other individuals that have affiliative bonds with the potential victim, and may risk losing future supportive coalitions with these individuals. Furthermore, in many primates with multi-male, multi-female social systems, both sexes copulate with many partners, thus reducing a male's certainty of paternity (Hrdy and Whitten 1987). Therefore, if an adult male attempts to kill and eat infants or juveniles within his group, the potential risk of him killing his own offspring may be high.

The factors related to social bonding among individuals represent a potential cost of cannibalism. Of course, such costs could equally apply to infanticidal individuals, but the benefits (usually reproductive) of infanticide that accrue to the killer may be so large that they exceed the costs on social bonding. The nutritional benefits of cannibalism would be unlikely to exceed this cost.

CANNIBALISM AMONG CHIMPANZEES

Observations of cannibalism

Cannibalism in chimpanzees has been observed in three different study sites, Budongo Forest in Uganda and Gombe National Park and Mahale Mountains National Park in Tanzania (Table 15.2). These observations are distributed over a long time and therefore cannibalism is unlikely to be an uncommon behaviour, peculiar to a particular locality or time. However, infanticide and cannibalism have not been observed in West African populations at Bossou, Guinea (Sugiyama 1984) and Tai National Park, Ivory Coast (Boesch and Boesch 1989), despite the long-term nature of these studies.

The frequency of cannibalism in chimpanzees is exceptionally high, compared with redtail monkeys, baboons, Japanese monkeys and gorillas; 12 out of 14 cases of direct observations of infanticide involved cannibalism of the victim infants (Table 15.2). In the two cases where the victim was not eaten, the carcass was not defended by the attackers after they killed it; one victim in Gombe was 'rescued' by another chimpanzee (Goodall 1977), and the body of the victim in Mahale was recovered by its mother immediately after it was killed (K. Masui, personal communication). Therefore, cannibalism of the victim infants following infanticide is likely to be the norm, as long as the attackers possess the body of the victim. Cannibal attacks are usually initiated by males; an adult male was the first attacker in eight of the 11 cases in which the identification of the attacker was known. The other three cases were initiated by a particular adult female, 'Passion', and her adolescent daughter 'Pom', of Gombe National Park (Goodall 1977).

Meat-sharing, in which the prey is usually shared among individuals including

Table 15.2 Infanticide and cannibalism in chimpanzees

Site	Date	Age of victim	Sex of victim	Occasion	Attacker†	Cannibals†	Degree of consumption	Remarks	Source
Budongo	Nov '67	Newborn	?	?	AM?	AM	at least one leg		Suzuki 1971
Gombe	Sep '71	1.5–2 yr	?	Inter	AM	3AMs	legs, arms & trunk	bizarre behaviour	Bygott 1972
Gombe	Aug '75	3 wk	f	Within	Af, AdF	2AFs, AdF, JM	head and 4 limbs		Goodall 1977
Gombe	Oct '75	1.5–2 yr	m	Inter	AM?	3AMs	2 legs,1 hand, all intestines, brain		Goodall 1977
Gombe	Nov '75	1.5–2 yr	f	Inter	AM	–	–	killed but not eaten	Goodall 1977
Gombe	Oct '76	3 wk	m	Within	Af, AdF	2AFs, AdF	head, 4 limbs, part of trunk, all intestines		Goodall 1977
Gombe	Nov '76	3 wk	f	Within	Af, AdF	AF, JM	not mentioned		Goodall 1977
Mahale	Apr '74	3 yr	m	Inter	?	3AMs	almost all		Nishida *et al.* 1979
Mahale	Jan '77	2.5 mo	m	Within	?	3AMs+	head, 1 leg, 2 arms		Norikoshi 1982
Mahale	Jun '79	1.5 mo	m	Within	?	4AMs, AF	almost all		Kawanaka 1981
Mahale	Jul '83	1 mo	m	Within	AM	4AMs, 3AFs	almost all		Takahata 1985
Mahale	Dec '83	3 mo	m	Within	AM?	5AMs, AF	almost all		Nishida and-Kawanaka 1985
Mahale	Mar '85	6 mo	m	Within	AM		–	killed but not eaten	Masui, pers. comm.
Mahale	Nov '89	6 mo	m	Within	AM		almost all		Hamai *et al.* 1992

† A, Adult; Ad, Adolescent; J, Juvenile; M, Male; F, Female

those who did not participate in hunting activities, is a special feature of chimpanzee predation (Teleki 1975; Takahata *et al.* 1984). This behaviour also occurs in cannibalistic episodes; the flesh of the victim infants is solicited by several individuals, who are attracted to the site by excited calls and is eventually distributed among them. Cannibals included adult males, adult females, and adolescent and juveniles of both sexes. Cannibalism by infants has never been observed. The mean number of cannibals per attack was three in Gombe and five in Mahale. The degree of consumption of the body varies from case to case; ranging from the consumption of one leg to almost the entire body. In four cases from Gombe, the rest of the body was abandoned after being partially eaten, but in all the cases from Mahale, almost nothing was left uneaten (Table 15.2).

Most reports of infanticide and accompanying cannibalism by chimpanzees have focused their discussion on the significance of infanticide rather than cannibalism. Nevertheless, Suzuki (1971) suggested that cannibalism should not be regarded as a mere accident, but rather should be considered as occurring 'in a particular mental state in which the whole group was in a vortex of excitement in their daily nomadic life'. Suzuki (1971) remarked that cannibalism might be a behaviour common to species of Hominoid stock, citing Roper's (1969) evidence of cannibalism among Pleistocene Hominids. Nishida *et al.* (1979) interpreted cannibalism as an extension of infanticide and immense aggression, while both Norikoshi (1982) and Nishida and Kawanaka (1985) suggest that cannibalism in Mahale was not any different from ordinary meat eating, 'except that they fed very slowly' (Nishida and Kawanaka 1985). Norikoshi's (1982) impression was that the behaviour of one of the adult males appeared to inhibit cannibalism among other individuals. Both Kawanaka (1981) and Takahata (1985) did not provide explanations for cannibalism.

A relationship between predatory habits and cannibalism in chimpanzees has been suggested by Goodall (1977, 1986) and Hiraiwa-Hasegawa (1988). Goodall (1977, 1986) stated that 'in view of the clear-cut predatory behaviour of chimpanzees, it should perhaps not be surprising that they, like humans, sometimes eat flesh of victims of their own species.' Hiraiwa-Hasegawa (1988) speculated that chimpanzees may kill conspecifics for food, since reports of infanticidal individuals cannibalizing their victims have been reported for many other carnivorous mammals, including lions *Panthera leo*, puma *Puma concolor*, brown bears *Ursus arctos*, polar bears *Ursus maritimus*, coatis *Nasua narica,* and golden jackals *Canis aureus* (see Packer and Pusey 1984). Similar behaviour has been reported for ground squirrels *Spermophilus beldingi* (Sherman 1981).

Before examining the relationships between carnivory and cannibalism in chimpanzees, it is necessary to summarize briefly the observations and explanations of infanticide in chimpanzees. Explanations of infanticide in chimpanzees are still controversial (Nishida and Kawanaka 1985; Takahata 1985; Goodall 1986; Hiraiwa-Hasegawa 1987). Almost all of the cases of infanticide in other primates have been explained in terms of male–male competition for the opportunity of siring offspring (Hausfater and Hrdy 1984; Hiraiwa-Hasegawa 1988). However, the benefits of infanticide in chimpanzees are less obvious.

Infanticide can be divided into 3 categories; between-group infanticide by males (Bygott 1972; Goodall 1977; Nishida *et al.* 1979), within-group infanticide by females (Goodall 1977), and within-group infanticide by males (Norikoshi 1982; Kawanaka 1981; Nishida and Kawanaka 1985; Takahata 1985; Hamai *et al.* 1992). Cannibalism occurs in all three categories (see Table 15.2).

Between-group infanticide was observed at the overlapping area of two neighbouring groups (= community or unit-group), when a mother–infant pair of one group encountered males of another group. However, in contrast to most cases of infanticide in other primate species, there was no immediate reproductive advantage to the killer males, because the mothers neither transferred to the killer's group nor mated with the killer.

Within-group infanticide by females was only observed in Gombe, and committed by one adult female, 'Passion' and her adolescent daughter, 'Pom'. On three occasions, this mother–daughter pair collaborated in surrounding another mother–infant pair, removing the infant from the mother and then killing and eating the infant. This behaviour might be interpreted as an attempt to exclude potential food competitors, because females that have lost infants, and cycle again, spend less time feeding in their core areas. Eventually the daughter, 'Pom', established her core area in that vicinity (Pusey 1983).

Within-group infanticide by males has been observed only at Mahale. All of the victims were infants of immigrant females whose duration of residency at that time in the group was less than five years. All of the females had mated with the males of the group into which they had migrated at least once prior to the infanticidal attack. Moreover, two of them had mated with these males during the time when the victimized infant might have been conceived. It is quite possible that the males killed their own infants (Kawanaka 1981; Takahata 1985). One interpretation of this behaviour is the elimination of infants with dubious paternity in relation to competition between males of neighbouring groups (Takahata 1985; Nishida and Kawanaka 1985; Hiraiwa-Hasegawa 1987), despite the behaviour incurring costs associated with social bonding (see above).

The integration process of immigrant females takes a long time, sometimes extending over five years. During this period, the immigrant females tend to spend much of their time in the periphery of the communal range, which provides them with opportunities to mate with males of another group. As a female chimpanzee experiences nearly three years of postpartum non-receptivity, the loss in reproductive opportunities for males that wait to mate, for the sake of the infants with even the slightest possibility of dubious paternity, may be high.

The other interpretation of within-group infanticide by males is elimination of future, possibly unrelated, male competitors (Hiraiwa-Hasegawa and Hasegawa in press). Although the sample size is still small, there seems to be a sex bias in the victims (Kawanaka 1981; Hiraiwa-Hasegawa 1987). In the M group of Mahale, six of the 12 male infants born to the 12 immigrant females as their first or second offspring after immigration, were killed or suspected of having been killed by males within the group. In contrast, none of the four female infants that were born to the same group of immigrant females as their first or second offspring were killed.

In the chimpanzee, females transfer between groups, but males typically remain in their natal group for life, sharing food resources and females within their communal range. In these circumstances, competition among unrelated males should be high, which selects for a female-biased sex allocation ratio (Clark 1978; Silk 1983). If the tendency for the victims of within-group infanticide to be male-biased is established (following an increased sample size), it may be possible to interpret this phenomenon as sex allocation ratio manipulation by the elimination of possibly unrelated males.

The nutritional value of infanticide victims

Chimpanzees are capable of digesting meat (Milton and Demment 1989), and as previously mentioned, the frequency of predation on mammals is exceptionally high (Butyinski 1982; Boesch and Boesch 1989; Takahata *et al.* 1984; Teleki 1973; Wrangham and Riss 1990; Uehara *et al.* 1992). The size of mammals eaten by chimpanzees ranges from mice to subadult bushbucks and adult female red colobus monkeys, but chimpanzees mainly prey on smaller monkeys and ungulates. Predation of vertebrates by primates may be taxonomically widespread; 38 species, representing nine families, have been observed to hunt and eat vertebrates in the wild, of which 18 species feed on mammals (Butynski 1982). However, predation of mammals seems to be extremely infrequent, because all records, except for those of baboons and chimpanzees, are based on only one or a few sightings (Butynski 1982).

The impact on prey populations by chimpanzee predation clearly illustrates the degree of carnivory in this species. Chimpanzees are estimated to be responsible for between 20 to 42 per cent annual loss of population of red colobus monkeys *Colobus badius,* in Gombe National Park, Tanzania (Wrangham and Riss 1990), and around four per cent of the red colobus population in Tai National Park, Ivory Coast (Boesch and Boesch 1989). These figures are comparable with the predation rates of large carnivores. For example, in Ngorongoro Crater, around 11 per cent of the wildebeest population and nine per cent of the zebra population were killed by hyenas (Kruuk 1972), and about six per cent of the zebra population was killed by lions in the Sergengeti (Schaller 1972). Apparently there is no other primate which exhibits, over the wide range of their distribution, such a high rate of predation on mammals on a regular basis.

Although it is difficult to estimate accurately the contribution of meat to the chimpanzee's overall diet, recent estimates from three different study sites give us a rough picture on the rates of predation (Boesch and Boesch 1989; Wrangham and Riss 1990; Uehara *et al.* 1992). In Tai National Park, Ivory Coast, a group of 79 chimpanzees was estimated to have made a minimum of 73 kills per year (Boesch and Boesch 1989). The corresponding figure in Gombe was 204 kills per year for 51 to 56 chimpanzees (Wrangham and Riss 1990), and 61 kills per year in Mahale for 100 chimpanzees (Uehara *et al.* 1992). However, the meat is not distributed equally among individuals in the group. Adult males eat considerably more than any of the other age and sex classes. Accounting for this bias, Wrangham and Riss (1990) estimate that adult males of Gombe consumed on average around 25 kg of meat per year, and the corresponding figures for adult females and offspring are 8 and 3 kg, respectively. The figures for Mahale and Tai will be less than half of those for Gombe.

Given this estimate of meat intake and the chimpanzee preference for meat, what is the value of the flesh of infanticide victims for cannibals? Victims of infanticide are mostly infants of less than one-year-old. There are no accurate data on the body weights of infant chimpanzees in the wild. However, the average birth weight for captive chimpanzees is 1.8 kg (Gavan 1953), and the youngest individual that was measured in Mahale, a six-year-old male, weighed 12.3 kg within three days of his death. Therefore, the average size of the victim infants is estimated to be around 3 kg. Since chimpanzees seldom kill wild mammals weighing more than 6 kg (Teleki 1973), infanticidal victims may represent a substantial meal.

Are conspecifics killed primarily for food?

Within-group cannibalism by the mother–daughter pair at Gombe could be explained as predation on a source of meat; the target of their aggression was the infant, who was eaten immediately after the kill, and the mother was ignored after they had caught the infant. Whether this behaviour is peculiar to this mother–daughter pair is unknown. It is possible that the daughter benefited from the subsequent absence of the victim's mother, but this does not explain why the infant was cannibalized. Aggression by adult females toward a mother–infant pair, apparently with the intention of killing the infant, and the mother expressing unusual fear, was also observed in Mahale (personal observation).

However, it is likely to be very difficult for a single female to attack another female and actually take the infant from her. 'Passion' was able to do this only with the help of her adolescent daughter, 'Pom', and they succeeded in taking the infant from a mother who was physically handicapped (Goodall 1977). 'Pom' ceased to attack the mothers after 'Passion's' death, indicating her inability to repeat this behaviour without her mother's help, although it is possible that cannibalistic attacks were no longer necessary, after she had gained her own territory (see above).

No categories of infanticide and accompanying cannibalism have been observed in West African populations, and within-group infanticide by males has not been observed in Gombe or in the K group of Mahale, despite high predation rates on mammals in all these study sites. The differences in the occurrence of cannibalism between the populations may be related to differences in the available protein in their diet. Perhaps protein is more readily available in the diet of the West African populations. Although vertebrate kill rates are slightly lower in the Tai population compared with Gombe, chimpanzees in the West African populations may obtain sufficient sources of protein from other sources, thus reducing the potential nutritional benefits of cannibalism. It would be interesting to compare the total protein intake for individuals from each of these populations.

The exploitation of infant flesh as food may be considered as one, but not the exclusive, cause of infanticide by males, since the selective advantages that may accrue to the killer remain ambiguous. Predation is mainly a male activity and the amount of meat intake by males considerably exceeds that taken by females (McGrew 1979). There is no evidence suggesting that the frequency of cannibalism is related to the availability of food. However, predation on mammals itself is opportunistic, and is not connected with scarcity of fruits (Takahata *et al.* 1984). Therefore male chimpanzees might prey upon conspecific infants whenever they can find a suitable potential victim, namely, an infant unrelated to the killer.

Interpretation of infanticide in chimpanzees is still a matter of contention. The adaptive significance of cannibalism of infanticide victims is most likely to be interpreted as an optional side-effect, bringing nutritional advantage to the killers. While we do not have a clear explanation for infanticide, exploitation of infant flesh as food remains a reasonable explanation.

ACKNOWLEDGEMENTS

I would like to thank Drs. T. H. Clutton-Brock, Robert Seyfarth, Mark Elgar, Bernie Crespi, Robert Elwood, Anne Pusey, T. Hasegawa, S. Uehara, and T. Nishida for their valuable comments on the earlier version of this chapter. Special thanks are due to the Government of Tanzania for permission to conduct research in the Mahale Mountains National Park; Japan International Cooperation Agency and the Ministry of Education, Science and Culture, Japan for their financial support; and Drs. T. Nishida, S. Uehara, K. Kawanaka, Y. Takahata, H. Hayaki, H. Takasaki, T. Hasegawa, and K. Masui for their contribution to collecting long-term data in Mahale.

REFERENCES

Angst, W. and Thommen, D. (1977). New data and a discussion of infant killing in old world monkeys and apes. *Folia Primatologica*, **27**, 198–229.

Boesch, C. and Boesch, H. (1989). Hunting behaviour of wild chimpanzees in the Tai National Park. *American Journal of Physical Anthropology*, **78**, 547–73.

Boggess, J. (1979). Troop male membership changes and infant killing in langurs (*Presbytis entellus*). *Folia Primatologica*, **32**, 65–107.

Butyinski, T.M. (1982). Vertebrate predation by primates: a review of hunting patterns and prey. *Journal of Human Evolution*, **11**, 421–30.

Bygott, D. (1972). Cannibalism among wild chimpanzees. *Nature*, **238**, 410–11.

Clark, A.B. (1978). Sex ratio and local resource competition in a prosimian primates. *Science*, **201**, 163–5.

Collins, D.A., Busse, C.D., and Goodall, J. (1984). Infanticide in two populations of savanna baboons. In *Infanticide: comparative and evolutionary perspectives*, (ed. G. Hausfater and S.B. Hrdy), pp. 193-216. Aldine, New York.

Curtin, R. and Dolhinow, P. (1978). Primate social behaviour in a changing world. *American Scientist*, **66**, 468–75.

Doyle, G.A., Pelletier, A., and Bekker, T. (1967). Courtship, mating and parturition in the lesser bushbaby (*Galago senegalensis moholi*) under semi-natural conditions. *Folia Primatologica*, **7**, 169–97.

Eisenberg, J. (1978). The evolution of arboreal herbivores in the class Mammalia. In *The ecology of arboreal folivores*, (ed. G.G. Montgomery), pp. 135-52. Smithsonian Institution Press, Washington, D.C.

Foerg, R. (1982). Reproduction in *Cheirogaleus medius*. *Folia Primatologica*, **39**, 49–62.

Fossey, D. (1984). Infanticide in mountain gorillas (*Gorilla gorilla beringei*) with comparative notes on chimpanzees. In: *Infanticide: comparative and evolutionary perspectives*, (ed. G. Hausfater and S.B. Hrdy), pp. 217–35. Aldine, New York.

Gavan, J.A. (1953). Growth and development of the chimpanzee: a longitudinal and comparative study. *Human Biology*, **5**, 93–143.

Goodall, J. (1977). Infant killing and cannibalism in free-living chimpanzees. *Folia Primatologica*, **28**, 259–82.

Goodall, J. (1986). *Chimpanzees of Gombe: patterns of behaviour*. Harvard University Press, Cambridge, MA.

Hamai, M., Nishida, T., and Takasaki, H. (1992). New records of within-group infanticide and cannibalism in wild chimpanzees of Mahale, Tanzania. *Primates*. (In press.)

Harding, R.S.O. (1981). An order of omnivores: nonhuman primate diets in the wild. In *Omnivorous primates*, (ed. R.S.O. Harding and G. Teleki), pp. 191–214. Columbia University Press, New York.

Hausfater, G. and Hrdy, S.B. (Ed.) (1984). *Infanticide: comparative and evolutionary perspectives*. Aldine, New York.

Hiraiwa-Hasegawa, M. (1987). Infanticide in primates and a possible case of male-biased infanticide in chimpanzees. In *Animal societies: theories and facts*, (cd. Y. Ito, J.L. Brown, and J. Kikkawa), pp. 125–39. Japan Scientific Societies Press, Tokyo.

Hiraiwa-Hasegawa, M. (1988). On the adaptive significance of infanticide in primates. *Trends in Ecology and Evolution*, **3**, 102–5.

Hiraiwa-Hasegawa, M. and Hasegawa, T. (1992). Infanticide in non-human primates: sexual competition or local resource competition? In *Protection and abuse of young in animals and man*, (ed. S. Parmigiani, F. vom Saal, and B. Svare). Harwood Scientific Publications, New York. (In press.)

Hrdy, S.B. (1979). Infanticide among animals: a review, classification, and examination of the reproductive strategies of females. *Ethology and Sociobiology*, **1**, 13–40.

Hrdy, S.B. and Whitten, P.L. (1987). Patterning of sexual activity. In *Primate societies*, (ed. B.B. Smuts, D.L. Cheney, R.M. Seyfarth, W.R. Wrangham, and T.T. Struhsaker), pp. 370–84. University of Chicago Press.

Izard, M.K. and Simons, E.L. (1986). Isolation of females prior to parturition reduces neonatal mortality in galago. *American Journal of Primatology*, **10**, 249–55.

Kawanaka, K. (1981). Infanticide and cannibalism in chimpanzees, with special reference to the newly observed case in Mahale Mountains. *African Study Monograph*, **1**, 69–99.

Kruuk, H. (1972). *The spotted hyena: a study of predation and social behavior*. University of Chicago Press.

McGrew, W.C. (1979). Evolutionary implications of sex differences in chimpanzee predation and tool use. In *The great apes*, (ed. D.A. Hamburg and E.R. McCown), pp. 441–64. Benjamin/Cummings, Menlo Park.

Manocha, S. and Long, J. (1977). Experimental protein malnutrition during gestation and breeding performance of squirrel monkeys, *Saimiri sciureus*. *Primates*, **18**, 923–30.

Milton, K. and Demment, M. (1989). Features of meat digestion by captive chimpanzee (*Pan troglodytes*). *American Journal of Primatology*, **18**, 45–52.

Nihonzaru. (1976). A case of cannibalism observed among wild Japanese monkeys. *Nihonzaru*, **1**, 123–9.

Nishida, T., Uehara, S., and Nyundo, R. (1979). Predatory behaviour among wild chimpanzees of the Mahale Mountains. *Primates*, **20**, 1–20.

Nishida, T. and Kawanaka, K. (1985). Within-group cannibalism by adult male chimpanzees. *Primates*, **26**, 274–84.

Norikoshi, K. (1982). One observed case of cannibalism among wild chimpanzees of the Mahale Mountains. *Primates*, **23**, 66–74.

Packer, P. and Pusey, A. E. (1984). Infanticide in carnivores. In *Infanticide: comparative and evolutionaryperspectives*, (ed. G. Hausfater and S. B. Hrdy), pp. 31–42. Aldine, New York.

Pusey, A.E. (1983). Mother–offspring relationships in chimpanzees after weaning. *Animal Behaviour*, **31**, 363–77.

Rhine, R.J., Norton, G.W., Wynn, G.M., Wynn, R.D., and Rhine, H.B. (1986). Insect and meat eating among infant and adult baboons (*Papio cynocephalus*) of Mikumi National Park, Tanzania. *American Journal of Physical Anthropology*, **70**, 105–18.

Roper, M.K. (1969). A survey of the evidence for intrahuman killing in the Pleistocene. *Current Anthropologist*, **10**, 427–59.

Rosenson, L.M. (1972). Observations of the maternal behaviour of two captive greater bushbabies (*Galago classicaudatus argentatus*). *Animal Behaviour*, **20**, 677–88.

Rothe, H. (1975). Influence of newborn marmosets' (*Callithrix jacchus*) behaviour on expression and efficiency of maternal and paternal care. In *Contemporary Primatology*, (ed. A. Ehara and S. Kondo), pp. 315–20. Karger, Basel.

Saayman, G.S. (1971). Behaviour of the adult males in a troop of free-ranging chacma baboons (*Papio ursinus*). *Folia Primatologica*, **15**, 36–57.

Sauer, E.G.F. (1967). Mother–infant relationships in galagos and the oral child-transport among primates. *Folia Primatologica*, **7**, 127–49.

Sauer, E.G.F. and Sauer, E.M. (1963). The South West African bushbaby of the *Galago senegalensis* group. *Journal of the South West African Scientific Society*, **16**, 5–36.

Schaller, G. B. (1972). *The Serengeti lion*. University of Chicago Press.

Sherman, P. W. (1981). Reproductive competition and infanticide in Belding's ground squirrels and other animals. In *Natural selection and social behaviour*, (ed. R. D. Alexander and D. W. Tinkle), pp. 311–31. Chiron Press, New York.

Silk, J.B. (1983). Local resource competition and facultative adjustment of sex ratios in relation to competitive abilities. *American Naturalist*, **121**, 56–66.

Smuts, B.B., Cheney, D.L., Seyfarth, R.M., Wrangham, W.R., and Struhsaker, T.T. (Ed.) (1987). *Primate societies*. University of Chicago Press.

Struhsaker, T.T. (1977). Infanticide and social organization in the redtail monkey (*Cercopithecus ascanius schmidti*) in the Kibale Forest, Uganda. *Zeitschrift für Tierpsychologie*, **45**, 75–84.

Strum, S.C. (1981). Processes and products of change: baboon predatory behaviour at Gilgil, Kenya. In *Omnivorous primates: gathering and hunting in human evolution*, (ed. R.S.O. Harding and G. Teleki), pp. 255–302. Columbia University Press, New York.

Sugiyama, Y. (1984). Population dynamics of wild chimpanzees at Bossou, Guinea, between 1976 and 1983. *Primates*, **25**, 391–400.

Suzuki, A. (1971). Carnivority and cannibalism observed among forest-living chimpanzees. *Journal of the Anthropological Society of Nippon*, **79**, 30–48.

Takahata, Y. (1985). Adult male chimpanzees kill and eat a male newborn infant: newly observed intragroop infanticide and cannibalism in Mahale National Park, Tanzania. *Folia Primatologica*, **44**, 161–70.

Takahata, Y., Hasegawa, T., and Nishida, T. (1984). Chimpanzee predation in the Mahale Mountains from August 1979 to May 1982. *International Journal of Primatology*, **5**, 213–33.

Teleki, G. (1973). *The predatory behaviour of wild chimpanzees*. Bucknell University Press, Lewisburg.

Teleki, G. (1975). Primate subsistence patterns: collector-predators and gatherer-hunters. *Journal of Human Evolution*, **4**, 125–84.

Uehara, S., Nishida, T., Hamai, M., Hasegawa, T., Hayaki, H., Haltman, M., Kawanaka, K., Kobayashi, S., Mitani, J., Takahata, Y., Takasaki, H., and Tsukahara, T. (1992). Characteristics of predation by the chimpanzees at Mahale and cross-population comparisons of chimpanzee hunting. In *The Symposium Proceedings of the XIII Congress of the International Primatological Society*, (ed. T. Nishida, W. C. McGrew, P. Marler, M. Pickford, and F. de Waal), The University of Tokyo Press. (In press.)

Vincent, J. (1969). Miscellaneous notes of natural history interest. 1. Murder. *Lammergeyer*, **10**, 99–100.

Walters, J.R. and Seyfarth, R.M. (1987). Conflict and cooperation. In *Primate societies*, (ed. B.B. Smuts, D.L. Cheney, R.M. Seyfarth, R.W. Wrangham, and T.T. Struhsaker), pp. 306–17. University of Chicago Press.

Wrangham, R.W. and Riss, E. van Zinnicq Bergmann (1990). Rates of predation on mammals by Gombe chimpanzees, 1972–1975. *Primates*, **31**, 157–70.

Zimmermann, E. (1989). Aspects of reproduction and behavioural and vocal development in Senegal bushbabies (*Galago senegalensis*). *International Journal of Primatology*, **10**, 1–16.

Author Index

Abler, T. S. 4
Abrams, J. 214
Abrams, P. A. 18, 30
Aguilera, E. 287, 289
Ahroon, J. A. 301
Alberts, J. R. 309, 310
Alden, R. H. 85, 87, 88, 89
Alexander, R. D. 4, 114, 120, 176, 178, 179, 185, 189, 191, 249, 277, 285
Alford, D. V. 216, 222, 223, 224
Alheit, J. 239
Alyakrinskaya, I. O. 113
Anderson, D. T. 132
Anderson, J. D. 258
Andreassen, E. M. 116
Angst, W. 323
Annett, C. 291
Antikainen, E. 279, 280
Aoki, K. 170
Aoki, S. 176
Appelbaum, S. 248
Apsit, V. J. 73
Archer, M. E. 188
Arens, W. 4
Armentrout, D. 257, 258, 261, 264
Armstrong, J. T. 2
Arnold, S. J. 262
Ashe, J. S. 182
Asheroff, A. F. 312
Ashmole, N. P. 280
Assem, J. van den 243, 245
Audesirk, G. 102, 119
Audesirk, T. 102, 119
Austad, S. N. 133, 142
Austin, A. D. 132
Axelrod, R. 78

Bachmann, M. D. 257, 258, 261
Bagh, R. 180
Bailey, K. H. 15, 16
Baird, A. M. 67, 71, 74, 75
Baldauf, R. 258
Balfour, E. 286
Ball, S. C. 259
Banks, C. J. 2, 157, 159, 161
Banks, E. M. 314
Bannerman, D. A. 286
Barber, F. G. 13
Barlow, G. W. 239, 248
Baroni, U. 193
Barth, E. K. 286
Bartke, A. 305
Bartlett, J. 157, 160, 171, 183
Bartram, C. 66
Batra, S. W. T. 185, 186, 216, 229
Baudvin, H. 286
Baur, A. 109, 111, 113, 114
Baur, B. 2, 8, 105, 107, 109, 111–20, 138, 249, 262, 264, 288
Bayne, C. 114, 117
Bazire-Bénazet, M. 193, 195
Beaver, R. Z. 158, 164
Beig, D. 199, 200, 230
Beilharz, R. G. 309
Bělar, K. 89
Bell, R. W. 314
Belles-Isles, J.-C. 38, 39, 57, 58, 60, 240, 242–6, 247, 250, 291
Bellman, R. E. 42
Ben Mordechi, Y. R. 217
Bengston, S. A. 117, 286
Bent, A. C. 280
Berger, D. 87
Berrio, I. 291
Berry, J. W. 132
Bhatnagar, S. P. 64, 65, 67
Birkhead, T. R. 129, 130, 137
Birney, E. C. 300
Bisazza, A. 239
Bishir, J. 74
Bishop, S. C. 258
Biswas, S. K. 107
Blackburn, T. M. 183
Blair, G. M. 106
Blair, W. F. 262
Blakely, H. M. 314
Blanke, R. 139
Blaustein, A. R. 264
Blest, A. D. 136
Blom, J. van der 216, 217, 219
Blum, M. S. 191, 196
Blumer, L. S. 38, 39, 40, 238, 239, 242, 251, 262
Bobisud, L. 19
Boesch, C. 329, 333
Boesch, H. 329, 333
Boggess, J. 323
Boldyrev, M. I. 180
Bonin, B. 286

Borland, J. R. 256
Bortolotti, G. R. 278, 281, 284–9
Boucher, D. H. 289
Bourke, A. F. G. 193
Bovee, E. C. 89
Boycott, A. E. 103
Bragg, A. N. 259, 260, 262–4, 267
Bragg, W. N. 262
Braid, J. D. 18
Breden, F. 157, 159, 160, 171
Breene, R. G. 137
Brennan, M. 65
Brian, M. V. 192, 194–7
Brigham, R. M. 290
Briscoe, D. A. 91
Bristowe, W. S. 131, 132, 134, 142, 143, 148, 149
Broida, J. 311
Bronson, F. H. 303, 313, 315
Brook, A. 286
Brooks, R. J. 306, 309, 311–15
Broom, D. M. 300
Brothers, D. H. 216, 229
Brothers, D. J. 185, 186
Brough, P. 71, 74
Brown, H. D. 157, 159, 161, 162, 180
Brown, J. L. 282, 284
Brown, M. E. 248
Brown, R. E. 309–11
Brown, R. G. B. 280, 282
Brown, R. Z. 315
Bruce, H. M. 313
Bruce, J. A. 143
Bruniquel, S. 193
Brunsting, A. M. H. 158, 159, 165
Brussard, P. F. 109
Buchholz, J. T. 72, 73, 75
Buhse, H. E. 85
Bull, J. J. 9
Burger, J. 280, 286
Burky, A. J. 104, 111
Burley, N. 65, 71, 74
Burlingame, L. L. 65
Burnet, F. M. 94
Burton, T. M. 68, 258, 260
Buskirk, R. E. 9, 129, 136, 137, 138, 150
Butler, C. G. 216
Butyinski, T. M. 333
Bygott, D. 330, 332

Cade, T. J. 281, 285
Cain, A. J. 102, 107
Caldwell, O. W. 71, 74
Calhoun, J. B. 315
Calow, P. 102
Cambridge, O. -P. 147
Cameron, S. A. 221
Campagno, L. J. V. 288
Caraco, T. 32
Carew, T. J. 118
Carico, J. E. 143
Carlisle, T. R. 39, 40, 43, 57
Carrick, R. 109, 114
Cavalier-Smith, T. 87
Cawston, J. M. 278
Cech, J. J., Jr. 249
Cei, J. M. 259, 260, 262, 268
Cerri, R. D. 25
Chamberlain, C. J. 71, 74
Chambers, L. A. 105
Chapman, R. N. 158
Chardine, J. W. 281
Charnov, E. L. 9, 22, 25, 32
Chase, R. 109
Chavelas, C. G. 288
Cheek, J. E. 261, 263, 269, 270
Chew, F. S. 177
Chia, F. -S. 107
Choi, J. -S. 66, 80
Choudhuri, D. 180
Christensen, H. 107
Christian, J. J. 315
Christie, S. L. 89
Cicirello, D. M. 305–7, 314
Clark, A. B. 323
Clark, C. W. 16, 32, 41, 42, 43, 47, 57, 60
Clarke, R. F. 258
Cleveland, L. R. 86, 89
Clevenger, G. A. 279
Clothier, G. W. 258
Cloudsley-Thompson, J. L. 128, 143
Coddington, J. 145
Cohen, D. 70
Coleman, J. R. 279
Coleman, R. M. 44
Colgan, P. W. 245
Colley, S. A. 258
Collins, D. A. 324
Collins, J. P. 261, 263, 269, 270
Collopy, M. W. 288
Conely, L. 314
Conover, M. R. 290
Cook, A. 106, 106
Coquelin, A. 313
Costantino, R. F. 13, 27, 158, 166, 167
Cottrell, C. B. 2, 4
Coulter, J. M. 65
Courtney, S. P. 177
Craig, C. L. 148
Craig, D. M. 166
Crane, S. 107
Crespi, B. J. 1, 2, 3, 14, 15, 23, 182, 223, 226, 249, 288, 289
Croll, R. P. 113
Crombie, A. C. 158, 164

Crook, J. R. 287
Crosland, M. W. J. 194
Crow, J. F. 170
Crowell, H. H. 113
Crowley, P. 2, 15, 17, 18, 19, 20, 24, 27
Crozier, R. H. 96, 185, 189, 222
Crozier, W. J. 102
Crump, M. L. 3, 8, 15, 18, 240, 241, 249, 256–60, 262, 264–8, 291
Cruz, Y. P. 176
Cumber, R. A. 225
Cunningham, J. 258, 260
Currey, J. D. 107
Curry, R. L. 283, 289
Curtin, R. 323
Cuthbertson, G. J. 291

Darwin, C. 143, 147
Davies, N. B. 17, 128
Davies, R. B. 27
Davis, J. W. F. 280, 282
Davis, W. J. 129, 130, 143
Dawson, J. A. 85, 87, 88, 89
Dawson, P. S. 156, 163, 168, 171
Day, C. S. 300–3
De Vita, J. 26
de Bruijn, L. L. M. 199–201, 204
de Montellano, B. R. O. 4
DeAngelis, D. 13, 18, 19, 27, 28
Degani, G. 258, 262
Dejean, A. 192, 193
Delage, B. 193
Delage-Darchen, B. 193
DeMartini, E. E. 40, 57, 247
Demment, M. 323
Dengler, R. E. 89, 91
DeSantis, D. T. 301, 307
Desharnais, R. A. 166, 167
Deusing, M. 280, 282, 291
Devi, R. V. 89, 90
Dewsbury, D. A. 300
Diamond, M. 301
Dickemann, M. 289
Dickinson, J. L. 280
Didier, B. 105
Diehl, S. 26, 31
Dimetry, N. A. 157
Dionne, M. 239, 240
Dirks, T. F. 187
Dixon, A. F. G. 157, 159, 161
Doak, C. C. 75
Dodson, G. N. 136
Dolhinow, P. 323
Dominey, W. J. 38, 39, 40, 238, 239, 242, 251, 262
Dong, Q. 1, 5, 10, 166, 214
Doorn, A. van 216, 222–4
Doutt, R. L. 183
Downes, J. A. 130, 140
Doyle, G. A. 65, 324, 325
Drummond, H. 288, 289
Duchateau, M. J. 186, 190, 216, 222–4
Dudley, D. 300
Duelli, P. 2
Duellman, W. E. 261
Duffy, K. T. I. 89, 92, 93
Dukas, R. 70
Duncan, C. J. 113, 118
Dunn, E. K. 280, 282
Dupler, A. W. 65
Durant, J. P. 249
Dushin, A. I. 260

Eberhard, W. G. 184
Edgerly, J. S. 184
Edmunds, M. 130, 134
Edwards, F. W., Jr. 130
Edwards, T. C., Jr. 130, 282, 288
Eggert, A. -K. 183
Eickwort, G. C. 181, 184, 214
Eickwort, K. R. 116, 157, 159, 177, 180, 184, 266
Eisenberg, J. 326
Eisenberg, R. M. 178
Elgar, M. A. 1, 2, 5, 6, 8, 14, 15, 23, 24, 118, 129, 131–3, 135, 137–42, 144, 145, 148–50, 249
Elmslie, L. J. 109
Elwood, R. W. 1, 3, 6, 7, 8, 249, 265, 300, 305–15
Emlen, S. T. 282
Engels, F. M. 67
Engels, W. 198, 200
Erdos, G. W. 89, 91
Errington, P. L. 286, 306
Evans, H. E. 183

Fajer, E. D. 285
Falconer, D. S. 171
Farley, R. D. 131
Favre-Duchartre, M. 64, 67, 70, 71, 75, 79
Ferguson, W. 279, 291
Fernandez, G. W. 67
Fidura 301
Field, L. H. 130
Filosa, M. F. 89, 91, 97
Fioroni, P. 104, 108–11, 120, 121
Fisher, R. M. 171, 186, 190, 279
FitzGerald, G. J. 3, 5, 8, 18, 38, 39, 45, 46, 48, 57–60, 241–7, 250, 265
Flanders, S. E. 205
Fletcher, D. J. C. 187, 191
Foelix, R. F. 142, 143

Foerg, R. 324, 325
Folkvord, A. 15, 16
Foltz, D. W. 118
Fordham, R. A. 280
Forester, L. C. 257
Forster, L. 133–5, 139, 140
Fossey, D. 324
Foster, S. A. 25, 245, 249-51
Fowler, D. P. 73
Fox, L. R. 2, 5, 10, 13, 15, 17, 19, 25, 26, 30, 119, 134, 156, 238, 239, 242, 264, 267
Frankel, J. 95
Franklin, E. C. 68, 73
Fraser, D. F. 16, 25, 26
Free, J. B. 186, 190, 216, 223
Freed, L. A. 287
Freeland, J. 193, 194
Fretter, V. 102, 118
Friedman, W. E. 66, 67, 74, 80
Fritts, T. H. 259, 260
Frömming, E. 103, 105–7, 109
Fuchs, S. 302
Fuhrman, F. A. 259
Fujioka, M. 289
Fukui, Y. 89, 92
Fukushima, K. 187

Gabriel, W. 27
Galatti, U. 259
Galef, B. G. 300–3
Gall, J. G. 86
Gallardo, C. S. 110
Gandelman, R. 301, 302, 307, 311
Gargett, V. 281, 285, 288
Garófalo, C. A. 216, 220, 221
Garrido, O. A. 110
Gavan, J. A. 334
Gebler, J. B. 48
Gelei, J. V. 85, 89, 90
Gerhardt, U. 133
Gerling, D. 217, 218
Gervet, J. 187, 190
Geyer, D. 106
Ghiraldi, L. L. 311
Ghose, K. C. 107, 109
Gianordoli, M. 67
Gibber, J. R. 309, 311
Gibbons, D. W. 281, 284
Giese, A. C. 84, 86–9
Giese, K. 111
Gilbert, H. A. 89, 286
Gilbert, J. J. 2
Gilliam, J. 16, 22, 25, 26, 32
Gilmer, R. W. 105, 118, 131
Glancey, B. M. 194
Glass, G. E. 238, 251
Godfray, H. J. C. 178, 183
Goethe, F. 281, 284
Goin, C. J. 267
Goldman, L. 301
Goldstein, H. 287
Goodall, J. 323, 324, 329, 330, 332, 334
Goodnight, K. E.
Goodwin, O. K. 266, 268
Goss-Custard, J. D. 15
Gould, J. L. 128, 143
Gould, S. J. 135, 138, 140
Goyens, J. 312
Grafen, A. 190
Graham, A. 102, 118
Grass, P. -P. 112
Grasshoff, M. 131, 139
Graves, J. A. 289, 290
Greene, A. 188
Greenwald, G. S. 307
Griffin, A. R. 68, 73
Grobler, J. H. 259, 260, 264
Gross, M. R. 39, 40, 43, 57, 239, 289
Grossu, A. V. 106, 107
Gruber, S. H. 288
Gubernick, D. J. 309, 310

Haefelfinger, H. -R. 107
Hagen, K. S. 162
Haig, D. 3, 66, 70, 74, 77, 80, 160, 289
Hamai, M. 332
Haime, J. 85, 89
Haines, R. J. 65, 67
Hall, D. 15
Hallas, S. E. A. 133, 136, 143
Hamai, M. 330
Hamilton, W. D. 115, 116, 156, 157, 168, 170, 177, 184, 185, 218, 224, 232, 245, 277, 299
Hamner, W. M. 104
Hampton, S. H. 262, 269
Hansson, L. 299
Harding, R. S. O. 328
Hardy, I. C. W. 183
Hare, H. 185, 189, 191, 195, 203, 222, 224
Harper, A. B. 4, 170
Harris, M. 4
Harris, R. N. 267
Hartung, T. G. 300
Harumoto, T. 87
Harvey, P. H. 145
Hasegawa, T. 332
Haskins, C. P. 193, 194
Haskins, E. F. 193, 194
Hassell, M. 13, 16, 17, 19, 26
Hastings, A. 13, 27, 166, 167
Hausfater, G. 6, 7, 15, 23, 323, 331
Havre, N. van 243, 245, 250
Hawbecker, A. C. 286
Hawkes, O. A. M. 161

Heatwole, H. 258
Hecht, T. 248
Heckmann, K. 90
Heessen, H. J. L. 158, 159, 165
Hefetz, A. 217
Heinsohn, R. G. 284
Heintzelman, D. S. 286
Henry, C. S. 178, 179
Hepper, P. G. 308
Heringa, J. 216, 222–4
Hermann, H. R. 187, 217
Hero, J. M. 259
Heusser, H. 259
Hewett, S. W. 88
Heyer, W. R. 257, 259, 267
Higgins, J. 13, 18, 27
Hiraiwa-Hasegawa, M. 1, 3, 5, 6, 262, 323, 331, 332, 334
Hirshfield, H. I. 88, 89
Hiwatashi, K. 96
Ho, F. K. 163, 168, 171
Ho, R. H. 70
Hoagland, K. E. 121
Hobbs, G. A. 216, 225
Hodasi, J. K. M. 109
Hoeck, H. N. 304
Hoelzer, G. 38, 40, 45, 57, 58, 60, 245–7
Hogeweg, P. 186, 190, 222
Hölldobler, B. 185, 192–7, 205
Holling, C. S. 19
Holomuzki, J. R. 257, 258, 268
Holthuijzen, A. M. A. 281
Honk, C. G. J. van 186, 190, 216, 222–4
Hoogland, J. L. 7, 289, 299, 305–7, 312, 314, 315
Horsfall, J. A. 289
Houston, A. I. 5, 16, 22, 32, 41, 42
Howard, L. S. 306, 309, 310
Hrdy, S. B. 6, 7, 15, 23, 238, 266, 277, 323, 329, 331
Huber, P. 224
Huck, U. W. 304, 309, 310, 313
Huffman, D. M. 89, 91
Hughes, R. N. 102–4, 119, 120
Hung, A. C. -F. 193, 195
Hunt, G. L., Jr. 280, 282
Hunt, M. W. 280
Hunter, J. R. 15, 16, 17
Hurd, L. E. 178
Hurst, L. D. 74
Huston, M. 30
Hutcherson, J. E. 258
Hyatt, K. D. 239, 245–7
Hyman, L. H. 102, 118
Hynes, H. B. N. 247

Iersel, J. J. A. van 242, 246
Imperatriz-Fonseca, V. L. 198, 200
Ingram, C. 285, 286
Inoue, T. 199, 200
Iperti, G. 157
Ishay, J. 188
Istock, C. A. 159, 165
Ivanic, M. 85, 86, 89
Iwahashi, O. 134, 139
Iwata, K. 219
Izard, M. K. 324, 325

Jackson, R. R. 132–4, 136, 143–5
Jakubowski, M. 310, 311, 313
Jamieson, I. G. 135, 138, 245
Janzen, D. H. 249, 284
Jasienska, G. 2
Jasienski, M. 2
Jay, S. C. 225
Jeanne, R. L. 185, 187, 188
Jensen, K. R. 104, 107
Jeppsson, B. 313
Jillson, D. 158
Johnsson, H. 68
Jones, D. 141
Joslin, D. 278
Joukowsky, D. 85
Jude, D. 13
Jungfer, K. H. 257

Kaddou, I. K. 156, 157, 159, 161
Kaplan, R. H. 257, 258, 266
Kaston, B. J. 129
Kaston, E. 129
Katayama, E. 186, 190, 216, 224
Kaur, D. 64, 65, 67
Kawai, A. 157, 159, 161, 162, 180
Kawanaka, K. 330–2
Keawjam, R. S. 104, 111
Keenleyside, M. H. A. 244
Keller, L. 193, 195
Kennedy, H. F. 309, 310, 313, 314
Kimble, D. P. 301
Kimbrell, C. A. 17
Kimsey, L. S. 221
King, C. E. 156
Kirkman, F. B. 280
Klekowski, E. J., Jr. 73
Kluge, A. G. 259
Koehl, M. A. R. 13, 26
Koenig, W. D. 3, 161, 249, 262, 282, 283, 290, 302
Koford, R. R. 284
Kohn, A. J. 102, 119
Konar, R. N. 65
Konig, B. 302, 308
Kosinska, M. 109

Kosinski, G. 72
Koski, V. 67, 68, 73
Kozlowski, J. 70
Krackow, S. 304
Krambrich, A. 286
Krebs, J. R. 13, 16, 17, 21, 22, 23, 26, 41, 128
Kristal, M. B. 305
Kruuk, H. 333
Kuhlmann, H. W. 90
Kukuk, P. F. 1–3, 6, 185, 186, 189, 190, 197, 200, 227–9, 249, 288
Kupfermann, I. 118
Kupferschmidt, J. 305, 306
Kurihara, Y. 2
Kusano, T. 257, 258, 268
Kynard, B. 247, 250

Labov, J. B. 300, 303, 304, 308, 309, 313
Lack, D. 120, 287
Lalli, C. M. 105, 119, 131, 138
Lampert, W. 27
Land, W. J. G. 64, 65, 71
Lande, R. 170
Landolt, P. J. 188
Landry, M. 17
Lane-Petter, W. 301
Lannoo, M. J. 257, 258, 261, 264, 265
Lapage, G. 85, 86, 89
Laverty, T. M. 186, 218
Lawn, R.M. 86
Lawrence, S. E. 129, 130, 134, 137
Lawson, A. A. 74, 75
LeCato, G. L. 163, 166
Lee, C. L. 67, 74
Lee, T. D. 63, 78
Lennartz, D. C. 89
Leonard, J. L. 104, 106, 107, 119
Lewis, W. M., Jr. 143
Lewis, W. H. 241
Lidicker, W. Z. 315
Lin, N. 184
Lindgren, D. 68, 73
Linkowski, T. B. 2
Lipinski, M. R. 2
Liske, E. 129, 130, 143
Liu, L. 166, 167
Lloyd, J. A. 315
Loadman, N. L. 239, 241
Loekle, D. M. 238, 265
Loflin, R. K. 284
Loiselle, P. V. 243, 265
Lomnicki, A. 185
Long, J. 324, 325, 326
Longley, A. J. 118
Longley, R. D. 118
Loveridge, A. 259
Low, C. B. 68
Lowenstein, J. M. 71
Lubin, Y. D. 132
Lukowiak, K. 104, 106, 107, 119
Lynch, C. B. 303, 304
Lyons, D. M. 314

Macdonald, D. 299
Macedo, R. 284
Maeta, Y. 216, 219
Maheshwari, P. 64
Main, B. Y. 129
Makin, J. W. 309
Malhotra, R. K. 75
Mallory, F. F. 306, 312, 313
Malsbury, C. W. 301
Mangel, M. 16, 32, 41–3, 47, 57, 60
Manning, A. 128, 143
Manocha, S. 324, 325, 328
Marchetti, K. 40, 57, 59, 245, 250
Marconato, A. 239
Markl, H. 302
Marques, D. M. 311
Marriner, T. F. 161
Marshall, C. J. 257–9
Marsteller, F. A. 303, 304, 315
Martens, P. 65
Martin, A. A. 259, 263
Martin, B. 18
Mason, C. F. 103
Mast, M. 301
Masui, K. 329, 330
Matheson, A. C. 73
Matthews, R. W. 189
Maupas, E. 85, 89
May, B. 228, 229
May, R. 13, 19, 26
Mays, C. E. 258
McCann, C. 260
McCarthy, M. M. 306, 307, 309–11
McCarty, R. 300
McCauley, D. E. 158–60, 165
McCauley, P. J. 314
McClure, P. A. 303, 304
McCoid, M. J. 259, 260
McCormick, S. 20
McCracken, G. F. 109
McFarland, D. 16, 41
McGrew, W. C. 325, 335
McLean, I. G. 312
McLoon, S. C. 280, 282
McNamara, J. M. 42
McNicholl, M. K. 278, 282, 285
McPhail, J. D. 245, 250
McWilliam, J. R. 68
Mebs, T. 286
Mecham, J. S. 266
Mecom, J. O. 2

Meffe, G. K. 17, 240, 241, 251, 265
Mehra, P. N. 75
Mennella, J. A. 305, 310
Merényi, L. 107
Mertz, D. B. 13, 18, 27, 156, 157, 163, 166–72
Metz, J. A. J. 30
Meyburg, B. -U. 285
Michener C. D. 176, 184–6, 190, 196, 199, 201, 202, 205, 214, 216–22, 226, 227, 229
Mikkola, L. 67, 68, 72
Miller, D. E. 66
Miller, W. R. 290
Mills, N. J. 161, 180
Milton, K. 333
Mitra, T. R. 107
Miyake, A. 87
Mock, D. W. 178, 277, 278, 285–90
Mockford, E. L. 179, 180
Molder, M. 67, 68, 72
Moltz, H. 305, 310
Montgomery, T. H., Jr. 133, 143, 149
Moodie, G. E. E. 245–7
Mora, G. 183, 184
Mordan, P. B. 103, 106
Morin, P. J. 257, 258, 260, 266
Morris, G. K. 135
Morris, R. D. 281
Morris, S. J. 76
Morse, D. H. 190, 216, 224
Morse, M. P. 105
Moseley, M. F. 71
Moyle, P. B. 249
Mrowka, W. 40, 44, 57, 58, 244
Muedeking, M. H. 257, 259
Muller, J. K 183
Mumme, R. L. 282, 283
Murdoch, W. W. 18, 19, 31
Murphy, E. C. 277, 290
Murphy, R. C. 280
Murray, J. J. 115
Myers, C. A. 3, 256
Myser, W. C. 216, 227

Nagai, Y. 258, 265
Namkoong, G. 74
Nash, D. R. 131, 134, 137, 140–2, 148, 149
Natarajan, A. V. 110
Naumann, M. G. 187
Neale, D. B. 76
Nelson, E. M. 95
Nentwig, W. 134, 145
Newman, J. A. 5, 14, 24, 129, 140, 147, 150
Newton, I. 281
Nickerson, M. A. 258
Nihonzaru 324
Nishida, T. 330–2
Noble, G. K. 257, 259, 262
Noirot, E. 301, 312
Nolan, W. J. 216, 227
Nonacs, P. 185
Norikoshi, K. 331, 332
Norris, D. O. 258
Norstog, K. J. 67, 74
Nummelin, M. 2, 8, 265
Nurse, P. 94
Nussbaum, R. A. 258
Nybakken, J. W. 119

O'Brien, W. 15
O'Connor, R. J. 277, 300, 302
O'Donnell, R. 159
O'Hara, R. K. 264
O'Neill, K. M. 183
Oberoi, Y. P. 65, 71
Ochi, H. 245
Ohmart, R. D. 281, 284, 289
Okamoto, K. 2
Olive, L. S. 89, 91
Oniki, Y. 201
Orians, G. H. 249, 284
Orr, B. K. 2, 13, 17, 18, 19, 20
Orton, G. L. 260, 262
Osawa, N. 156, 157, 158, 162
Osborn, T. G. B. 65
Oster, G. F. 41, 43, 44
Ostermeyer, M. C. 300, 305, 306, 308, 310–12, 314, 315
Owen, G. 112
Owen, R. E. 186, 224
Owens, J. N. 67, 68, 72, 76
Owiny, A. M. 109

Packer, L. 216, 229
Packer, P. 331
Padmavathi, P. B. 89
Page, R. E., Jr. 185, 227
Pagel, M. 145
Paine, R. T. 104, 106, 119, 131
Paller, M. H. 241
Palmer, R. S. 280
Paludan, K. 280, 291
Pardi, L. 187
Park, T. 156, 158, 162, 166, 168, 169
Park, Y. S. 73
Parker, G. A. 136, 140, 142, 178, 288
Parmelee, D. F. 286
Parmigiani, S. 309, 312
Parsons, J. 280, 282, 291
Passera, L. 192–5, 197
Paul, L. 305, 306, 308
Paullin, D. G. 278
Paulus, H. F. 221

Pawson, P. A. 109
Paynter, R. A. 280
Pearce, J. B. 110
Peckham, E. G. 133, 145
Peckham, G. W. 133, 145
Peeters, C. 185, 222
Perrigo, G. 301, 303, 309–11, 314
Perron, F. E. 120
Persson, L. 26, 31
Peters, L. C. 305
Petersen, C. W. 245, 250
Peterson, C. L. 40, 57, 59
Petranka, J. W. 256
Pettingill, O. S., Jr. 280
Pfennig, D. W. 262, 264, 265, 267, 269, 270
Pickering, J. 178
Pickwell, G. 285
Picman, J. 291
Pienaar, U. de V. 259, 260
Pienkowski, R. L. 157, 161, 180
Pierce, B. A. 95, 261, 269
Pierce, E 89
Pierotti, R. 277, 290, 291
Pilz, W. R 286
Pitcher, T. J. 249
Plant, J. D. 221
Plath, O. E. 186, 216
Platnick, N. 143
Plowright, R. C. 186, 216, 224, 225
Plummer, J. M. 109
Plym Forshell, C. 68
Poldi, B. 194
Polis, G. A. 1–3, 5, 10, 13–20, 23–6, 30, 31, 114, 119, 131, 134, 138, 143, 144, 156, 163, 166, 214, 238, 239, 249, 256, 264, 266, 267, 284
Pollard, S. D. 132
Pomeroy, L. V. 259, 260, 262, 264, 269
Pomeroy, N. 186, 190, 216
Porter, R. D. 281, 291
Porter, R. H. 309
Poulin, R. 242
Poulsen, B. A. 132
Powders, V. N. 258, 260
Prakash, I. 299
Preece, J. 303
Pressley, P. H. 39, 40, 43, 57
Price, R. A. 71
Procter, D. L. C. 280
Pukowski, E. 182
Purchon, R. D. 102
Pusey, A. E. 331, 332
Pyke, G. 13, 22, 23, 26

Queller, D. C. 78, 185

Raboud, C. 112, 115, 116
Radovich, J. 13
Raper, K. B. 97
Rapport, D. 16
Rasmussen, P. C. 286
Ratneiks, F. L. W. 185, 203, 225, 226, 228, 232
Raut, S. K. 107
Rechten, C. 245, 250
Reebs, S. G. 242
Rees, W. J. 109
Reese, J. G. 285
Reig, O. A. 269
Reynolds, L. G. 71
Reynolds, R. D. 290, 301
Rhine, R. J. 328
Ribbands, C. R. 228
Ricci, N. 88
Rice, B. 78
Richards, K. W. 216, 224
Richards, L. 16
Richards, M. P. M. 307
Richmond, K. 286
Richmond, N. D. 259
Ricker, W. E. 13, 27
Ridgway, M. 245, 250
Ridley, M. 145, 245, 250
Rigby, C. 192, 194
Ringler, N. H. 239, 245–7
Rinkevich, B. 3
Riss, E. 333, 334
Robbins, R. K. 177
Robertson, F. J. 287
Robinson, B. 131, 132, 140, 143, 147, 148
Robinson, G. E. 216, 227
Robinson, M. H. 131, 132, 140, 143, 147, 148
Roeder, K. D. 129
Roest, A. I. 279, 285
Rohwer, S. 38–40, 56–60, 238, 245, 246, 247, 250, 256, 277, 282, 287, 290
Roland, C. 133
Rollo, C. D. 105, 106
Romagnano, L. C. 291
Rondelaud, D. 103, 105, 119
Roper, M. K. 331
Rose, F. L. 257, 258, 261, 264
Röseler, P. -F. 186, 222–5
Rosenberg, K. M. 311
Rosenblatt, J. S. 307
Rosenson, L. M. 312, 324, 325
Ross, I. K. 91
Ross, J. 132
Ross, K. G. 133, 143, 187, 189, 194, 195
Rothe, H. 324, 325
Rothstein, S. I. 306
Rovner, J. S. 133, 143
Rowell, T. E. 300, 301
Rowley, I. 280
Rubenstein, D. I. 19, 133

Rudin, D. 68
Rugh, R. 256
Ruibal, R. 260, 264
Russell, P. 94
Rutowski, R. L. 105, 118, 131, 138
Rypstra, A. L. 9, 183

Saayman, G. S. 324
Safina, C. 281, 286
Saga, Y. 89, 91
Saila, S. B. 13
Sakagami, S. F. 185, 187, 196–201, 203, 204, 216, 217, 219, 220, 226, 230
Salfert, I. G. 245–7
Salthe, S. N. 266
Samuels, O. 311
Sanday, P. R. 4
Sandlant, G. R. 130
Sandon, H. 85
Sargent, R. C. 3, 5, 14, 39–41, 43–8, 56, 57, 59, 238, 243, 289
Sarvas, R. 64–72, 74
Sasaki, T. 132, 134, 139
Sauer, E. G. F. 324, 325
Sauer, E. M. 325
Savage, R. M. 267
Savory, T. 129, 144
Saxton, W. T. 71, 75
Schaller, G. B. 333
Scharf, W. C. 286
Schick, R. X. 180
Schlosser, I. 25
Schmaltz, L. W. 301, 307
Schmekel, L. 110, 120, 121
Schnarf, K. 67
Schoener, T. W. 13, 16, 20, 22, 31, 32
Schüz, E. 281, 284
Schwagmeyer, P. L. 313
Schwanck, E. 244
Schwarz, M. P. 230
Schwarzkopf, L. 309, 311
Scott, M. P. 182
Seapy, R. R. 104, 106, 119
Sederoff, R. R. 76
Semler, D. E. 247
Seyfarth, R. M. 329
Shear, W. A. 145
Sherman, A. R. 282
Sherman, G. F. 311
Sherman, P. W. 257, 258, 266, 299, 304, 312
Sherry, T. W. 291
Shields, W. M. 287
Shine, R. 239
Siebert, L. K. 286
Siebt, U. 180
Siegel, H. I. 307
Siegenthaler, S. 281, 284
Sierwald, P. 143
Sih, A. 16, 25, 26
Silk, J. B. 333
Simmons, L. W. 136, 140, 142
Simoes, D. 187
Simon, M. P. 256, 257, 259, 267, 268, 302, 307
Simons, E. L. 324, 327
Singh, H. 64, 71
Skaife, S. H. 190
Skeel, M. 107
Skutch, A. F. 284
Sladen, F. W. L. 216, 223, 224
Slater, P. J. B. 251
Smart, J. L. 303
Smith, B. J. 104, 108, 110, 121
Smith, C. C. 68, 73
Smith, R. L. 133, 143
Smith, R. S. 243
Smuts, B. B. 328
Smyly, W. J. P. 130
Snyder, N. F. R. 286
Solem, A. 102
Solter, D. 77
Sommeijer, M. J. 199–201, 204
Sorensen, F. C. 73, 74
Soroker, V. 309, 310
Southwick, C. H. 300, 301
Southwood, T. R. E. 16, 17
Spanier, E. 104
Spellerberg, I. F. 280
Sperry, D. G. 262
Spight, T. M. 111, 120
Spradbery, J. P. 187, 188
Springer, S. 288
Sprunt, A., Jr. 280
Stacey, P. B. 282
Stanback, M. T. 3, 161, 249, 262, 281, 285, 302
Stark, R. E. 216, 218, 226, 232
Starr, C. K. 185, 189, 204
Starrett, P. 260
Stearns, S. C. 70, 72
Stenseth, N. C. 315
Stephens, D. W. 13, 16, 19, 21–3, 26, 41
Stephenson, A. G. 63, 78
Stevens, L. 1–3, 5, 8, 15, 17, 18, 157, 169–71, 180
Stinson, C. H. 285
Stouffer, P. C. 291, 306
Stowe, M. K. 132
Strand, M. R. 183
Strassmann, J. E. 189
Strenna, L. 286
Strong, D. 20
Struhsaker, T. T. 324
Strum, S. C. 328
Stuart, R. J. 193
Stutchbury, B. J. 287
Sugiyama, Y. 329

Summers, K. 259, 267
Suzuki, A. 330, 331
Svare, B. B. 305, 311, 312
Swanson, H. 301
Sweet, G. B. 69, 70
Sweet, M. H. 137
Synder, R. J. 250
Szabo, J. 105, 107
Szabo, M. 105, 107
Szmidt, A. E. 76

Taborsky, M. 40, 57
Tago, K. 258
Tait, D. E. N. 289
Takahashi, K. 180
Takahata, Y. 330–2, 334, 335
Taki, A. 193–5
Tallamy, D. W. 181
Tamaki, G. 158, 165
Tang, W. 67
Taylor, J. D. 102
Taylor, M. H. 71, 74
Teleki, G. 331, 333, 334
Teleky, S. B. 158, 166
Terkel, J. 309–11, 314
Test, F. H. 258
Thiele, H. -U. 158, 165
Thomas, E. 260
Thomas, J. A. 300
Thommen, D. 323
Thompson, T. E. 105
Thornhill, R. 130, 133, 136, 142
Thorson, G. 110, 111, 120
Thulin, I. J. 69
Tilley, S. G. 257, 258, 266, 267
Tinbergen, N. 280
Tompa, A. S. 113, 117
Torossian, C. 193, 194
Townsend, D. S. 257, 259, 266, 267
Trail, P. W. 279, 280, 281, 284
Traniello, J. F. A. 182
Trivers, R. L. 39, 185, 189, 191, 195, 203, 222, 224, 246, 266, 304
Trost, C. H. 290
Trueb, L. 261
Trumbo, S. T. 183
Tschinkel, W. 158, 165
Tulchin, N. 88, 89
Turner, J. 16

Ueda, H. 260
Uehara, S. 333, 334
Unger, L. M. 45, 46, 47
Urrutia, L. P. 289

Vail, V. A. 109
Valdes, E. S. 18, 27, 239
Valerio, C. E. 179, 180
Valerio, M. 239, 248
van den Bosch, F. 18, 27
Vargo, E. L. 194, 195
Vasil, V. 65
Vaught, K. C. 104
Vehrencamp, S. L. 284
Velestein, E. S. 311
Velthuis, H. H. W. 186, 216, 217, 222–4
Vickery, W. 238, 242, 243, 245, 250
Villa, P. 4
Vincent, J. 286, 324
Visscher, P. K. 216, 227, 228
Vogel, G. 92, 93
Vollrath, F. 147
Volpe, E. P. 262, 269
vom Saal, F. S. 305, 306, 309–11, 313
Voss, S. H. 194–6

Waddell, D. R. 2, 4, 5, 8, 19, 89, 92–5, 249, 291
Wade, M. J. 157, 159, 160, 163, 170–2
Wagner, D. B. 76
Wagner, T. L. 158, 164, 165
Waldman, B. 264
Walsh, T. P. 281
Walters, J. R. 329
Warburg, M. R. 257, 258, 268
Ward, G. 242
Ward, H. L. 280, 284
Wassersug, R. J. 257, 262
Watanabe, T. 87, 96
Watanuki, Y. 278, 280, 291
Watson, C. N. 134
Weissman, I. L. 3
Wellington, W. G. 105, 106
Wells, K. D. 266
Werner, E. 15, 22, 25, 32, 291
West, D. L. 110
West, M. J. 178, 179
West-Eberhard, M. J. 187, 192, 195, 196, 198, 200, 203, 204, 226
Westoby, M. 66, 70, 74, 77, 78, 80
Weyer, G. 89, 193
Weygoldt, P. 257, 266
Whatley, J. M. 76
Wheeler, W. M. 183
White, C. M. 285
White, T. L. 69
Whiten, A. 289, 290
Whitham, T. G. 67
Whitmore, M. J. 284
Whitten, P. L. 329
Whoriskey, F. G. 3, 5, 8, 18, 38, 39, 59, 60, 243, 245, 247, 250, 265
Wickler, W. 180

Wicklow, B. J. 89, 90, 95, 96
Wiemeyer, S. N. 281, 291
Wiens, J. A. 20
Wilbur, H. M. 18, 266
Wilde, W. H. A. 180
Wiley, J. W. 286
Wilkins, M. A. 279
Willard, D. E. 304
Williams, G. C. 39, 57
Williams, N. E. 85
Willoughby, E. J. 281
Willson, M. F. 65, 70, 71, 74
Wilson, C. M. 91
Wilson, E. O. 41, 43, 44, 176, 181, 185, 192–7, 205, 214
Winston, M. L. 216, 221, 222
Wolda, H. 107, 116
Wolff, J. O. 305–7, 312, 314
Wood, F. E. 258, 260
Wood, J. T. 258, 260, 266, 268
Wood, T. K. 181
Woodward, I. 280
Woolfenden, G. E. 279, 280
Worgan, J. P. 43
Wourms, J. 3, 5, 248, 249
Woyciechowski, M. 185
Woyke, M. 216, 227, 228
Wrangham, R. W. 333, 334
Wright, S. 170, 172
Wynne-Edwards, V. C. 277

Yamane, S. 216, 226
Yanigsawa, K. 89, 91
Yom-Tov, Y. 106, 107, 109, 279, 280, 282, 289
Young, C. M. 3, 8, 9
Young, E. C. 280, 281

Zack, S. 105, 118
Zahavi, A. 284
Zimmermann, E. 257, 325
Zimmermann, H. 256
Zucchi, R. 198, 199, 216, 224

Species Index

Abies balsamea 71
Acanthocyclops viridis 130
Accipiter cooperii 286, 290
A. gentilis 290
A. nisus 281
Achatina achatina 109
A. fulica 109, 112
Acomys cahirinus 306, 307
Adalia bipunctata 157, 161
A. decempunctata 157
Adonia undecimnotata 157
A. variegata 157
Aegopinella 103
A. nitidula 103, 106
Aeolidia papillosa 105, 118
Agathis 64, 67
Agonum assimile 158, 165
Alouatta palliata 327
A. seniculus 323, 327
Amanita 92
Ambystoma annulatum 258
A. tigrinum 257, 261–4, 270
A. t. mavortium 258, 261
A. t. nebulosum 258, 261, 264, 268
A. t. tigrinum 258, 261, 264, 265
Ameles decolor 13
Amoeba gjorgjevici 85, 89
A. verrucusa 85, 86, 89
A. vespertilio 85, 89
Amonna 192
Ancylus fluviatilis 105
Andrias japonicus 258
Anurogryllus muticus 178, 179–82
Aphaenogaster cockerelli 193
A. subterranea 193
Aphelocoma coerulescens 279–80
A. ultramarina 279–81
Apis 186, 227, 228, 232
A. cerrana indica 227, 228
A. mellifera 227, 228
Aquila verreauxi 281, 288
Araneus 139
A. diadematus 131, 134, 135, 137
A. pallidus 131, 139
A. quadratus 131
Araucaria 67
Archachatina marginata 109
Archipsocus gurneyi 180, 181
Arctocebus calabarensis 327
Ardea herodias 285, 286
Argiope 139, 148
A. aemura 132, 134, 139
A. aetheria 132
A. argentata 132
A. aurantia 132
A. bruennichi 131
A. cuspidata 131
A. florida 131
A. ocyaloides 131
A. picta 132
A. radon 132
A. reinwardti 132
Arianta arbustorum 107, 109, 111–17, 119
Arion 107
A. subfuscus 105
Arvicola terrestris 313
Ascaloptynx furciger 179, 181
Asio flammeus 286
Asplanchana sieboldi 90
Atta 196
A. cephalotes 193
A. laevigata 193, 195
A. rubopilosa 193
A. sexdens 193

Balea perversa 117
Baryphyma pratense 136
Basiceros 197
Basiceros manni 193
Belonogaster 187
Bezzia modocensis 130
Biota 71
Blepharisma 87, 88, 89, 95
Blepharisma americanum 89
B. undulans 85, 87
Bombina bombina 259
B. orientalis 259
Bombus 185, 186, 188–90, 202, 205, 222–7, 229, 231, 232
B. affinis 190
B. atratus 190, 224
B. huntii 225
B. hypnorum 225
B. hypocrita 190
B. lapidarius 190
B. ruderatus 190
B. terrestris 190, 222–4
Bothriurus asper 144
B. bonariensis 131, 144

Bothriurus asper (*cont*)
B. flavidus 144
Bracon hebetor 183
Bubo virginianus 286
Buccinum 111
Buccinum undatum 108, 111
Bufo arenarum 258
B. boreas halophilus 258, 260
B. vulgaris 258, 265
Buscycon spiratum 104
Buteo buteo 286
B. jamaicensis 279
B. lagopus 286
B. platypterus 286
B. swainsoni 286
Buthus occitanus 131, 144

Calliopaea oophaga 104, 109
Callithrix 327
C. jacchus 324, 325
Canis aureus 331
Carabus auratus 130
Cardiapoda placenta 104
Carinaria cristata 104
Cassidaria echinophora 108
Catasigerpes occidentalis 130
Catharacta maccormicki 280, 281
Cathartes burrovianus 278
Cebuella pygmaea 327
Cebus apella 327
C. capucinus 327
Cepaea hortensis 113
C. nemoralis 109, 112, 113, 116, 117
Cephalotrigona 199
Cepphus columba 286
Ceratina 231
C. flavipes 220
C. japonica 220
Ceratophrys 257, 263, 267, 268
Ceratophrys cornuta 259
C. ornata 259
C. pierottii 259
Cercopithecus albigena 327
C. ascanius 323, 327
C. campbelli 327
C. mitis 327
C. pogonias 328
Chacophrys 260, 263, 267
Chamydophrys schaudinni 89
Cheilomenes lunata 157, 161
Cheirogaleus major 327
C. medius 324, 325, 327
Chirixalus eiffingeri 260
Chromodoris zebra 105
Cichlasoma citrinelli 248
C. nigrofasciatum 244
Ciconia ciconia 281
Circus aeruginosus 286
C. cyaneus 286
Clarias gariepinus 248
Coccinella axiridis 157
C. septempunctata 157, 161
Coleomegilla maculata 161
C. maculata lengi 157
Colobus badius 328, 333
C. guereza 328
Colus islandicus 110
Conus 119
Cophixalus parkeri 259
Coptopteryx viridis 130
Corcorax melanorhamphos 284
Corvus corone 279, 280
C. monedula 279, 280
Coryphella stimpsoni 105
Cottus gobio 239
Crepidula cerithicola 108
C. plana 108
Cricetus cricetus 299
Crotophaga ani 284
C. sulcirostris 284
Cryptobranchus alleganiensis 258
Cupressus 75
Cynomys ludovicianus 299, 304, 306, 307
Cyphoderris 138
Cyrtophora cicatrosa 139
C. citricola 132
C. moluccensis 132

Daudebardia 107
Daudebardia langi 106, 107
D. rufa 106
Dendrobates auratus 259, 266, 267
D. pumilio 266
Deroceras agreste 109, 114
D. caruanae 106
D. laeve 106
D. sturanyi 109
Desmognathus 260
D. fuscus fuscus 258
D. monticola 258
D. ochrophaeus 258
Dicamptodon ensatus 258, 260
Dictyna calcarata 132
Dictyostelium caveatum 85, 89, 92–5, 96, 97, 98
D. discoideum 89, 91, 92, 94, 97
D. mucoroides 97
Dineutes 165
D. assimilis 159
D. horni 159
D. nigrior 159
Dolichoderus quadripunctatus 193
Dolichovespula 188, 189
D. arenaria 188

D. maculata 188
Dolomedes fimbriatus 133
D. triton 133
Duckeola, see *Trigona*
Dysdera crocata 132

Eciton 192
Elaphrothrips tuberculatus 182, 184
Eleutherodactylus coqui 259, 266
Embletonia fuscata 105
Eobania vermiculata 109
Ephedra 64, 65
Eriophora fuliginea 132
E. transmarinus 132
Euglandina rosea 105, 106
Euglossa 231
Euglossa cordata 220–2, 231
Eumicrota 182
E. corruscula 182
E. socia 182
Euscorpius 144
Exoneura 230
E. bicolor 230
E. cordata 231

Falco mexicanus 281
F. peregrinus 290
F. sparverius 281, 284, 286
F. tinnunculus 281, 286
Fasciolaria tulipa 104, 108
Formica pergandei 193, 195
F. rufa 193
Fregata magnificens 280
Friesella, see *Plebeia*
Frieseomelitta, see *Trigona*
Fringilla coelebs 279
Frontinella pyramitela 133
Frontonia 90
F. leucas 89, 90
Fulica americana 278

Galago alleni 327
G. classicaudatus 324, 325, 327
G. demidovii 327
G. elegantulus 327
G. garnetti 324, 325, 327
G. senegalensis 324, 325, 327
Gallinula chloropus 278, 281
Gallus gallus 291
Gambusia affinis 239
Gasterosteus aculeatus 38, 48, 240
Gastrostyla steinii 89
Gea eff 132
Geococcyx californianus 281, 284
Gerbillus dasyurus 299
Ginkgo 64, 67, 71, 74
Glaucus atlanticus 105, 119, 131
Gnetum 65
Goniozus nephantides 183
Gorilla gorilla 328
G. g. beringei 323
Guira guira 284
Gymnorhina tibicen 279, 280

Halictus ligatus 229
Harmonia axyridis 157, 158, 161, 162
H. quatturodecimpunctata 157, 161
Harpagoxenus americanus 193
Helix aspersa 109
H. pomatia 109, 112–14, 117
Hermissenda 138
H. crassicornis 105, 118, 131
Herpyllus blackwalli 132
Hestiasula brunneriana 130
Heterometrus scaber 131, 144
Heteropoda venatoria 132
Hierodula membranacea 130, 137
Hippodamia quinquesignata 157
Hirundo rustica 282
Hoplophryne rogersi 257, 259
Humbertiella similis 130
Hyla pseudopuma 259, 265, 268
H. rosenbergi 259
Hynobius nebulosus 257, 258, 268
Hypoclinea quadripunctata 192, 193
Hypotrigona 202

Ips 164
I. avulsus 158, 164
I. calligraphus 158, 164, 165
Iridomyrmex humilis 193, 195
Isometrus maculatus 131, 144

Jacksonoides queenslandica 134
Juniperus communis 7, 69
J. monosperma 67

Labidomera clivicollis 157, 159, 180
Lanius collurio 281
Larus 279
L. argentatus 280–2
L. dominicanus 280
L. fuscus 280, 282
L. glaucescens 280
L. occidentalis 291
L. ridibundus 280
L. schistisagus 280, 291

Lasioglossum 185, 188–90
L. versatum 229
L. zephyrum 185, 190, 229
Lasius 193, 195
Latrodectus 137, 139
L. hasselti 133, 134, 139, 140
L. hesperus 133
L. katipo 140
L. mactans 137
Lechriodus fletcheri 259, 263
Leiurus quinquestriatus 131, 144
Lembadion 90
L. lucens 90
Leontopithecus rosalia 327
Lepidobatrachus 260, 263, 264, 267, 268
L. asper 259
L. laevis 265
L. llanensis 259, 265, 268
Lepomis macrochirus 44
Leptinotarsa decemlineata 158, 165
L. juncta 158, 165
Leptodactylus knudseni 259
L. pentadactylus 259
Leptothorax recedens 193
Lestrimelitta 202
L. limao 203
Leucage 133
Leurotrigona, see *Trigonisca*
Limax maximus 106
Limicolaria martensiana 109
Linyphia litigiosa 134
Lioadalia flavomaculata 157, 161
Loris tardigradus 327
Lycosa rabida 133, 143
L. scutulata 133
Lymnaea stagnalis 107

Macaca fuscata 323, 328
M. mulatta 328
Macrochlamys indica 107
Macrozamia 71
Mallochohelea nitida 130
Manorina melanocephala 284
Mantis religiosa 130, 135, 137
Marpissa marina 133
Melanerpes formicivorus 81, 282
Melipona 199–201, 226, 230
M. nigra schencki 197
M. scutellaris trinitatis 200
Meliponula 200, 201
Meriones unguiculatus 300, 301
Mesocricetus auratus 301
Messor capitatus 193
M. semirufus 193
Meta segmentata 133
Metapolybia 187
M. azteca 187
M. aztecoides 187
Metasequoia 67
Microcebus coquereli 327
M. murinus 327
Microcycas 71, 74, 75, 76
Miomantis caffra 130
M. paykullii 130
Miopithecus talapoin 328
Mischocyttarus 190
M. drewseni 187
Motacilla alba 278, 281
Murex brandaris 108, 110
M. incarnatus 108
M. trunculus 104
M. virgineus 110
Mus musculus 301, 303, 305, 306
Myrmecia 197
M. forceps 193
M. gulosa 192, 194
M. simillima 194
M. rubra 192, 194–6
Myrmica 197

Nasua narica 331
Natica unifasciata 103, 104, 120
Navanax inermis 104, 106, 107, 118, 119, 131
Nebo hierichonticus 144
Necturus maculosus 258
Neotoma floridana 303
Nephila clavipes 132
N. edulis 132
Neptunea antiqua 110
Nesovitrea hammonis 103
Nicrophorus 161, 171, 182–4
N. orbicollis 183
N. tomentosus 182
N. vespilloides 157, 160, 183
Nicticebus concang 327
Nothomyrmecia macrops 194
Notonecta 18
Notophthalmus 260
N. viridescens 258, 268
N. viridescens dorsalis 267
N. viridescens viridescens 258
Nucella crassilabrum 108
N. lapillus 108

Odontaspis tauris 288
Oecophylla 95, 196
O. longinoda 194
O. smaragdina 194
Ondatra zibethica 306
Onychodromus grandis 85, 89
O. quadricornutus 89, 90, 96
Onychomys 299
Opeas gracile 107

Opistophthalmus latimanus 144
Orthodera ministralis 130
Oryzaephilus surinamensis 158, 164
Osteopilus septentrionalis 260
Oxychilus alliarius 105, 107
  O. cellarius 107
  O. draparnaudi 103, 119
Oxylebius pictus 57
Oxypilus hamatus 130, 134
Oxytricha 85
  O. hymenostoma 87, 89

Palpomyia 130
Pan troglodytes 323, 328
Pandinus imperator 131, 144
Panthera leo 331
Papio cynocephalus 328
  P. c. anubis 323, 324
  P. c. ursinus 323, 324
Parabuthus planicauda 144
Paramecium 85, 87, 90
Paratrechina flavipes 194, 195
Paravespula vulgaris 188
Paruroctonus 144
  P. luteolus 131
  P. mesacensis 137
  P. mesaensis 131
Pelecanus occidentalis 280
Perforatella incarnata 113
  P. rubiginosa 109
Pernis apivorus 286
Peromyscus californicus 300
  P. leucopus 305
  P. maniculatus 301, 303, 306, 307
Pheidole 194
  P. pallidula 194, 197
Phiddipus audax 133
  P. clarus 133
  P. johnsoni 133, 134, 137
  P. morsitans 133
  P. rufus 133
Philanthus basilaris 183
Philonthus decorus 159, 165
Phodopus sungorus 307
Phonognatha graeffii 138
Pica nuttalli 290
  P. pica 281, 290
Picea 64, 67, 68, 70, 73
  P. abies 66, 68, 69, 70, 72
Pila 104, 111
Pimephales promelas 47, 57
Pinaroloxias inornata 291
Pinus 64, 67, 68, 70
  P. monticola 72
  P. sylvestris 66, 68, 70, 71
Pisaura mirabilis 133, 142
Plagiodera versicolora 8, 157, 159
Plagiolepis pygmaea 194, 195
Plebeia 199, 200, 202, 204
Plebia, see *Plebeia*
Plethodon cinereus 258
  P. glutinosus 258, 260
Pleurobranchaea californica 104, 119
Pleurodema nebulosa 259
Pleuroploca australasia 108
Pleurotricha laceolata 85
Poeciliopsis 251
  P. occidentalis occidentalis 251
Pogonomyrmex badius 192, 194
Polistes 185–90
  P. annularis 187
  P. fedwigae 187
  P. fuscatus 187
  P. gallicus 187
Polybia 187
Polyspilota 130
Pomacea 111
  P. urceus 104, 111
Portia 143
  P. labiata 133
  P. schultzi 133
Precuthona peachii 105
Presbytis cristatus 328
  P. entellus 323, 328
Prionopelta 197
  P. amabilis 194
Probezzia concinna 130
Propyleae, see *Harmonia*
Prosierola bicarinata 183
Protopolybia 187, 190
  P. exigua 187
  P. sedula 187
Pseudocrenilabrus multicolor 57, 244
Pseudotsuga menziesii 66
Pterostichus nigrita 158, 165
  P. oblongopunctatus 158, 165
Puma concolor 331
Punctum pygmaeum 117, 118
Pyxicephalus 260, 263, 264, 267
  P. adspersus 259, 260

Quercus 64

Rana 260
  R. esculenta 260
  R. ridibunda 260
  R. rugulosa 260
  R. tigrina 260
Rhinophrynus dorsalis 260
Rhizopertha 164
Rhizopertha dominica 158, 164
Rissoa membranacea 108
Rotaria 90

Rynchops niger 280, 286
Saguinus fascicollis 327
S. geofroyi 327
S. oedipus 327
S. tamarin 327
Saimiri oerstedii 327
S. sciureus 324, 325, 327, 328
Salamandra atra 261
S. salamandra 257, 258
Scaphiopus 15, 260–3, 267, 269
S. bombifrons 259
S. hammondi 259
S. holbrooki holbrooki 259
S. h. hurteri 259
S. multiplicatus 259, 262, 264, 265, 269, 270
Scaptotrigona 199–203, 226
S. postica 199, 200, 230
Scaura, see *Plebeia*
Schizodactylus monstrosus 180
Schwarziana, see *Plebeia*
Scleroderma macrogaster 183
Scoloderus cordatus 132
Scolytus multistriatus 158, 164
S. scolytus 158, 164
Semiadalia, see *Adonia*
Serromyia femorata 130
Simaetha paetula 133
Solenopsis 195
S. invicta 191, 194–6
Spermophilus beldingi 299, 304, 306, 331
S. parryii 312
Sphincterochila boissieri 106, 107, 109
Sphodromantis lineola 130, 134
Stagmonantis carolina 130
Stegastes rectifraenum 38, 244
Stenopelmatus fuscus 130
S. nigrocapitatus 130
Stentor coeruleus 85, 89, 90, 92
Sterna fuscata 280, 284
S. paradisaea 280
Strix aluco 281
S. nebulosa 278
Sturnus vulgaris 291
Stylonychia 87, 90
S. curvata 85, 89
Succinea putris 105, 107, 117
Supunna picta 132
Synoeca 187

Tapinoma erraticum 194
Tarachodes afzelii 130, 134
Taraxacum 64
Taricha torosa 257, 258
Tarsius spectrum 327
Taxus 65
Telmatodytes palustris 291
Temnothorax recedens 192
Tenodera aridifolia 130, 143
Testacella haliotidea 106
Tetragnatha 142
Tetragona, see *Trigona*
Tetragonula, see *Trigona*
Tetrahymena 85, 88
T. patula 85
T. vorax 85
Tetramorium caespitum 194
Thais emarginata 110, 111
T. lamellosa 120
Theodoxus fluviatilis 108
Tilapia 244
Tityus 144
T. tfasciolatus 131, 144
T. trinitatis 144
Tomicus piniperda 158, 164
Torreya 65
T. taxifolia 65, 71
Tribolium 15, 18, 166–71
T. brevicornis 158
T. castaneum 158, 163, 165–7
T. confusum 157, 158, 162, 163, 166, 168, 169
T. madens 158
Trichonympha campanula 86, 89
Trigona 199–202, 226, 230
Trigonisca 199, 201, 226, 230
Triodopsis albolabris 109
Tritonia hombergi 105
Troglodytes aedon 291
Turdoides squamiceps 284
Turdus merula 279–81
Tyto alba 281, 284, 286

Ululodes mexicana 179
Urophonius brachycentrus 131, 144
Ursus arctos 331
U. maritimus 331

Vallonia costata 117, 118
Vejovoidus longiunguis 131, 144
Vertigo pusilla 118
Vespa orientalis 188
Vespula 188
V. atropilosa 188
Volutopsis norvegicus 108

Welwitschia 65

Xema sabini 280
Xenopus 260
X. laevis 259

Xylocopa 217, 220, 226, 231, 232
X. pubescens 217
X. sonorina 218
X. sulcatipes 217, 218, 220
X. virginica 217
Xysticus 134
X. cristatus 142
Zacryptocerus varians 194
Zamia 74
Zilla sp. Wau # 1 132
Z. sp. Wau # 2 132
Zonitoides nitidus 105
Zophobas atratus (= *rugipes*) 158, 165
Zygopachylus albomarginatus 183, 184

Subject Index

abortion 64, 66–70, 73
of ovules 67–8
of seeds 68, 77–9
assymetry in size, importance for cannibalism 2, 15, 87, 91, 103, 106, 145–7, 166, 248, 287, 288
avoiding cannibalism, *see* defense

behaviour
bizzare 203
unnatural, abnormal or vicious 1, 256, 278, 300–1, 323
benefits of cannibalism
acquisition of gametes 38–61, 245–8, 266, 277, 282, 290, 304, 306
avoiding misdirected parental care 304, 306
growth 97
mate choice 140–1
not obvious 241
nutritional 110, 137, 141, 157, 161, 163, 167–8, 200, 214, 221, 226, 231, 239–41, 245, 248, 263, 265, 278, 289, 299, 301–5, 333–5
reduces competition for resources 6, 219, 222–4, 229, 230, 231, 240, 242, 266, 304, 306
survival 97, 115
brood reduction 69, 71, 285–7
as cannibalism 3, 63, 161
Bruce effect 313

cannibalism
as unnatural, abnormal or vicious behaviour 1, 256, 278, 300–1, 323
ambiguities of interpretation 3, 66, 85
among humans, not relevant 4
and competition 24
and courtship behaviour 9
and diet choice 19–21, 22, 113, 140–2
and food limitation 5, 15, 90, 92, 94, 106, 164
and sexual dimorphism 9
as ensuring paternity 140
as female choice 140–1
as foraging decisions 22–4, 140–2
as indiscrimate predation 88, 119, 106, 135–6, 156, 158–9, 162, 260
benefits, *see* benefits of cannibalism
bizzare behaviour 203
by different life-stages 103–11, 156, 257–61, 278–9
complicity by victim 96, 138–40, 228
costs, *see* costs of cannibalism
defined 2, 3, 6, 7, 66, 129, 176, 214, 239, 256, 278
frequency in population 86, 90, 134, 160, 162, 167, 182, 227, 283, 329, 330
genetic evidence 90, 93, 169–71, 227, 251
of kin, *see* sibling and filial cannibalism
of kin and misleading distinctions 7
on different life-stages 103–11, 156
risk of 14-21
seasonal changes 246–7
taxonomic distribution 2–3, 89, 104–6, 129–34, 157, 158–9, 178–9, 181–4, 185–8, 193–4, 215–16, 258–60, 280–1, 286, 323–6
cannibalistic polyphenisms 8, 9, 261–2, 269, 291
environmental basis 9, 261, 262
genetical basis 9, 261
carnivory
associated with cannibalism 326–8, 331
carrion 107, 119
cohort selection 163
competition
intraspecific 6, 24, 72, 162, 162, 166–7, 188, 229, 231, 242–3
for resources 6, 219, 222–4, 229, 230, 231, 240, 242, 266, 304, 306
reproductive 219, 222–4, 229, 242, 284
complicity by cannibal victim 96, 138–40, 228
conelet abscission 68–9
conflicting demands 43, 58
and trade-offs 44, 46, 58, 60, 96–8
conflicting interests 5, 6, 7, 9, 66, 128, 150, 177, 184, 188–90, 200, 204, 214, 222, 224, 233, 290
consequences of cannibalism
for individuals 1, 8–10, 26–30, 147–9, 165, 181
for populations 1, 10, 14, 26, 28–9, 119, 162, 166–7, 238, 267, 314
for social behaviour 181, 183, 188, 315
constraints on cannibalism
dietary 262, 287, 326–8
morphological 117, 262, 287
oviparity 288
social 328–9

costs of cannibalism 1, 7–8, 23, 97, 136, 263–5, 299–300, 328–9
  injury 263
  loss of inclusive fitness 23, 95, 177, 243, 264, 299–300
  loss of sexual partners 141–2
  smaller group size 95, 160
courtship behaviour
  and cannibalism 1, 9, 142–5, 147–9

defense against cannibalism
  behaviour 138, 142–5, 249–50, 267–8, 307–9, 312–14
  morphological traits 148–9
  parental care 182, 184, 287, 307–10, 312–14
  patch selection 24–6, 245
  spatial segregation 25, 163, 245, 267
  temporal segregation 25
definition of cannibalism 2, 3, 6, 7, 66, 129, 176, 214, 239, 256, 278
density effects
  on cannibalism 1, 5, 10, 14, 17–19, 31, 119, 164, 263, 302, 315
developmental selection 72–4, 75, 77–9
  in angiosperms 78

eating males 131–4
egg dispersal
  importance for cannibalism 117, 162
embryo
  cannibalism of, *see in utero* cannibalism
ensuring paternity 140
eusociality 184–204
evolution
  of cannibalism 4–7, 116–18, 136–42, 157
  of sociality 181, 183, 188,
  of trophic eggs 181, 191–2, 195–6, 226
evolutionary stable strategy 77–8

female choice 140–1
filial cannibalism 23, 39, 38–60, 157, 160, 183, 238, 244–8, 257, 284–5, 288, 299, 300–4
filicide 3, 63
food availability
  and cannibalism 5, 15, 90, 92, 94, 106, 164, 263, 289
food limitation 5, 15, 90, 92, 94, 106, 164
foraging decisions
  and diet choice 19–21, 22, 113, 140–2
  and sexual cannibalism 22–4, 140–2
frequency of cannibalism 86, 90, 134, 160, 162, 167, 182, 227, 283, 329, 330

gamete acquisition
  and cannibalism 38–61, 245–8, 266, 277, 282, 290, 304, 306
genetic
  basis of cannibalism 9, 90, 93, 169–71, 227, 251, 261
  relatedness 7, 159, 160
gerontophagy 3, 90, 328
  see also carrion
giant cells
  and cannibalism 86, 87, 89, 95
group selection 165

Hamilton's Rule 115–16, 157, 168, 177, 184, 218, 232, 245, 277, 299
hatching asynchrony
  influence on cannibalism 113, 114, 120, 121, 288
heterocannibalism 59
human cannibalism, no relevance 4

in utero cannibalism 5, 248, 260
  not cannibalism 260
inclusive fitness 115, 157, 159, 190, 245
  loss through cannibalism 23, 95, 177, 243, 264, 299–300
indiscrimate predation 88, 119, 106, 135–6, 156, 158–9, 162, 260
infanticide, 7, 278, 306–10, 312, 323, 326, 330, 332–3
  distinct from cannibalism 7, 278, 305, 326
  without cannibalism 326-8
intraspecific competition 6, 24, 72, 162, 162, 166–7, 188, 229, 231, 242–3

kin selection 168, 214, 218, 228, 230, 277, 300

mate choice 140–1
mating 87
  see also sexual cannibalism
misdirected parental care 304, 306
multiple mating 159
  influence on cannibalism 115, 118, 185, 190, 225, 227

nest raiding 245–6
nurse eggs, *see* trophic eggs
nutrition
  increased by cannibalism 110, 137, 141, 157, 161, 163, 167–8, 200, 214, 221, 226, 231, 239–41, 245, 248, 263, 265, 278, 289, 299, 301–5, 333–5

oophagy 9, 107, 111–18, 120, 161, 176–205, 214–29, 257
and breeding system 109, 190, 232
as reproductive suppression 189, 218, 220
between clutches 111, 115
followed by oviposition 188, 197
not considered cannibalism 3, 257
within clutches 111, 116
see also cannibalism
optimal foraging theory 13
oral replacement
and cannibalism 85
oviposition
following cannibalism 188, 197
rituals 197–9, 200–1

parasites and disease
largely ignored 7
risk of infection by cannibalism 1, 7, 24, 290
parental care 47–59, 184
and cannibalism 182, 287, 307–10, 312–14
parental manipulation
and cannibalism 3, 4, 114, 184, 188, 194
parental resource allocation 67, 69–70, 77, 181, 289, 302
and cannibalism 6, 39–40, 50–1, 56, 59, 70–1, 120, 221, 244–8, 289, 301–4
paternal investment
by eating males 9, 136–40
phagocytosis 90, 92, 94
polyembryony 65, 66, 70–7
population dynamics
and cannibalism 1, 10, 14, 26, 28–9, 119, 162, 166–7, 238, 267, 314

recognition of kin 185, 243
and cannibalism 185, 264–5, 308
no evidence of 115, 243, 309
recognition of self 86, 88, 93–4, 96, 97
reproductive
strategies 38–60, 74–6
suppression 189, 218, 220
risks of cannibalism 7–8, 14–21, 24, 290
see also costs

scavenging 3, 278, 289–90
sexual cannibalism 9, 118–19, 128–50
and mistaken identity 135
importance of timing 134
sexual size dimorphism
influenced by cannibalism 147–9
siblicide 3, 63, 230
sibling cannibalism 23, 86, 112, 159, 177–8, 248, 285, 288
spatial segregation 25, 163, 245, 267
social behaviour
and cannibalism 181, 183, 188, 315

taxonomic distribution
of cannibalism 2–3, 89, 104–6, 129–34, 157, 158–9, 178–9, 181–4, 185–8, 193–4, 215–16, 258–60, 280–1, 286, 323–6
temporal segregation 25
theoretical models
of cannibalism 5, 10, 21–4, 26–31, 32, 39, 43–56, 136–7, 140–2, 163, 166, 167, 238, 242–3
trade-offs 44, 46, 58, 60, 96–8
trophic eggs
as cannibalism 108, 110, 120, 121, 176–205, 283, 288
not considered cannibalism 3, 257

variation in cannibalism
by age 112, 257
by density 1, 5, 10, 14, 17–19, 31, 119, 164, 263, 302, 315
by food availability 5, 15, 90, 92, 94, 106, 164, 263, 289
by genetic relatedness 178,
by habitat 113–14
by sex 300, 311
vulnerability of cannibalism 30, 87, 91, 157

worker policing
and cannibalism 203, 225, 226, 228, 232
Wright's shifting balance theory 170–1